AQA Mathematics

for GCSE

Exclusively endorsed and approved by AQA

Series Editor
Paul Metcalf
Series Advisor
David Hodgson
Lead Author
Steven Lomax

June Haighton
Anne Haworth
Janice Johns
Andrew Manning
Kathryn Scott
Chris Sherrington
Margaret Thornton
Mark Willis

HIGHER

Linear 1

Nelson Thornes
a Wolters Kluwer business

Published in 2006 by:
Nelson Thornes Ltd
Delta Place
27 Bath Road
CHELTENHAM
GL53 7TH
United Kingdom

07 08 09 10 / 10 9 8 7 6 5 4 3

A catalogue record for this book is available from the British Library.

ISBN 978 0 7487 9751 6

Cover photograph: Salmon by Kyle Krause/Index Stock/OSF/Photolibrary
Illustrations by Roger Penwill
Page make-up by MCS Publishing Services Ltd, Salisbury, Wiltshire

Printed in Great Britain by Scotprint

Acknowledgements

The authors and publishers wish to thank the following for their contribution:
David Bowles for providing the Assess questions
David Hodgson for reviewing draft manuscripts

Thank you to the following schools:
Little Heath School, Reading
The Kingswinford School, Dudley
Thorne Grammar School, Doncaster

The publishers thank the following for permission to reproduce copyright material:

Explore photos
Diver – Corel 55 (NT); Astronaut – Digital Vision 6 (NT);
Mountain climber – Digital Vision XA (NT); Desert explorer – Martin Harvey/Alamy.

Archery – AAS; Plant – Stockbyte 29 (NT); Lighthouse – Corel 502 (NT);
Discs – Nick Koudis/Photodisc 37 (NT); Eden Project – Philippe Caron/Sygma/Corbis;
London Eye – Peter Adams/Digital Vision BP (NT); Students in exam – Digital Stock 10 (NT);
Beach – Corel 777 (NT); Factory worker – Digital Stock 7 (NT); Vitruvian man – Corel 481 (NT);
Solar System Montage – NASA; Stars in the Tarantula Nebula – NASA, The Hubble Heritage Team, STScl, AURA; Coins – Corel 590 (NT); Electricity pylons – Photodisc 4 (NT); Crufts dog show – Homer Sykes/Alamy; Shopping centre – Corel 641 (NT); Telephone survey – Photodisc 55 (NT);
Supermarket – Stockbyte 9 (NT); Concert hall – Digital Stock 13 (NT);
Puppy – G/V Hart/Photodisc 50 (NT); Seagulls – Photodisc 6 (NT);
Rangoli – Dinodia/Alamy; Goldfish – Alex Homer; Duke on the Crater's Edge – NASA, John W. Young; An Ancient Storm in the Jovian Atmosphere – NASA, The Hubble Heritage Team, STScl, AURA, Amy Simon Cornell; Bacteria – Photodisc 72 (NT); House sale – Photodisc 76 (NT);
Tiger – Nat Photos/Digital Vision AF (NT); Formula One car – Corel 231 (NT).

The publishers have made every effort to contact copyright holders but apologise if any have been overlooked.

Contents

Introduction v

1 Statistical measures 1

Learn 1 Mean, mode, median and range for a frequency table 2
Learn 2 Mean, mode and median for a grouped frequency table 6
ASSESS 10

2 Integers 11

Learn 1 Factors and multiples 12
Learn 2 Prime numbers and prime factor decomposition 14
Learn 3 Reciprocals 16
ASSESS 17

3 Rounding 19

Learn 1 Estimating 20
Learn 2 Finding minimum and maximum values 23
ASSESS 25

4 Use of symbols 27

Learn 1 Expanding brackets 28
Learn 2 Expanding linear expressions 31
Learn 3 Factorising expressions 34
Learn 4 Factorising quadratic expressions 36
Learn 5 Simplifying algebraic fractions 40
ASSESS 41

5 Decimals 43

Learn 1 Multiplying decimals 44
Learn 2 Dividing decimals 46
Learn 3 Fractions and decimals 48
Learn 4 Recurring decimals and fractions 50
ASSESS 52

6 Area and volume 55

Learn 1 Perimeters and areas of triangles and parallelograms 58
Learn 2 Areas of compound shapes 61
Learn 3 Circumferences of circles 65
Learn 4 Areas of circles 68
Learn 5 Volumes of cubes, cuboids, prisms and cylinders 71
Learn 6 Surface areas of prisms and cylinders 74
Learn 7 Volumes and surface areas of pyramids, cones and spheres 78
Learn 8 Lengths of arcs and areas of sectors and segments 84
ASSESS 87

7 Fractions 91

Learn 1 Adding and subtracting fractions 92
Learn 2 Multiplying and dividing fractions 94
ASSESS 96

8 Surds 97

Learn 1 Simplifying surds 98
Learn 2 Rationalising the denominator of a surd 101
ASSESS 104

9 Representing data 105

Learn 1 Stem-and-leaf diagrams 108
Learn 2 Frequency diagrams, line graphs and time series 111

Learn 3 Cumulative frequency diagrams 117
Learn 4 Box plots 123
Learn 5 Histograms 128
ASSESS 132

10 Scatter graphs 135

Learn 1 Scatter graphs 136
Learn 2 Lines of best fit 140
ASSESS 144

11 Properties of polygons 147

Learn 1 Diagonal properties of quadrilaterals 148
Learn 2 Angle properties of polygons 151
ASSESS 154

12 Indices and standard form 157

Learn 1 Rules of indices 158
Learn 2 Standard form 162
ASSESS 165

13 Sequences 167

Learn 1 The *n*th term of a sequence 168
ASSESS 171

14 Coordinates 173

Learn 1 Coordinates and lines 174
Learn 2 The midpoint of a line segment 177
Learn 3 Coordinates in three dimensions 179
ASSESS 183

15 Collecting data 185

Learn 1 Collecting data 186
Learn 2 Sampling methods 191
ASSESS 195

16 Percentages 197

Learn 1 Increasing or decreasing by a given percentage 198
Learn 2 Using successive percentage changes 200
Learn 3 Compound interest 202
Learn 4 Expressing one quantity as a percentage of another and finding a percentage increase or decrease 204
Learn 5 Reverse percentage problems 206
ASSESS 208

17 Equations 209

Learn 1 Equations where the unknown appears on both sides 210
Learn 2 Equations with brackets 213
Learn 3 Equations with fractions 214
Learn 4 More complex equations with fractions 215
ASSESS 216

18 Reflections and rotations 217

Learn 1 Drawing reflections and rotations 219
Learn 2 Describing reflections and rotations 224
Learn 3 Combining reflections and rotations 227
Learn 4 Reflection symmetry in 3-D solids 229
ASSESS 231

19 Ratio and proportion 233

Learn 1 Finding and simplifying ratios 234
Learn 2 Using ratios to find quantities 238
Learn 3 Ratio and proportion 240
Learn 4 Calculating proportional changes 243
Learn 5 Direct and inverse proportion 246
ASSESS 249

Glossary 253

Introduction

This book has been written by teachers and examiners who not only want you to get the best grade you can in your GCSE exam but also to enjoy maths.

Each chapter has the following stages:

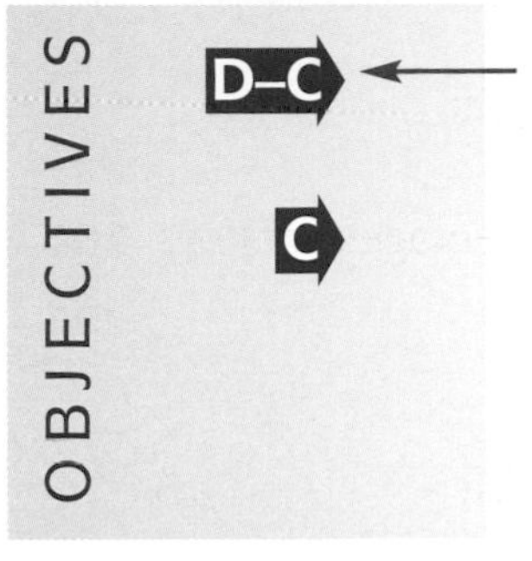

The objectives at the start of the chapter give you an idea of what you need to do to get each grade. Remember that the examiners expect you to perform well at the lower grade questions on the exam paper in order to get the higher grades. So, even if you are aiming for an A grade you will still need to do well on the D grade questions on the exam paper.*

Learn 1 — *Key information and examples to show you how to do each topic. There are several Learn sections in each chapter.*

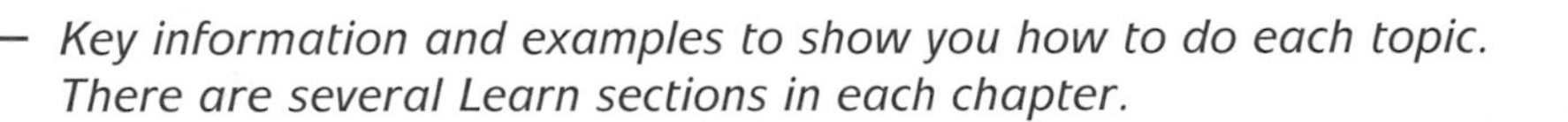

Apply 1 — *Questions that allow you to practise what you have just learned.*

Means that these questions should be attempted with a calculator.

Means that these questions are practice for the non-calculator paper in the exam and should be attempted without a calculator.

Get Real! — *These questions show how the maths in this topic can be used to solve real-life problems.*

1 — *Underlined questions are harder questions.*

Explore — *Open-ended questions to extend what you have just learned.*

ASSESS — *End of chapter questions written by an examiner.*

Some chapters feature additional questions taken from real past papers to further your understanding.

1 Statistical measures

OBJECTIVES

D **Examiners would normally expect students who get a D grade to be able to:**

Calculate the mean for a frequency distribution

Find the modal class for grouped data

C **Examiners would normally expect students who get a C grade also to be able to:**

Find the mean for grouped data

Find the median class for grouped data

What you should already know ...

- Multiplication and division of a set of numbers without the use of a calculator
- Find the mean, median, mode and range for a set of numbers
- Compare the mean and range of two distributions
- Understand the inequality signs $<$, $>$, $\leqslant$ and $\geqslant$

VOCABULARY

Frequency table or **frequency distribution** – a table showing how many times each quantity occurs, for example:

Number in family	2	3	4	5	6	7	8
Frequency	2	3	8	4	2	0	1

Average – a single value that is used to represent a set of data

Mean – found by calculating

$$\frac{\text{the total of all the values}}{\text{the number of values}}$$

Mode – the value that occurs most often

Modal class – the class with the highest frequency

Median – the middle value when all the values have been arranged in order of size; for an even set of numbers, the median is the mean of the two middle values

Range – a measure of spread found by calculating the difference between the largest and smallest values in the data, for example, the range of 1, 2, 3, 4, 5 is $5 - 1 = 4$

Data – information that has been collected

Discrete data – data that can only be counted and take certain values, for example, number of cars (you can have 3 cars or 4 cars but nothing in between, so $3\frac{1}{2}$ cars is not possible)

Continuous data – data that can be measured and take any value; length, weight and temperature are all examples of continuous data

Grouped data – data that has been grouped into specific intervals

Learn 1 Mean, mode, median and range for a frequency table

Example:

From this table work out the mean, mode, median and range of the number of goals scored.

Goals scored (x)	Frequency (f)
0	3
1	6
2	5
3	3
4	2
5	1

Completing the table:

Goals scored (x)	Frequency (f)	Frequency × goals scored (fx)
0	3	$0 \times 3 = 0$
1	6	$1 \times 6 = 6$
2	5	$2 \times 5 = 10$
3	3	$3 \times 3 = 9$
4	2	$4 \times 2 = 8$
5	1	$5 \times 1 = 5$
Total	$\Sigma f = 3 + 6 + 5 + 3 + 2 + 1 = 20$	$\Sigma fx = 0 + 6 + 10 + 9 + 8 + 5 = 38$

Σ means 'the sum of' so Σf means the sum of the frequencies and Σfx means the sum of the (frequency × goals scored)

The mean $= \dfrac{\Sigma fx}{\Sigma f} = \dfrac{38}{20} = 1.9$ goals

The mode is the value that occurs the most often.

Mode = 1 goal — In a frequency table, this is the value with the highest frequency. The highest frequency is 6 so the mode is 1

The median is the middle value when all the values have been arranged **in order of size.**

The middle value is the $\dfrac{\Sigma f + 1}{2}$ th value $= \dfrac{20 + 1}{2}$ th value = 10.5th value.

So the median lies between the 10th and 11th value.

To find the 10th and 11th terms do a quick running total of the frequencies.

Goals scored (x)	Frequency (f)	Running total
0	3	3
1	6	$3 + 6 = 9$
2	5	$9 + 5 = 14$
3	3	
4	2	
5	1	

The 10th and 11th terms are in this class so the median = 2 goals

Range — Largest number = 5
Smallest number = 0

The range is the difference between the largest and smallest numbers = 5 – 0 = 5

Apply 1

1 The shoe sizes of ten people are shown in the table.

Shoe size (x)	Frequency (f)	Frequency × shoe size (fx)
3	3	
4	3	
5	4	

a Copy and complete the table.

b Find the mean shoe size.

2 A dice is thrown 100 times. The scores are shown in the table.

Score (x)	Frequency (f)	Frequency × score (fx)
1	18	
2	19	
3	16	
4	12	
5	15	
6	20	

Copy and complete the table and find the mean score.

3 Ten people were asked to give the ages of their cars. Their answers are shown in the table.

Age of car (x)	Frequency (f)
1	2
2	3
3	4
4	1

Tom says that the mean age of the cars is 6 years.

a Find the mean.

b What do you think Tom did wrong?

4 **Get Real!**
Andrew has five people in his family. He wondered how many people there were in his friends' families. He asked 20 of his friends and put his results in a table.

Number in family	2	3	4	5	6	7	8
Frequency	2	3	8	4	2	0	1

From the data calculate:

a the median

b the mode

c the mean

d the range.

e Which average is the most appropriate to use? Why?

5 Get Real!

A group of 30 teenagers were learning archery and were allowed ten shots each at the target. The instructor counted the number of times they hit the target and recorded the following results.

2 6 2 0 2 3 8 7 2 5 6 1 7 4 8
8 6 1 0 3 2 2 1 3 8 3 2 6 7 5

a Copy the table and use the figures to complete it.

Number of hits	0	1	2	3	4	5	6	7	8
Frequency									

b Use the table to calculate:

i the mode **ii** the median **iii** the mean **iv** the range.

c Which average is the most appropriate to use with this data? Why?

6 Get Real!

Rachel is investigating the number of letters in words in her reading book. Her results are shown in the following table.

Number of letters in a word	Frequency
1	5
2	6
3	10
4	4
5	7
6	8
7	4
8	2

She says, 'The mode is 10, the median is 4 and the mean is 4.2'.
Is she right? Give a reason for your answer.
Which average do you think best represents the data? Why?

7 Get Real!

Mr Farrington, a technology teacher, has estimated that the average length of a box of bolts is 60 mm, to the nearest millimetre.
The class measured the lengths of 20 boxes and recorded their results in a table.

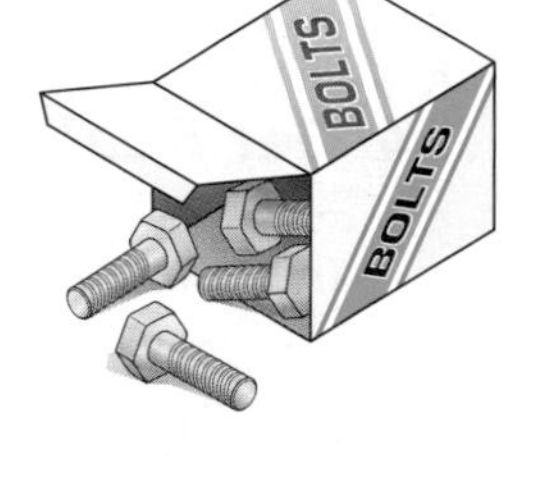

Length (mm)	Frequency	Frequency × length
58	3	
59	4	
60	2	
61	2	
62	8	
63	1	
Total		

Copy and complete the table.
Work out the mean length.
Was Mr Farrington correct in his estimation?

8 Get Real!

These marks were obtained by a class of 28 students in a science test. The maximum mark possible was 25.

20 17 21 15 16 15 12 14 15 19 21 17 20 22
16 19 21 13 22 15 14 18 20 13 18 16 15 19

By drawing a frequency table:

a Calculate the mean and work out the median and mode.

b Work out the range of the marks.

c What would you give as the pass mark? Explain your answer.

9 The mean of five integers is 4 and the median is 3. Find three different groups of numbers that fit this statement.

10 Get Real!

In five schools, the heights of a sample of the students were recorded and the mean calculated. The results are shown in the table.

School	Sample size	Mean height
A	150	1800 mm
B	200	1770 mm
C	400	1810 mm
D	350	1830 mm
E	400	1810 mm

a What is the mean height of the complete sample over all the schools?

b How many schools show mean heights below the overall mean?

Explore

A DIY shop sells a selection of large letters that can be used to design name boards for houses

The shopkeeper wants to buy 1000 letters but realises that some letters will be more popular than other letters

For example

needs two Es, two Os, two Ts and one of each of the other letters

How many of each letter should the shopkeeper buy?

Investigate further

Explore

- Draw a straight line on a sheet of paper
- Place a ruler some distance away from the line
- Let several people look at the line and the ruler
- Ask each one to estimate the length of the line
- Put your results into a frequency table and calculate the mean
- How does this mean compare with the actual length of the line?

Investigate further

Learn 2 Mean, mode and median for a grouped frequency table

Example:

The weights of 50 potatoes are measured to the nearest gram and shown in the table below.

From this table work out the mean, modal class and the class containing the median.

Weight in grams	Frequency
75–79	3
80–84	3
85–89	3
90–94	10
95–99	7
100–104	7
105–109	5
110–114	4
115–119	2
120–124	4
125–129	1
130–134	1

The weight of potatoes is continuous

The class 90–94 means any weight between 89.5 and 94.5
The lower bound is 89.5 and the upper bound is 94.5
The midpoint of the class is

$$\frac{89.5 + 94.5}{2} = 92$$

Classes can be written in different ways:
for example 124.5 up to 129.5
$124.5 \leqslant x < 129.5$
$124.5 < x \leqslant 129.5$
etc

Completing the table:

Weight in grams	Frequency (f)	Midpoint (x)	Frequency × midpoint (fx)
75–79	3	77	$77 \times 3 = 231$
80–84	3	82	$82 \times 3 = 246$
85–89	3	87	$87 \times 3 = 261$
90–94	10	92	$92 \times 10 = 920$
95–99	7	97	$97 \times 7 = 679$
100–104	7	102	$102 \times 7 = 714$
105–109	5	107	$107 \times 5 = 535$
110–114	4	112	$112 \times 4 = 448$
115–119	2	117	$117 \times 2 = 234$
120–124	4	122	$122 \times 4 = 488$
125–129	1	127	$127 \times 1 = 127$
130–134	1	132	$132 \times 1 = 132$
Totals	$\Sigma f = 50$		$\Sigma fx = 5015$

Start by finding the midpoint of each class and then continue as a frequency table

$\Sigma fx = 231 + 246 + 261 + 920 + 679 + 714 + 535 + 448 + 234 + 488 + 127 + 132$

The mean $= \dfrac{\Sigma fx}{\Sigma f} = \dfrac{5015}{50} = 100.3$ g

Remember that this is only an estimate as you have used the midpoints

The modal class is the class that occurs the most often.

Modal class = 90–94 g

In a frequency table, this is the class with the highest frequency

The median is the middle value when all the values have been arranged **in order of size.**

The middle value is the $\dfrac{\Sigma f + 1}{2}$ th value $= \dfrac{50 + 1}{2}$ th value = 25.5th value.

So the median lies between the 25th and 26th value.

To find the 25th and 26th terms do a quick running total of the frequencies.

Weight in grams	Frequency (f)	Running total
75–79	3	3
80–84	3	$3 + 3 = 6$
85–89	3	$6 + 3 = 9$
90–94	10	$9 + 10 = 19$
95–99	7	$19 + 7 = 26$
100–104	7	
105–109	5	
110–114	4	
115–119	2	
120–124	4	
125–129	1	
130–134	1	

The 25th and 26th values are in this class so the class containing the median = 95–99 g

Apply 2

1 Get Real!

The table shows the wages of 40 staff in a small company.

Wages (£)	Frequency
$50 \leqslant x < 100$	5
$100 \leqslant x < 150$	13
$150 \leqslant x < 200$	11
$200 \leqslant x < 250$	9
$250 \leqslant x < 300$	0
$300 \leqslant x < 350$	2

Find:

a the modal class

b the class that contains the median

c an estimate of the mean.

2 Get Real!

The scores obtained in a survey of reading ability are given in this table.

Reading scores (x)	Frequency
$0 \leqslant x < 5$	15
$5 \leqslant x < 10$	60
$10 \leqslant x < 15$	125
$15 \leqslant x < 20$	260
$20 \leqslant x < 25$	250
$25 \leqslant x < 30$	200
$30 \leqslant x < 35$	90

a What is the modal class?

b Calculate an estimated mean reading score.

3 Get Real!

The lengths, to the nearest millimetre, of a sample of a certain type of plant are given below.

51 51 58 54 59 60 52 52 55 49 51 53
55 60 58 57 51 57 56 50 53 58 59 57

a Calculate the mean length.

b Calculate an estimate of the mean length by grouping the data in class intervals of 47–49, 50–52, etc.

c Comment on your findings. What do you notice?

4 For each of these sets of data, work out the:

a mean **b** modal class **c** class containing the median.

i

Mark	Frequency
21 up to 31	1
31 up to 41	1
41 up to 51	3
51 up to 61	9
61 up to 71	8
71 up to 81	6
81 up to 91	2

ii

Daily takings ($)	Frequency
480–499	2
500–519	3
520–539	5
540–559	7
560–579	11
580–599	13
600–619	6
620–639	4
640–659	0
660–679	1

5 The table shows the weights of ten letters.

Weight (grams)	$0 \leqslant x < 20$	$20 \leqslant x < 40$	$40 \leqslant x < 60$	$60 \leqslant x < 80$	$80 \leqslant x < 100$
Number of letters	2	3	2	2	1

Calculate an estimate of the mean weight of a letter.

6 A survey was made of the amount of money spent at a supermarket by 20 shoppers. The table shows the results.

Amount spent, A (£)	$0 \leqslant A < 20$	$20 \leqslant A < 40$	$40 \leqslant A < 60$	$60 \leqslant A < 80$
Number of shoppers	1	7	8	4

Calculate an estimate of the mean amount of money spent by these shoppers.

Explore

- Sarah has collected some data from students
- She has found that the 'average' student is male, has brown eyes and hair, and is 165 cm tall
- Is this true for the students in your class?
- What else can you say about students in your class?

Investigate further

Statistical measures

ASSESS

The following exercise tests your understanding of this chapter, with the questions appearing in order of increasing difficulty.

1 The table shows the number of trainers sold in one day in a sports shop.

Size	5	$5\frac{1}{2}$	6	$6\frac{1}{2}$	7
Frequency	10	15	9	3	1

Find the mode, median, mean and range of this data.

2 J A Williams has a dental surgery. The information below shows the waiting times for patients during one day.
Find the class intervals that contain the mode and median and calculate an estimate of the mean waiting time at J A Williams' surgery.

Waiting time, x (minutes)	$0 \leqslant x < 3$	$3 \leqslant x < 6$	$6 \leqslant x < 9$	$9 \leqslant x < 12$	$12 \leqslant x < 15$	$15 \leqslant x < 18$
Number of patients	9	15	12	8	4	2

3 Debbie asks some of the students in her class how many brothers and sisters they have. She puts the information in a table.

Number of brothers \ **Number of sisters**	**0**	**1**	**2**	**3**	**4**	**5**
5		1				
4	0	2				
3	2	2	1			
2	4	3	3	1		
1	5	3	2	1	1	
0	5	6	2	1		1

a How many students have no sisters?

b How many students have only one brother?

c How many students have equal numbers of brothers and sisters?

d How many students did Debbie survey altogether?

e What is the modal number of sisters?

f Calculate the mean number of brothers.

4 The information below shows the speeds of 60 white vans passing a speed camera. Find the class intervals that contain the mode and median and calculate an estimate of the mean.

Speed, x (mph)	$30 \leqslant x < 40$	$40 \leqslant x < 50$	$50 \leqslant x < 60$	$60 \leqslant x < 70$	$70 \leqslant x < 80$	$80 \leqslant x < 90$	$90 \leqslant x < 100$
Frequency	2	10	18	16	11	2	1

5 An office manager monitored the time members of staff took on 'private' telephone calls during working hours. Calculate an estimate of the mean length of the telephone calls, giving your answer to the nearest minute.

Time (nearest min)	3–5	6–8	9–12	13–16	17–20	21–25	26–30	31–40
Frequency	67	43	28	13	7	4	3	2

2 Integers

OBJECTIVES

C **Examiners would normally expect students who get a C grade to be able to:**

Recognise prime numbers

Find the reciprocal of a number

Find the least common multiple (LCM) of two simple numbers

Find the highest common factor (HCF) of two simple numbers

Write a number as a product of prime factors

B **Examiners would normally expect students who get a B grade also to be able to:**

Find the least common multiple (LCM) of two or more numbers

Find the highest common factor (HCF) of two or more numbers

What you should already know ...

- Understand the four rules of number
- Understand place value
- Understand the inequality signs <, >, ⩽ and ⩾
- Know the meaning of 'sum' and 'product'
- Apply the four rules to positive and negative numbers
- Change a decimal into a fraction
- Change a mixed number into a top-heavy fraction

VOCABULARY

Integer – any positive or negative whole number or zero, for example, −2, −1, 0, 1, 2 ...

Factor – a natural number which divides exactly into another number (no remainder), for example, the factors of 12 are 1, 2, 3, 4, 6 and 12

Multiple – the multiples of a number are the products of its multiplication table, for example, the multiples of 3 are 3, 6, 9, 12, 15 ...

Least common multiple (LCM) – the lowest multiple which is common to two or more numbers, for example,

the multiples of 3 are 3, 6, 9, 12, 15, 18, 24, 27, 30, 33, 36 ...
the multiples of 4 are 4, 8, 12, 16, 20, 24, 28, 32, 36 ...
the common multiples are 12, 24, 36 ...
the least common multiple is 12

Common factor – factors that are in common for two or more numbers, for example,

the factors of 6 are 1, 2, 3, 6
the factors of 9 are 1, 3, 9
the common factors are 1 and 3

Highest common factor (HCF) – the highest factor that two or more numbers have in common, for example,

the factors of 16 are 1, 2, 4, 8, 16
the factors of 24 are 1, 2, 3, 4, 6, 8, 12, 24
the common factors are 1, 2, 4, 8
the highest common factor is 8

Prime number – a natural number with exactly two factors, for example, 2 (factors are 1 and 2), 3 (factors are 1 and 3), 5 (factors are 1 and 5), 7, 11, 13, 17, 23, ... 59 ...

Index notation – when a product such as $2 \times 2 \times 2 \times 2$ is written as 2^4, the 4 is the index (plural **indices**)

Prime factor decomposition – writing a number as the product of its prime factors, for example, $12 = 2^2 \times 3$

Reciprocal – any number multiplied by its reciprocal equals one; one divided by a number will give its reciprocal, for example, the reciprocal of 3 is $\frac{1}{3}$ because $3 \times \frac{1}{3} = 1$

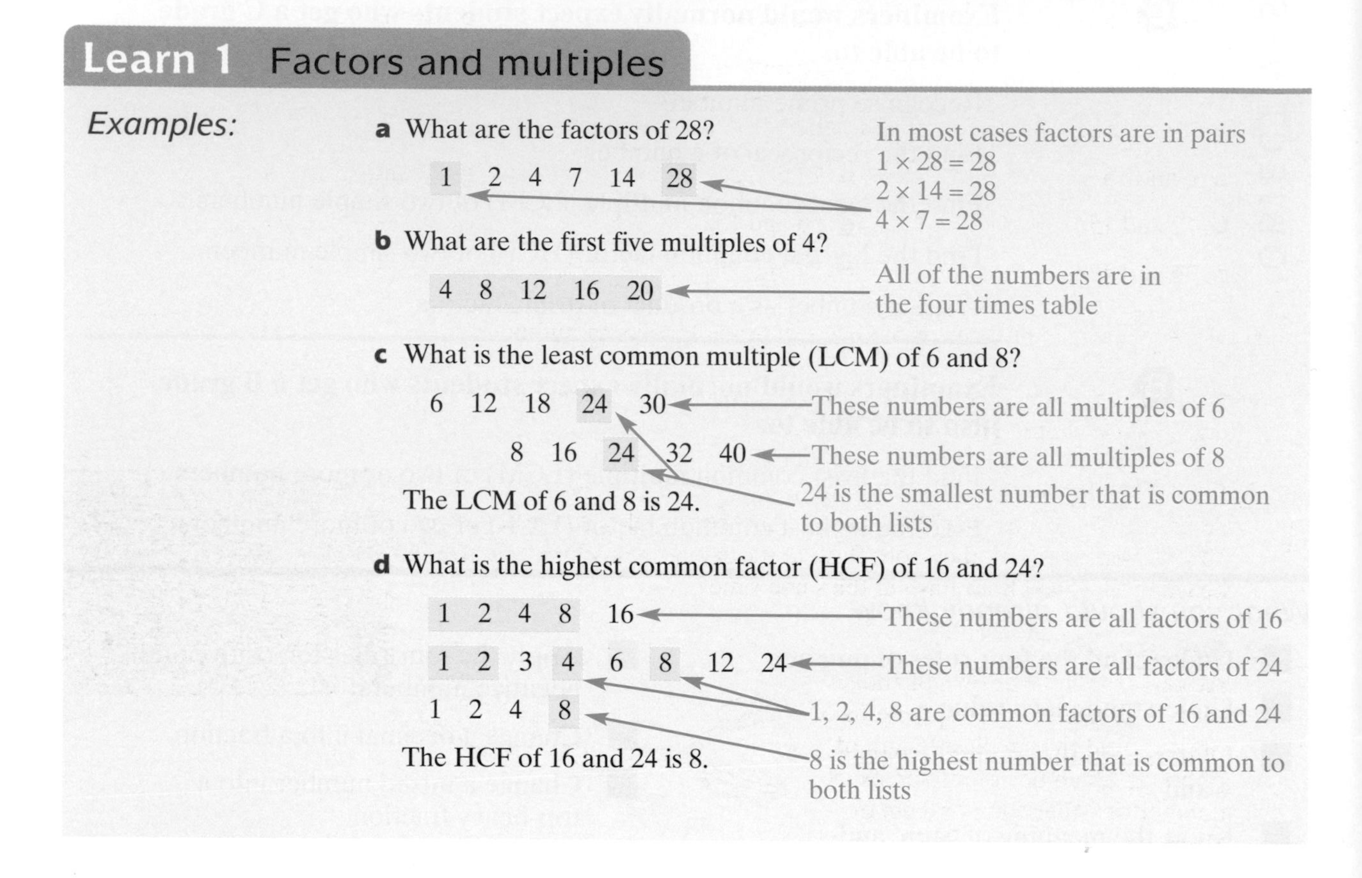

Learn 1 Factors and multiples

Examples:

a What are the factors of 28?

1 2 4 7 14 28

In most cases factors are in pairs
$1 \times 28 = 28$
$2 \times 14 = 28$
$4 \times 7 = 28$

b What are the first five multiples of 4?

4 8 12 16 20

All of the numbers are in the four times table

c What is the least common multiple (LCM) of 6 and 8?

6 12 18 24 30 – These numbers are all multiples of 6

8 16 24 32 40 – These numbers are all multiples of 8

The LCM of 6 and 8 is 24.

24 is the smallest number that is common to both lists

d What is the highest common factor (HCF) of 16 and 24?

1 2 4 8 16 – These numbers are all factors of 16

1 2 3 4 6 8 12 24 – These numbers are all factors of 24

1 2 4 8 – 1, 2, 4, 8 are common factors of 16 and 24

The HCF of 16 and 24 is 8.

8 is the highest number that is common to both lists

Apply 1

1 Write down all the factors of:

a 15 **d** 10 **g** 32

b 64 **e** 40 **h** 72

c 48 **f** 36 **i** 84

2 Write down the first five multiples of:

a 2 **d** 6 **g** 11

b 5 **e** 9 **h** 8

c 7 **f** 12 **i** 13

3 Find the LCM of these sets of numbers.

a 6 and 15
b 12 and 6
c 5 and 7
d 3 and 8
e 4 and 6
f 4, 10 and 12
g 3, 5 and 6
h 6, 8 and 32

4 What are the common factors of:

a 6 and 15
b 9 and 48
c 4 and 64
d 10 and 16
e 25 and 40
f 24 and 36?

5 Find the HCF of these pairs of numbers.

a 6 and 15
b 12 and 15
c 32 and 48
d 24 and 36
e 56 and 152
f 84 and 70
g 27 and 36

6 The HCF of two numbers is 5. Give five possible pairs of numbers.

7 Write down all the factors of 20 and 24. Hence find the common factors and write down the HCF of 20 and 24.

8 Get Real!
A lighthouse flashes every 56 seconds. Another lighthouse flashes every 40 seconds. At 9 p.m. they both flash at the same time. What time will it be when they next both flash at the same time?

9 Get Real!
Alison is making her own birthday cards. She needs to cut up lengths of ribbon. Find the smallest length of ribbon that can be cut into an exact number of either 5 cm or 8 cm or 12 cm lengths.

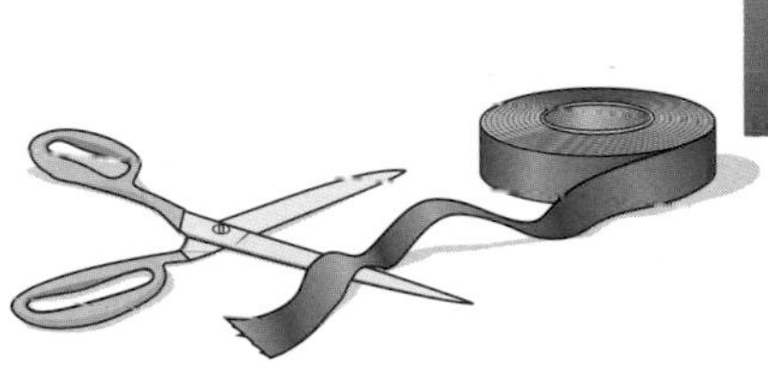

10 Get Real!
One political party holds its annual conference at Eastbourne every four years. Another holds its annual conference there every six years. They both held their conference in Eastbourne in 2006. When will they next be there in the same year?

11 Get Real!
Nick visits his grandparents every two days. Briony visits them every three days. If they both visit on a Monday, on which day do they next both visit?

12 Richard always goes to his local café on a Saturday afternoon. Zoë works there on every third afternoon. She serves Richard one Saturday afternoon. How many weeks will it be before she will serve him again?

13 A farmer has 24 cows and 30 sheep. She decides to divide these equally between her sons.

a What is the greatest number of sons she could have?

She increases her livestock to 40 cows and 72 sheep.

b What is the greatest number of sons she could now have?

14 Emily sets her alarm clock using the town hall clock. The problem is that the town hall clock gains one hour a day. How long will it be until both clocks next show the same time?

Explore

- Write down all the numbers between 1 and 30
- Work out the number of factors for each number
- Can you work out a rule for numbers that have:
  - **a** two factors only
  - **b** an odd number of factors?

Investigate further

Explore

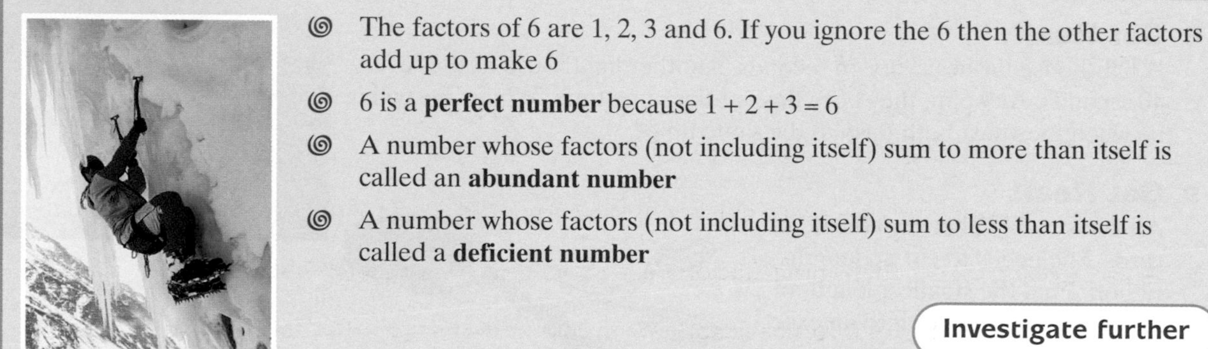

- The factors of 6 are 1, 2, 3 and 6. If you ignore the 6 then the other factors add up to make 6
- 6 is a **perfect number** because $1 + 2 + 3 = 6$
- A number whose factors (not including itself) sum to more than itself is called an **abundant number**
- A number whose factors (not including itself) sum to less than itself is called a **deficient number**

Investigate further

Learn 2 Prime numbers and prime factor decomposition

Example: Find the prime factors of 40.

Start by finding the smallest prime number that divides into 40.
Continue dividing by successive prime numbers until the answer becomes 1.

2	40
2	20
2	10
5	5
	1

Prime numbers are numbers with exactly two factors, for example, 2, 3, 5, 7, 11, 13, ...

The prime factors of 40 are 2, 2, 2, 5.
40 written as a product of prime factors is $2 \times 2 \times 2 \times 5$.
This can be written as $2^3 \times 5$.

This is called 'index notation'. The index tells you how many times the factor 2 occurs

Apply 2

1 Sam describes a number in the following way:
'It is a prime number. It is a factor of 21. It is not a factor of 12.'
What number is she describing?

2 Chris describes a number in the following way:
'It is below 100. It is a multiple of 6. It is also a multiple of 4.
The sum of its digits is a prime number.'
What number is he describing?

3 Make up two number descriptions of your own.

4 Write each of the following numbers as a product of prime factors.

a 20	**c** 36	**e** 90	**g** 63	**i** 84
b 18	**d** 66	**f** 100	**h** 48	**j** 96

5 Express each number as a product of its prime factors.

a 24 **b** 72 **c** 45

6 Express each number as a product of its prime factors.
Write your answers using index notation.

a 220	**c** 136	**e** 720	**g** 390	**i** 624
b 144	**d** 300	**f** 480	**h** 450	**j** 216

7 Clare says that one must be a prime number. Is she correct? Explain your answer.

8 Write 2420 as a product of its prime factors. Write your answer using index notation.

9 Write 9240 as a product of its prime factors. Write your answer using index notation.

10 Write 8820 as a product of its prime factors. Write your answer using index notation.

11 If $1080 = 2^x \times 3^y \times 5^z$, what are the values of x, y and z?

Explore

You will need a 100 square

- Cross out the number 1
- Put a circle round the number 2 and then cross out all of the other multiples of 2
- Put a circle round the next number after 2 that has not been crossed out
- Cross out all of the other multiples of that number
- Put a circle round the next number not crossed out and cross out every multiple of that number
- Continue until you run out of numbers in the 100 square

What do you notice about the numbers that are left?

Investigate further

Learn 3 Reciprocals

Examples: Find the reciprocal of: **a** 5 **b** $\frac{1}{4}$ **c** 0.3 **d** $2\frac{1}{2}$

a The reciprocal of 5 is $\frac{1}{5}$ ← You can write 5 as $\frac{5}{1}$. The reciprocal of $\frac{5}{1}$ is $\frac{1}{5}$

b The reciprocal of $\frac{1}{4}$ is $\frac{4}{1} = 4$ ← It is better to write $\frac{4}{1}$ as 4

c The reciprocal of 0.3 is the same as the reciprocal of $\frac{3}{10}$ ← Write 0.3 as a fraction

The reciprocal of $\frac{3}{10}$ is $\frac{10}{3} = 3\frac{1}{3}$ ← It is better to write $\frac{10}{3}$ as $3\frac{1}{3}$

d The reciprocal of $2\frac{1}{2}$ is the same as the reciprocal of $\frac{5}{2}$ ← Write $2\frac{1}{2}$ as $\frac{5}{2}$

The reciprocal of $\frac{5}{2}$ is $\frac{2}{5}$

The reciprocal of 0 is not defined and your calculator will give an error for $\frac{1}{0}$

Apply 3

1 Write down the reciprocal of:

a 4 **b** 6 **c** 8 **d** 10 **e** 7 **f** 0.25

2 Find the reciprocal of:

a $\frac{1}{2}$ **b** $\frac{1}{5}$ **c** $\frac{1}{7}$ **d** $\frac{1}{8}$ **e** $\frac{1}{12}$ **f** 0.8

3 Find the reciprocal of:

a $\frac{2}{7}$ **b** $\frac{3}{5}$ **c** $\frac{2}{3}$ **d** $\frac{5}{6}$ **e** $\frac{3}{11}$

4 Write down the reciprocal of:

a 0.3 **b** 0.7 **c** 0.4 **d** 0.125 **e** $0.\dot{3}$

5 Find the reciprocal of:

a $2\frac{1}{4}$ **b** $3\frac{1}{2}$ **c** $1\frac{3}{4}$ **d** 1.25 **e** 3.6 **f** $1.\dot{6}$

<u>6</u> Find the reciprocals of the numbers 2 to 12, as decimals. If they are not exact, write them as recurring decimals. Which of the numbers have reciprocals that:

a are exact decimals

b have one recurring figure

c have two recurring figures?

Explore

- Write down the reciprocals of 2, 1, $\frac{1}{2}$, $\frac{1}{4}$, $\frac{1}{8}$ ...
- Continue the pattern
- What do you notice?

Investigate further

Integers

ASSESS

The following exercise tests your understanding of this chapter, with the questions appearing in order of increasing difficulty.

1 At a party it was discovered that Siobhan, Gareth, Nathan and Ulrika had birthdays on the 6th, 15th, 27th and 30th of the month.
Sven joined the group and it was discovered that his birthday was a factor of everyone else's.
If Sven was not born on the 1st, on what day of the month was Sven born?

2 Find the prime factors of:

a 420

b 13 475

3 Amrit says that the value of the expression $n^2 + n + 41$, where $n = 0, 1, 2, 3, \ldots$ always gives prime numbers.

a Show this is true for $n = 0$ to 5.

b Without any calculation, name one value of n that disproves Amrit's theory.

4 Find the reciprocals of:

a 5

b -8

c $\frac{5}{8}$

d -0.2

e $0.\dot{2}$

5 **a** What is the only number that is the same as its reciprocal?

b What is the only number that has no reciprocal?
Explain your answer.

6 Sam says he knows a different way of finding the HCF of two numbers. He says:

1. Write down the numbers side by side.
2. Cross out the largest number and write underneath it the difference between the two numbers.
3. Repeat step 2 until the two numbers left are the same.
4. The remaining number is the HCF of the two original numbers.

Try Sam's method with the following numbers:

a 16 and 36

b 60 and 225

c 456 and 640

d Does the method work with three numbers? Investigate.

Try some real past exam questions to test your knowledge:

7 Tom, Sam and Matt are counting drum beats.

Tom hits a snare drum every 2 beats.
Sam hits a kettle drum every 5 beats.
Matt hits a bass drum every 8 beats.

Tom, Sam and Matt start by hitting their drums at the same time.
How many beats is it before Tom, Sam and Matt **next** hit their drums at the **same** time?

Spec A, Higher Paper 1, June 04

8 **a** Express 144 as the product of its prime factors.
Write your answer in index form.

b Find the highest common factor (HCF) of 60 and 144.

Spec B, Mod 3 Intermediate, June 03

3 Rounding

OBJECTIVES

D **Examiners would normally expect students who get a D grade to be able to:**

Estimate answers to calculations such as $\frac{22.6 \times 18.7}{5.2}$

C **Examiners would normally expect students who get a C grade also to be able to:**

Estimate answers to calculations such as $\frac{22.6 \times 18.7}{0.52}$

Find minimum and maximum values

B **Examiners would normally expect students who get a B grade also to be able to:**

Round to a given number of significant figures

What you should already know ...

- Round numbers to the nearest 1000, nearest 100, nearest 10, nearest integer, significant figures, decimal places ...
- Estimate calculations involving decimals
- Estimate square roots

VOCABULARY

Round – give an approximate value of a number. Numbers can be rounded to the nearest 1000, nearest 100, nearest 10, nearest integer, significant figures, decimal places ... etc

Significant figures – the digits in a number; the closer a digit is to the beginning of a number then the more important or significant it is; for example, in the number 23.657, 2 is the most significant digit and is worth 20, 7 is the least significant digit and is worth $\frac{7}{1000}$; the number 23.657 has 5 significant digits

Decimal places – the digits to the right of a decimal point in a number, for example, in the number 23.657, the number 6 is the first decimal place (worth $\frac{6}{10}$), the number 5 is the second decimal place (worth $\frac{5}{100}$) and 7 is the third decimal place (worth $\frac{7}{1000}$); the number 23.657 has 3 decimal places

Upper bound – this is the maximum possible value of a measurement, for example, if a length is measured as 37 cm correct to the nearest centimetre, the upper bound of the length is 37.5 cm

Lower bound – this is the minimum possible value of a measurement, for example, if a length is measured as 37 cm correct to the nearest centimetre, the lower bound of the length is 36.5 cm

Degree of accuracy – the accuracy to which a measurement or a number is given, for example, to the nearest 1000, nearest 100, nearest 10, nearest integer, 2 significant figures, 3 decimal places

Estimate – find an approximate value of a calculation; this is usually found by rounding all of the numbers to one significant figure, for example, $\frac{20.4 \times 4.3}{5.2}$ is approximately $\frac{20 \times 4}{5}$ where each number is rounded to 1 s.f.; the answer can be worked out in your head to give 16

Learn 1 Estimating

Example: Estimate the answer to 3.86×2.14

Round all the numbers to one significant figure, then work out the approximate answer in your head.

$3.86 \times 2.14 \approx 4 \times 2 = 8$

This curly equals sign means 'is approximately equal to'

It is easy to make mistakes when using decimals, so it is a good idea to estimate to find the approximate size of the answer so that you can see if you are right.

Apply 1

1 For each question, decide which is the best estimate.

		Estimate A	Estimate B	Estimate C
a	2.89×9.4	2.7	18	27
b	1.2×29.4	3	30	300
c	$9.17 \div 3.2$	3	4	18
d	48.5×9.8	5	50	500
e	$4.2 \div 1.9$	1	2	3
f	22.4×6.1	12	120	180
g	$7.8 \div 1.2$	8	78	80
h	$2.1 \times 3.1 \div 4.2$	1	1.5	2
i	$20.9 \div 6.9 \times 4.1$	10	11	12

2 Estimate the answers to these calculations by rounding to one significant figure. You may wish to use a calculator to check your answers.

a $2.9 + 3.2$

b $7.9 \div 2.2$

c $67.8 + 22.1$

d $\dfrac{20.4 \times 7.7}{5.2}$

e 0.2×5.4

f $4.3 - 3.7$

g 5.3×8.2

h $\dfrac{20.4 \times 7.7}{0.52}$

i $\dfrac{75.5 \times 2.7}{0.12}$

j $\dfrac{28.5 + 53}{64.1 - 53.7}$

k $\dfrac{102.4 + 8.7}{0.22}$

l $\dfrac{102.4 \times 8.7}{0.22}$

m $21.3(7.56 + 3.89)$

n $21.3(7.56 - 3.89)$

HINT Be careful when dividing by numbers less than one.

3 For each pair of calculations, estimate the answers to help decide which will have the bigger answer.
Use a calculator to check your answers

a 5.2×1.8 or 3.1×2.95

b $28.4 \div 5.9$ or 2.03×3.78

c $9.723 + 4.28$ or $39.4 \div 2.04$

d 39.5×21.3 or 81.3×7.8

4 The square root of 10 lies between the square root of 9 and the square root of 16. The square root of 9 is 3 and the square root of 16 is 4, so the answer lies between 3 and 4.

Use this method to estimate the square root of:

a 12 **b** 20 **c** 50 **d** 3 **e** 1000

5 Estimate 5.92×3.82 by rounding to the nearest whole number. Explain why the answer is an over-estimate of the exact answer.

6 Sam says, 'I estimated the answer to $16.7 - 8.6$ as 8 by rounding up both numbers, so the answer is an over-estimate.' Show that Sam is not correct.

7 Ali says, '35 divided by 5 is 7, so 35 divided by 0.5 is 0.7'. Is Ali right? Give a reason for your answer.

8 Give one example that shows that dividing can make something smaller and another to show that dividing can make something bigger.

9 Hannah says:

- When I rounded the numbers in a calculation I got $\frac{110}{0.2}$
- Then I multiplied the top and the bottom both by 10 to give $\frac{1100}{2}$
- So the answer is 550.

Is Hannah correct? Give a reason for your answer.

10 a Find five numbers whose square roots are between 7 and 8.

b Find two consecutive whole numbers to complete this statement: 'The square root of 60 is between ... and ... '

11 Get Real!

Estimate the total cost of three books costing £3.99, £5.25 and £10.80

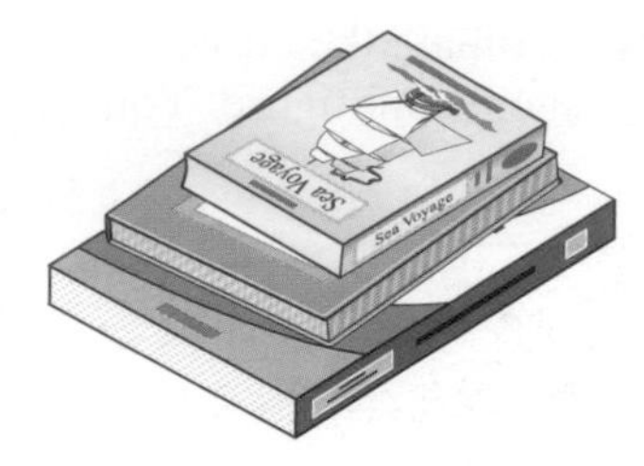

12 Get Real!

Estimate the length of fencing needed for this field.

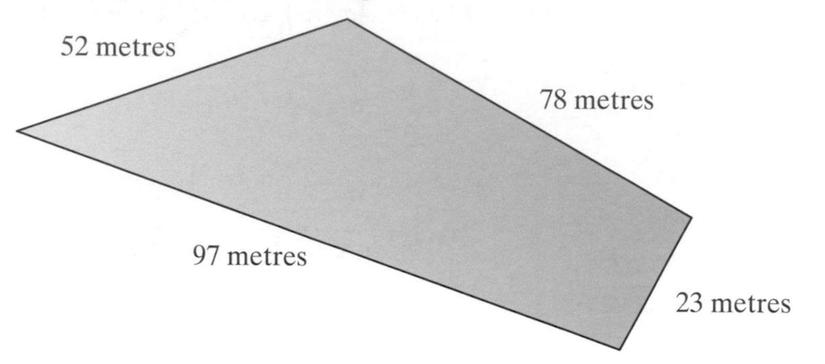

13 Get Real!

Anne's car goes 6.2 miles on every litre of petrol. Estimate how far she can drive if her fuel tank has 24.5 litres in it.

14 Get Real!

A group of 18 people wins £389 540 on the lottery. Estimate how much each person will get when the money is shared out equally.

15 Get Real!

It's Harry's birthday! He asks his mum for a cake like a football pitch. She makes a cake that is 29 cm wide and 38 cm long.

a She wants to put a ribbon round the cake. She can buy ribbon in various lengths: 1 m, 1.5 m, 2 m, 2.5 m or 3 m.
Estimate the perimeter of the cake and say which length ribbon she should buy.

b She can buy ready-made green icing for the top of the cake. The icing comes in packs to cover 1000 cm^2.
Estimate the area of the top of the cake and decide whether one pack will be enough.

Explore

- You know that $20 \times 30 = 600$
  So the answers to all these calculations will be close to 600, as the numbers are close to 20×30:
  a 19.4×28.7 **b** 23.4×30.2 **c** 18.8×29.6 **d** 21.2×30.4 **e** 29.8×20.4 **f** 31.4×18.7 **g** 23.4×33.4 **h** 22.3×29.8
- Can you decide which answers will be less than 600, and which will be more than 600?
- Check your predictions with a calculator
- When can you be sure an estimate is lower than the actual answer?
- When can you be sure an estimate is higher than the actual answer?

Investigate further

Explore

- You know that $8 \div 2 = 4$
  Use a calculator to divide numbers close to 8 by numbers close to 2
  For example, you might try $8.2 \div 1.9$ and $7.8 \div 1.8$
- Can you decide which answers will be less than 4, and which will be more than 4?
- Check your predictions with a calculator
- When can you be sure an estimate is lower than the actual answer?
- When can you be sure an estimate is higher than the actual answer?

Investigate further

Learn 2 Finding minimum and maximum values

Example: The length of a table is measured as 60 cm, correct to the nearest centimetre. What are the minimum and maximum possible lengths of the table?

All these numbers are nearer to 60 than they are to 59 or to 61

60.5 rounds up to 61

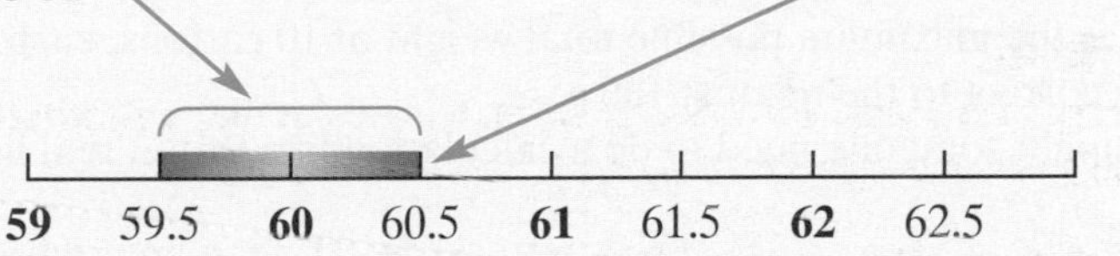

Any length given to the nearest centimetre could be up to half a centimetre smaller or larger than the given value.

A length given to the nearest centimetre as 60 cm could be anything between 59.5 cm and 60.5 cm. The minimum and maximum possible lengths are 59.5 cm and 60.5 cm or $59.5 \leqslant x < 60.5$

The length cannot actually be 60.5 cm, as this measurement rounds up to 61 cm – but it can be as close to 60.5 cm as you like, so 60.5 cm is the top limit (or **upper bound**) of the length

Apply 2

1 Each of these quantities is rounded to the nearest whole number of units. Write down the minimum and maximum possible size of each quantity.

a 54 cm
b 5 kg
c 26 m
d 17 mℓ
e £45
f 175 g

2 Jane says,
'If a length is 78 cm to the nearest centimetre, then the maximum possible length is 78.49 cm.'
Is Jane right? Explain your answer.

3 The volume of water in a tank is given as 1500 litres.

a Decide if the volume has been rounded to the nearest litre, nearest 10 litres or nearest 100 litres if the minimum possible volume is

i 1450 ℓ **ii** 1499.5 ℓ **iii** 1495 ℓ

b If the actual volume is V litres, complete this statement in each case: ... $\leqslant V <$ ....

c Explain why there is a 'less than or equal to' sign before the V but a 'less than' sign after the V.

d How do you know that a volume written as 1500 litres has not been measured to the nearest tenth of a litre?

4 Get Real!
ChocoBars should weigh 40 grams with a tolerance of 5% either way.
If the bars weigh 40 grams correct to the nearest 10 grams, will they be within the tolerance? Show how you worked out your answer.
Why do you think that manufacturers have a 'tolerance' in the sizes of their products?

5 Get Real!
What is the maximum possible total weight of 10 cartons, each weighing 1.4 kg correct to the nearest 100 g?
Why might someone need to do a calculation like this in real life?

Explore

- Write a note to explain to someone else how to find the maximum and minimum possible ages of a person whose age is given as a whole number of years, for example, 8 years
- Write a note to explain to someone else how to find the maximum and minimum possible amounts of money when the quantity is given to the nearest pound, for example £18

Investigate further

Rounding

ASSESS

The following exercise tests your understanding of this chapter, with the questions appearing in order of increasing difficulty.

1 The average person's heart beats about once a second. Estimate how many times it beats during a year.

2 The (movement) energy of an athlete of mass 43 kg running at a velocity of 9.7 m/s can be found by working out $43 \times 9.7 \times 9.7 \div 2$. Use appropriate approximations for 43 and 9.7 and estimate the athlete's energy.

3 Ngugi lives on the equator, which is a circle of diameter 12 756 km. George lives in the UK on a circle of latitude with diameter 7854 km. To calculate the distance each boy moves in one day due to the Earth's rotation we multiply each diameter by 3.14

Write the diameters given above correct to **the nearest thousand** and the multiplier to the nearest whole unit, and hence estimate how much further Ngugi travels than George in one day.

4 Estimate the value of **a** $\dfrac{24.8 \times 3.2}{0.54}$ **b** $\dfrac{29.9 + \sqrt{0.918}}{(16.2 - 6.15)^2}$

5 **a** The length of a rectangle is given as 27 m correct to the nearest m. Write down the minimum and maximum possible lengths it could be.

b A different length is given as 5.0 cm correct to nearest mm. Write down the minimum and maximum possible lengths it could be.

6 Copy and complete the table below.

Starting number	To two significant figures	To three significant figures	To one decimal place	To two decimal places
186.487		186		
3.14159	3.1			
0.51627			0.5	
0.0080990				
8				

7 Estimate the value of $25 \times 4.2 + \dfrac{5(93.1 - 32 \times 2.4)}{11.3 + 9.1}$

Try some real past exam questions to test your knowledge:

8 **a** Work out 600×0.3

b Work out $600 \div 0.3$

c You are told that $432 \times 21 = 9072$
Write down the value of $9072 \div 2.1$

d Find an approximate value of $\dfrac{2987}{21 \times 49}$

You **must** show all your working.

Spec A, Int Paper 1, Nov 03

9 This is a true statement.

Write down:

a the minimum age that Kylie could be

b the maximum age that Kylie could be.

Spec B, Int Paper 1, Mar 04

4 Use of symbols

OBJECTIVES

D **Examiners would normally expect students who get a D grade to be able to:**

Multiply out expressions with brackets such as $3(x + 2)$ or $5(x - 2)$

Factorise expressions such as $6a + 8$ and $x^2 - 3x$

C **Examiners would normally expect students who get a C grade also to be able to:**

Expand (and simplify) harder expressions such as $x(x^2 - 5)$ and $3(x + 2) - 5(2x - 1)$

B **Examiners would normally expect students who get a B grade also to be able to:**

Expand (and simplify) quadratic expressions such as $(x + 4)(x - 2)$, $(2x + y)(3x - 2y)$ and $(x + 2)^2$

Factorise quadratic expressions such as $4x^2 + 6xy$ and $x^2 - 8x - 16$

Simplify rational expressions such as $\dfrac{2(x + 1)^2}{x + 1}$

A **Examiners would normally expect students who get an A grade also to be able to:**

Factorise harder quadratic expressions such as $a^2 - 16b^2$ and $5x^2 + 13x - 6$

A* **Examiners would normally expect students who get an A* grade also to be able to:**

Factorise harder quadratic expressions such as $x^2 - 10x + a$, writing them in the form $(x + b)^2 + c$

Simplify harder rational expressions such as $\dfrac{x^2 + 2x}{x^2 - 4}$

What you should already know ...

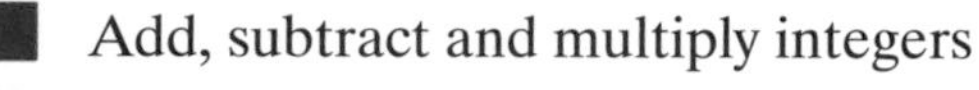

- Add, subtract and multiply integers
- Multiply a two-digit number by a single-digit number
- Simplify expressions with more than one variable such as $2a + 5b + a - 2b$

VOCABULARY

Variable – a symbol representing a quantity that can take different values such as x, y or z

Term – a number, variable or the product of a number and a variable(s) such as 3, x or $3x$

Algebraic expression – a collection of terms separated by + and – signs such as $x + 2y$ or $a^2 + 2ab + b^2$

Product – the result of multiplying together two (or more) numbers, variables, terms or expressions

Collect like terms – to group together terms of the same variable, for example, $2x + 4x + 3y = 6x + 3y$

Consecutive – in sequence

Simplify – to make simpler by collecting like terms

Expand – to remove brackets to create an equivalent expression (expanding is the opposite of factorising)

Factorise – to include brackets by taking common factors (factorising is the opposite of expanding)

Linear expression – a combination of terms where the highest power of the variable is 1

Linear expressions	Non-linear expressions
x	x^2
$x + 2$	$\frac{1}{x}$
$3x + 2$	$3x^2 + 2$
$3x + 4y$	$(x + 1)(x + 2)$
$2a + 3b + 4c + \ldots$	x^3

Quadratic expression – an expression containing terms where the highest power of the variable is 2

Quadratic expressions	Non-quadratic expressions
x^2	x
$x^2 + 2$	$2x$
$3x^2 + 2$	$\frac{1}{x}$
$4 + 4y^2$	$3x^2 + 5x^3$
$(x + 1)(x + 2)$	$x(x + 1)(x + 2)$

Rational expression – a fraction, for example,

$$\frac{x^2 - 9}{x + 3}$$

Coefficient – the number (with its sign) in front of the letter representing the unknown, for example:

$4p - 5$ — 4 is the coefficient of p

$2 - 3p^2$ — −3 is the coefficient of p^2

Equation – a statement showing that two expressions are equal, for example, $2y - 7 = 15$

Formula – an equation showing the relationship between two or more variables, for example, $E = mc^2$

Identity – two expressions linked by the ≡ sign are true for all values of the variable, for example, $3x + 3 \equiv 3(x + 1)$

Learn 1 Expanding brackets

Examples:

a Expand $5(2y - 1)$.

	$2y$	-1
5	$10y$	-5

-5 ← $5 \times -1 = -5$

$5(2y - 1) = 10y - 5$ ← Write $10y - 5$ not $10y + -5$

$10y - 5$ is a linear expression because the highest power of the variable (y) is 1.

b Expand $2p(p^2 - 5)$.

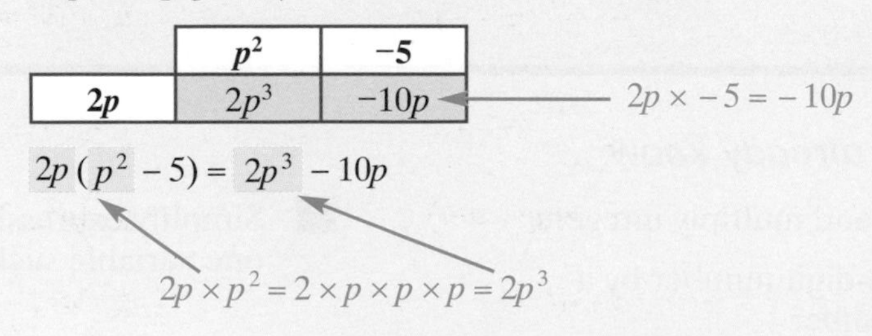

c Expand and simplify $3(x-2)-5(2x-1)$.

Treat this as two separate algebraic expressions, $3(x-2)$ and $-5(2x-1)$ and merge the answers together at the end

Step 1 Expand $3(x-2)$.

	x	-2
3	$3x$	-6

$3(x-2) = 3x - 6$

Step 2 Expand $-5(2x-1)$.

	$2x$	-1
-5	$-10x$	$+5$

$-5 \times -1 = +5$

$-5(2x-1) = -10x + 5$

Step 3 Merge the two answers by collecting like terms:

$3(x-2) - 5(2x-1) = \underline{3x} \; (-6) \; \underline{-10x} \; (+5)$
$= -7x - 1$

Underlining or circling like terms (including their sign) helps when collecting them:
$3x - 10x = -7x$ and $-6 + 5 = -1$

Apply 1

1 Multiply these out:

a $4(x+2)$
b $6(y+3)$
c $3(3+2y)$
d $7(g-4)$
e $5(2d-3)$
f $8(5f-1)$
g $\frac{1}{2}(4b+6)$
h $\frac{1}{4}(16f-4)$
i $\dfrac{35h+10}{5}$
j $\frac{3}{4}(12a-28)$
k $-4(q+2)$
l $-5(2m-3)$
m $-7(-4a-1)$

2 Sam thinks the answer to $5(3x-2)$ is $15x-2$. Hannah says he is wrong. Who is correct and why?

3 Expand:

a $p(p+3)$
b $b(b-4)$
c $2a(a+5)$
d $x(x^3+3)$
e $4d(1-2d)$
f $\dfrac{m}{3}(m-9)$
g $\dfrac{2h}{3}(3h+6)$
h $w(ig+am)$
i $t(t^2-1)$
j $x^2(4+x^3)$
<u>k</u> $y^2(y^5+4y^3)$
<u>l</u> $2y^2(4y-2y^3)$
<u>m</u> $4p^2q(3pq+2q)$
<u>n</u> $2ab\left(\dfrac{4}{a}+\dfrac{1}{2b}\right)$

4 The answer is $12y - 36$.
Write down five questions of the form $a(by + c)$ with this answer.
(a, b, and c are integers – positive or negative numbers.)

5 Expand and simplify:

a $4(x + 2) + 2(x + 3)$
b $2(p + 3) + 3(2p - 4)$
c $6x - (2 - x)$
d $3(m - 1) - 4(m - 2)$
e $\frac{1}{2}(6y - 3) + \frac{1}{4}(12 - 4y)$
f $4x - (x + 2)$
g $4(t - 2) - 2(t + 1)$

6 Simplify:

a $4(2m - 3) + 3(m - 6)$
b $3(2a - 1) - 3(4 - a)$
c $5(6x - 3) + 2(3 - 2x)$
d $4(2y - 1) - 4(3y - 5)$
e $5(2t - 4) - 7(2 - 3t)$
f $\frac{x}{2}(3x - 4) + \frac{x}{4}(2x - 8)$
g $2x(2x + 3) - 5x(3x + 4)$

7 Josie thinks the answer to $3(2m - 1) - 4(m - 2)$ is $2m - 11$.
Explain what she has done wrong.

8 Find the integers a and b if $4(x - a) - b(x - 1) = 2x - 14$.

9

Expand and simplify:

a 2A + 3C **b** C − 2D **c** 2A − B

d Work out a combination of two cards that gives the answer 13.

10 Get Real!
The diagram shows an L-shaped floor with dimensions as shown.
The floor is to be covered with tiles, each measuring 1 m by 1 m.

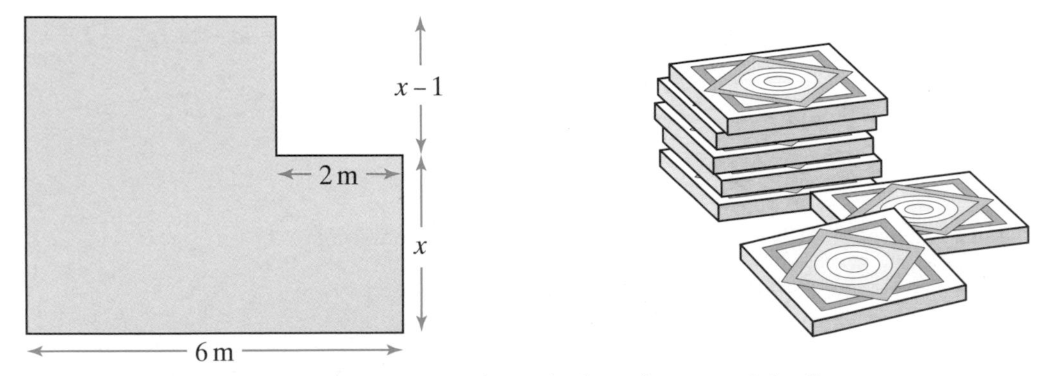

a By splitting the floor into two rectangles, calculate the area of the floor.

b By splitting the floor into two different rectangles, calculate the area of the floor.

Are your answers the same each time?
Give a reason for your answer.

11 The answer is $2y + 3$.
Find five questions of the form $a(by + c) \pm d(ey + f)$ with this answer.
(a, b, c, d, e and f are all integers.)

Explore

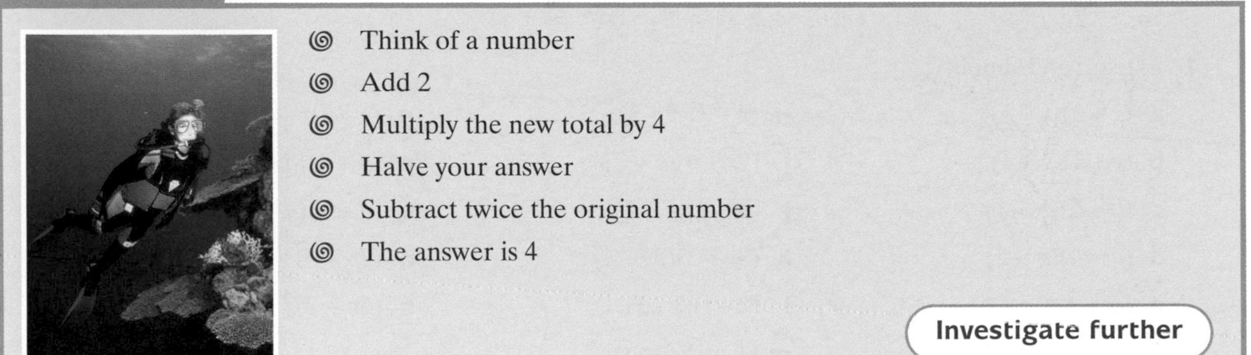

- Think of a number
- Add 2
- Multiply the new total by 4
- Halve your answer
- Subtract twice the original number
- The answer is 4

Investigate further

Explore

- Pick a blue card
- Double the number
- Add 1 to the new number
- Multiply the new number by 5
- Pick a white card
- Add this number to the previous result
- Subtract 5
- What do you notice?

Blue cards: 5 6 7 8

White cards: 1 2 3 4

Investigate further

Learn 2 Expanding linear expressions

Examples:

a Expand and simplify $(x + 4)(x - 2)$.

	x	$+4$
x	x^2	$+4x$
-2	$-2x$	-8

Remember: $-2 \times 4 = -8$

$$(x + 4)(x - 2) = x^2 + 4x - 2x - 8$$
$$= x^2 + 2x - 8$$

Underlining or circling like terms ($+4x$ and $-2x$ in this example) helps when simplifying your answer

b Expand and simplify $(2x + y)(3x - 2y)$.

	$2x$	$+y$
$3x$	$6x^2$	$+3xy$
$-2y$	$-4xy$	$-2y^2$

$2x \times -2y = -4xy$ because $2 \times -2 = -4$ and $x \times y = xy$

$$(2x + y)(3x - 2y) = 6x^2 + 3xy - 4xy - 2y^2$$
$$= 6x^2 - xy - 2y^2$$

Write $3xy - 4xy$ as $-xy$ rather than $-1xy$

Apply 2

1 Expand and simplify:

a $(x+3)(x+2)$
b $(y+3)(y+4)$
c $(b+5)(b+1)$
d $(a+4)(a-5)$
e $(m+2)(m-6)$
f $(h-4)(h+8)$
g $(3-x)(x+7)$
h $(x-3)(x+3)$
i $(2x+4)(x+2)$
j $(p-4)(p-3)$
k $(2y+3)(3y-4)$
l $(2-t)(t-3)$
m $(4-a)(5-a)$
n $(x+2)^2$
o $(y-3)^2$
p $(d+5)(d-5)$
q $(d-1)(d-2)$
r $(2p+4)^2$
s $(5k-2)^2$

2 Katya thinks the answer to $(x-2)(x+3)$ is x^2+x+1.
Becky thinks the answer is x^2+x-6.
Who is correct and why?

3 Paul thinks the answer to $(x+3)^2$ is x^2+9. Is he correct?
Give a reason for your answer.

4 **a** The answer is x^2+8x+c where c is an integer.

Find five questions in the form $(x+a)(x+b)$ where a and b are integers.

b The answer is $x^2+dx+48$ where d is an integer.

Find five questions in the form $(x+e)(x+f)$ where e and f are integers.

c The answer is $60+gx-x^2$ where g is an integer.

Find five questions in the form $(h-x)(k+x)$ where h and k are integers.

5 Calculate 102^2-98^2 without a calculator.

6 Find the integers p and q if $(x-p)(x+q)=x^2-3x-10$.

HINT Use the answer from question **1**, part **h** to help you.

7

A $y+2$	**B** $y-3$	**C** $y-1$	**D** $y+3$

Expand and simplify:

a **AC** **b** **CD** **c** **DC** **d** $\mathbf{B^2}$

e Work out a product of two cards that gives the answer y^2-9.

8 Show that $(n-2)^2+2(n-2)+2n=n^2$.

9 Get Real!

Farmer Forgetful can't remember the exact dimensions of his field.
He has found a scrap of paper with the following sketch:

$x + 15$

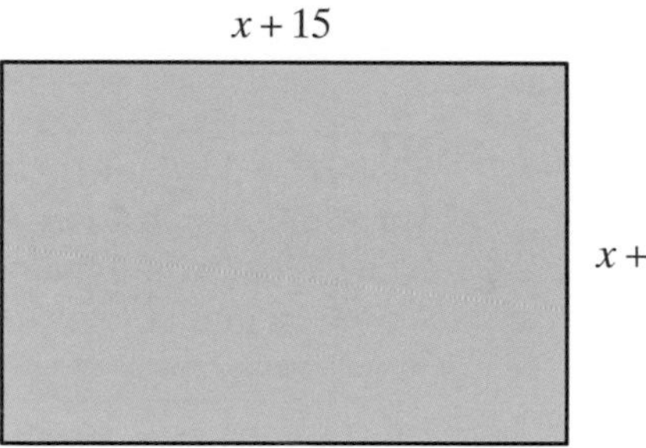

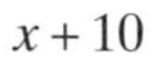

$x + 10$

a Write an expression for the area of the field.

Farmer Forgetful has forgotten he needs two separate fields, so he divides his field like this:

$x + 15$

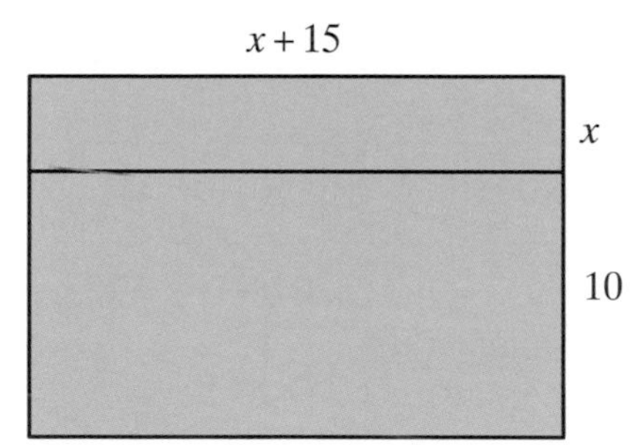

x

10

b Write an expression for the area of each new field.

c Write an expression for the total area of the two fields.

d By expanding and simplifying the expressions for parts **a** and **c**, show that the total area of land is the same in each case.

10 The answer is $3x^2 + gx + 12$ where g is an integer.
Find five questions in the form $(mx + n)(x + p)$ where m, n and p are integers.
State the value of g in each case.

11 Expand and simplify:

a $(3x + 2y)(x + y)$

b $(2p + q)(3p + 2q)$

c $(2a - b)(4a + b)$

d $(3p + 2q)(2p - 2q)$

e $(3x - y)(x - y)$

f $(b - 3c)(b + 3c)$

g $(2x - 4)(2x + 4)$

h $(6t - 2s)(3t - 4s)$

i $(c + d)^2$

j $(3x + y)^2$

k $(5p - q)^2$

l $(4w - 3v)^2$

12 The answer is $6x^2 + fx + 12$ where f is an integer.
Find five questions in the form $(mx + n)(px + q)$ where m, n, p and q are integers.
State the value of f in each case.

Explore

- Pick three consecutive even numbers
- Multiply them together
- Is the total a multiple of 8?

HINT How do you write an expression in n that is always even? What would the next even number be?

Investigate further

Explore

- Pick two cards from the list below:

1	2	3	4	5	6	7	8	9	10

- The cards are designed so that the number on the back is the complement in 20, that is, on the back of [2] the number is [18]
- Multiply the top number of card 1 by the reverse number of card 2
- Multiply the reverse number of card 1 by the reverse number of card 2
- Multiply the top number of card 2 by 20
- Add the three totals
- Try again with two other cards

HINT If the top number of a card is x, what can you say about the reverse number?

Investigate further

Learn 3 Factorising expressions

Examples:

a Factorise $5a - 10$

$5a - 10 = 5(a - 2)$

$5a = (5) \times a$ and $10 = 2 \times (5)$
Both terms have 5 as a common factor

b Factorise $12s^2t - 16s$

$12s^2t - 16s = 4s(3st - 4)$

$12s^2t = 3 \times (4 \times s) \times s \times t$ and $16s = 8 \times 2 \times s$
Or $12s^2t = 6 \times 2 \times s \times s \times t$ and $16s = 4 \times (4 \times s)$
Both terms have s as a common factor and 4 as the highest common factor

HINT You can always check if you have factorised correctly by multiplying the bracket out to make sure you get the question.

Treat factorising as the 'reverse' of expanding.

Apply 3

HINT Two of these expressions cannot be factorised.

1 Factorise the following expressions:

a $2x + 6$
b $3y + 12$
c $7y - 63$
d $8y - 40$
e $14y + 20$
f $6t - 18$
g $21 - 42t$
h $44t - 55c$
i $13a + 23b$
j $20b + 35a + 15c$
k $-12a - 16$
l $-42ab + 70cd + 14ef$
m $18f - 36g + 12h$
n $20xy + 60tu - 80pq$
o $2f - 11g + 33h$

2 Sandra thinks that $12a + 18$ factorised completely is $3(4a + 6)$.
Is she correct? Explain your answer.

3 The answer is $4(\ldots p + \ldots)$ where $\ldots p + \ldots$ is an expression.
Find five expressions that can be factorised completely with answers in this form.

4 Rashid thinks he has factorised $2x + 3y$ correctly as $2(x + 1\frac{1}{2}y)$.
Rana thinks the expression cannot be factorised.
Who is correct and why?

5 Factorise the following expressions:

a $2ab + 5b$
b $3xy + 5xz$
c $p^2 + 2p$
d $x^2 - 3x$
e $4x^2 + 6xy$
f $a^2 - a$
g $6xy + 7xyz$
h $y^3 - 2y^2$
i $wig - wam$
j $bugs + bunny$
k $hocus - pocus$
l $silly + billy$
m $wonkey - donkey$
n $2p^2q^3 + 5pq^2$
o $7m^2n^3 + 11m^3n^2p + 3m^4n^3q$
p $m^2uf^2c^3 + 3mu^3f^4c$

6 Get Real!
Sanjit and Anna are playing snooker. They notice that the red balls fit inside the triangle in the following pattern:
1, 3, 6, 10, 15, ...

Sanjit uses a method of differences and finds the nth term to be $\frac{1}{2}n^2 + \frac{1}{2}n$.

The expression gives the 5th triangular number as 15 (test it!) which is correct.

Factorise Sanjit's expression to find another expression for the nth triangular number.

7 Kenny thinks that when $4x^2y + 6xy^2$ is factorised completely the answer is $xy(4x + 6y)$. Is he correct? Justify your answer.

8 Factorise completely the following expressions:

a $4xy + 6x$
b $3cd + 12d^2$
c $24gh - 4g$
d $3pq^2r + 12p^2q$
e $28f^2g^2h^2 - 21f^2gh^3$
f $72p^2q + 32pqr^2 - 48q^2rs^2$
g $5x^2yz + 15xy^3z^2$
h $13s^2t^2u^2 + 91s^2t^2u$
i $12ab^2c^3d^4 - 14a^4b^3c^2d$
j $6a^2bc - 9ab^2c^2d$
k $3(a + b)^2 + 4(a + b)$
l $6(x + y)^2 - 4(x + y)$
m $(2p + q)^3 + 5(2p + q)^2$
n $3(2x + y) - 8x - 4y$

9 Copy these two tables.
Match the expression with the correct factorisation buddy.
Fill in the missing buddies.

Expression
$2x^2 + 8x$
$6x^2y - 3xy$

Factorisation
$2x(x + 4)$
$3xy(2y - 1)$
$x(x + 8)$

10 Charles thinks that $xy + 2x + 2y + 4$ cannot be factorised completely.
Chica says it can. Who is correct? Justify your answer.

11 Factorise the following expressions completely:

a $ab - 2a + 3b - 6$

b $2x - 6 + yx - 3y$

c $8fg + 12f + 6g + 9$

d $4ab - 8a - 3b + 6$

e $2p^2q + 2p^2t - q^2 - qt$

f $12wy + 9y + 4w + 3$

Explore

- Pick four consecutive odd numbers
- Add them together
- Is the answer a multiple of 8?
- Investigate further by picking four other consecutive odd numbers

HINT How do you write an expression in n that is always odd? What would the next odd number be?

Investigate further

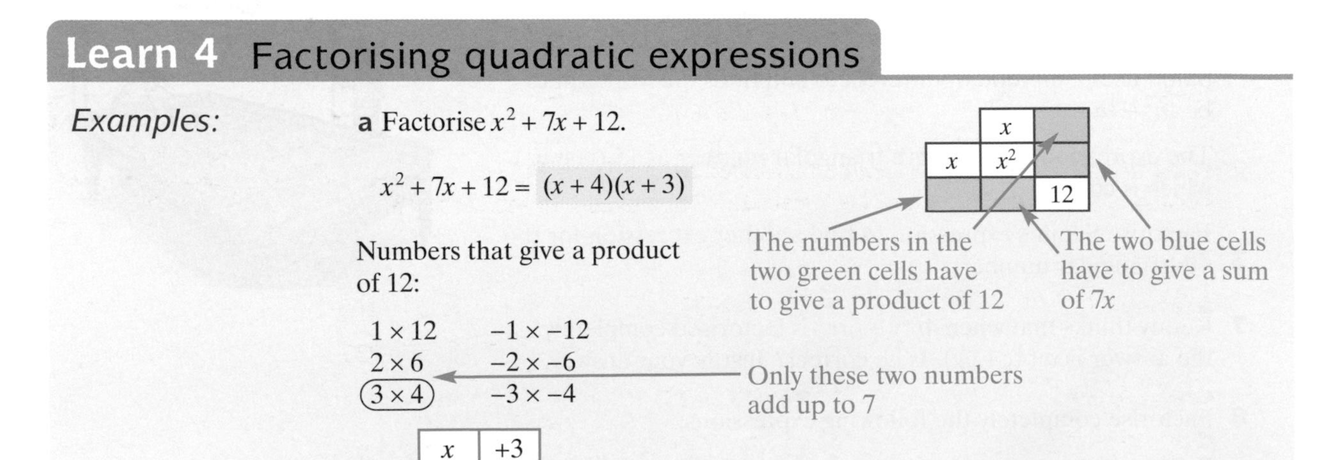

Learn 4 Factorising quadratic expressions

Examples:

a Factorise $x^2 + 7x + 12$.

$x^2 + 7x + 12 = (x + 4)(x + 3)$

	x	
x	x^2	
		12

Numbers that give a product of 12:

1×12 -1×-12
2×6 -2×-6
3×4 -3×-4

Only these two numbers add up to 7

	x	$+3$
x	x^2	$+3x$
$+4$	$+4x$	12

Sum $= 7x$

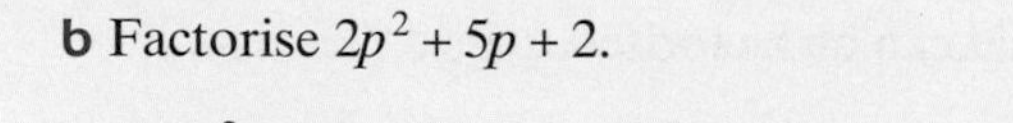

b Factorise $2p^2 + 5p + 2$.

$$2p^2 + 5p + 2 = (2p + 1)(p + 2)$$

Check:

- $2p \times p = 2p^2$
- $1 \times 2 = 2$
- $2p \times 2 + p \times 1 = 5p$

$(2p + 1)(p + 2)$

Note: $(2p + 2)(p + 1)$ would give $2p \times 1 + p \times 2 = 4p$ not $5p$

HINT You can always check if you have factorised correctly by multiplying the bracket out to make sure you get back to the question.

$2p^2 + 5p + 2$ is a quadratic expression because the highest power of the variable, p, is 2.
The coefficient of p^2 is 2, and the coefficient of p is 5.

c Factorise $x^2 - 10x + 3$ and write it in the form $(x + b)^2 + c$.

$$(x + b)^2 + c = x^2 + 2bx + b^2 + c$$

Expanding:

	x	$+b$
x	x^2	$+bx$
$+b$	$+bx$	$+b^2$

So comparing:

$$x^2 + 2bx + b^2 + c = x^2 - 10x + 3$$

$+2bx = -10x$ $\qquad b^2 + c = 3$

$2b = -10$ $\qquad 25 + c = 3$ ← As $b = -5$

$b = -5$ $\qquad c = -22$

So $x^2 - 10x + 3 = (x - 5)^2 - 22$

Factorising is the 'reverse' of expanding.

Apply 4

1 Factorise completely:

a $x^2 + 5x + 6$

b $x^2 + 8x + 7$

c $x^2 + 8x + 12$

d $y^2 + 14y + 49$

e $b^2 + 8b + 15$

f $p^2 + 22p + 21$

g $x^2 + 17x + 42$

h $d^2 + 13d + 42$

i $x^2 + 11x + 18$

j $w^2 + 2w + 8$

k $x^2 + 8x$

HINT One of these expressions cannot be factorised.

2 Billie thinks that $x^2 + 7x + 8$ can be factorised to $(x + 7)(x + 1)$ because $7 \times x$ gives the $7x$ and $7 + 1$ gives the 8 at the end. Is she correct?

3 Factorise:

a $y^2 + y - 6$

b $t^2 - t - 6$

c $p^2 - 5p - 6$

d $x^2 - 8x - 9$

e $b^2 - 12b - 13$

f $x^2 + 2x - 35$

g $x^2 - 5x - 24$

h $d^2 + 10d - 11$

i $y^2 + 4y - 12$

j $x^2 - 25$

k $p^2 - 100$

l $y^2 - 225$

m $x^2 - a^2$

n $6 - x - x^2$

o $15 + 2x - x^2$

4 Calculate the following, without using a calculator.

a $78^2 - 22^2$

b $59^2 - 41^2$

c $8.25^2 - 1.75^2$

d $0.62^2 - 0.38^2$

HINT Use the answer to **3j** to help you.

5 The answer is $(x+2)(x-a)$ where a is an integer.
Find five quadratic expressions that can be factorised to give this answer.

6 Are $(x-6)$ and $(x+2)$ the factors of $x^2+4x-12$?
If they are not, find the correct factors.

7 Factorise:

a $x^2-10x+25$

b $y^2-14y+13$

c $d^2-14d+48$

d $p^2-9p+20$

e $t^2-13t+36$

f $a^2-7a+12$

g $x^2-16x+28$

h $b^2-28b+27$

i $x^2-11x+30$

j $k^2-18k+56$

k y^2-y+5

HINT One of these expressions cannot be factorised.

8

A x^2+6x+8	**B** x^2+4x+4	**C** x^2+3x+2	**D** x^2-4

a What factor do all four quadratic expressions have in common?

b Which quadratic expression is the difference of two squares?

c Which quadratic expression has a repeated factor?

9 Factorise:

a $x^2+13x+40$

b $y^2-17y+60$

c $b^2-10b+21$

d $p^2+12p+11$

e x^2+3x

f y^2-27y

g x^2-49

h x^2+36

i $y^2-3y-18$

j $b^2-8b+16$

HINT One of these expressions cannot be factorised.

10 Copy the tables.
Match the quadratic expression with the correct factorisation buddy.
Fill in the missing buddies.

Quadratic expression
$x^2+13x+40$
x^2-81
x^2+16
$x^2-13x+40$

Factorisation
$(x-9)^2$
$(x+4)^2$
$(x+8)(x+5)$
Cannot be factorised

11 Copy and fill in the gaps.

a $x^2+\square x-12=(x-3)(x+\square)$

b $x^2+5x-\square=(x-3)(x+\square)$

c $x^2+14x+\square=(x+\square)^2$

d $x^2-\square x+16=(x-\square)^2$

12 Get Real!

Bob and Pam are investigating the number sequence of piles of cannonballs: 1, 4, 10, 20, ...
Using a method of differences, Bob finds the expression for the nth term to be $\frac{1}{6}n^3 + \frac{1}{2}n^2 + \frac{1}{3}n$.
Pam searches the internet and discovers that the numbers are called tetrahedral numbers and the nth term is $\frac{1}{6}n(n+1)(n+2)$.
Bob prefers Pam's expression.
Factorise Bob's expression into Pam's expression.

HINT Write all the fractions with the same denominator.

13 $x^2 + bx + a = (x + c)^2$ where a, b and c are integers.
Find five different sets of values for a, b and c to make the statement correct.

14 Christopher notices that the coefficient of x^2 in quadratic expressions is not always one. He reckons he can factorise other quadratic expressions, for example:

a $2x^2 + 7x + 6 = (2x + 2)(x + 3)$ because $2 \times 3 = 6$

b $3x^2 - x - 2 = (3x + 2)(x - 1)$ because $2 \times -1 = -2$

c $5x^2 - 14x + 8 = (5x - 2)(x - 4)$ because $-2 \times -4 = 8$

d $6x^2 + 16x + 8 = (2x + 2)(3x + 4)$ because $2 \times 4 = 8$

Expand Christopher's expressions and say whether he has factorised the quadratic expressions correctly. If he has not, say which coefficients are incorrect.

15 Factorise completely:

a $2x^2 + 7x + 3$

b $3x^2 + 14x + 8$

c $5x^2 + 8x + 3$

d $2x^2 - x - 1$

e $7x^2 - 27x - 4$

f $5x^2 - 17x + 6$

g $4x^2 - 9$

h $16x^2 - 4$

i $2x^2 - 18$

j $20x^2 - 80$

k $6x^2 - x - 1$

l $8x^2 + 10x + 3$

m $6x^2 - x - 5$

n $10x^2 - 2x - 12$

o $12x^2 - 14x - 6$

p $9x^2 - 18x + 5$

q $6x^2 + 8x - 8$

r $4x^2 - 10x - 6$

s $15x^2 - 39x - 18$

t $6 + x - 2x^2$

u $15 + 4x - 3x^2$

v $12 - 2x - 4x^2$

16 Copy the tables.
Match the quadratic expression with the correct factorisation.
Fill in the missing buddies.

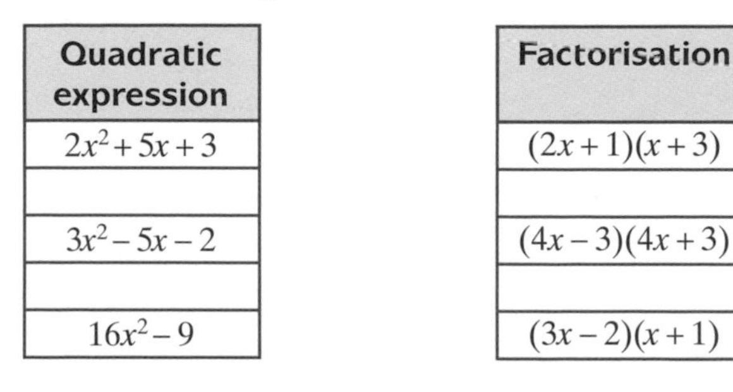

Quadratic expression
$2x^2 + 5x + 3$
$3x^2 - 5x - 2$
$16x^2 - 9$

Factorisation
$(2x + 1)(x + 3)$
$(4x - 3)(4x + 3)$
$(3x - 2)(x + 1)$

Learn 5 Simplifying algebraic fractions

You can simplify fractions by factorising and dividing by common factors.

Examples:

a Simplify $\frac{x^2 - 2x}{x - 2}$ ← This is a rational expression (that is, a fraction) It has a numerator and a denominator

$\frac{x^2 - 2x}{x - 2} = \frac{x(x - 2)}{x - 2}$ ← Factorise the numerator

$= x$ ← After dividing the numerator and denominator by $(x - 2)$

b Simplify $\frac{x^2 + 6x + 5}{x + 5}$

$\frac{x^2 + 6x + 5}{x + 5} = \frac{(x + 5)(x + 1)}{x + 5}$ ← Dividing the numerator and denominator by $(x + 5)$

$= x + 1$

c Simplify $\frac{3x^2 + 2x - 1}{x^2 - 1}$

$\frac{3x^2 + 2x - 1}{x^2 - 1} = \frac{(3x - 1)(x + 1)}{(x - 1)(x + 1)}$ ← Factorise the numerator and the denominator

$= \frac{3x - 1}{x - 1}$ ← Dividing the numerator and denominator by $(x + 1)$

Apply 5

1 Simplify these expressions.

a $\frac{x^2 + 3x}{x + 3}$

b $\frac{x^2 + 7x + 10}{x + 2}$

c $\frac{x^2 + 4x + 3}{x + 1}$

d $\frac{x^2 + 2x}{x^2 - 4}$

e $\frac{x^2 + 6x + 5}{x^2 + 8x + 15}$

f $\frac{x^2 - 9}{x + 3}$

g $\frac{2(x + 1)^2}{x + 1}$

2 Simplify these expressions.

a $\frac{2x^2 + 5x - 3}{x + 3}$

b $\frac{3x^2 + 7x - 6}{x + 3}$

c $\frac{4x^2 + 2x}{2x^2 - 5x - 3}$

d $\frac{5x^2 + 7x - 6}{2(x + 2)^2}$

e $\frac{3x^2 - 11x + 6}{2x^2 - 5x - 3}$

f $\frac{3x^2 - 10x + 8}{18x^2 - 32}$

g $\frac{2x^2 - x - 15}{4x^2 - 25}$

Explore

This is a proof that $1 = -1$

Let	$a = 1 \quad (*)$
Square both sides:	$a^2 = 1$
Subtract 1 from both sides:	$a^2 - 1 = 0$
Factorise:	$(a + 1)(a - 1) = 0$
Divide by $(a - 1)$:	$a + 1 = 0$
Subtract 1 from both sides:	$a = -1$
but from $(*)$	$a = 1$
	so $1 = -1$

Obviously, this cannot be true

Investigate further

Use of symbols

ASSESS

The following exercise tests your understanding of this chapter, with the questions appearing in order of increasing difficulty.

1 Remove the brackets from the following expressions.

a $2(a + 3)$
b $5(3a - 1)$
c $-3(4a + 5)$
d $-7(3a - 6)$
e $7(a + 2b)$
f $5(6a - 3b)$
g $-3(3a + 2b)$
h $-7(3a - 2b)$
i $-6(4a - 3b)$
j $-2(a - 2b + 3)$

2 Factorise the following expressions.

a $2a + 10$
b $10b - 12$
c $16 - 4c$
d $5d + 20e + 35f$
e $6g - 9h + 12i$
f $10j + 15k - 20l$
g $30p - 45q - 75r$
h $5x + 15y$
i $2ab - 3a$
j $x^2 + 7x$
k $4a^2 - 6a$
l $bil + ben$
m $abra + cadabra$
n $12cat + 3sat - 6mat$

3 Remove the brackets from the following expressions and simplify.

a $2c + 3(c + 5)$
b $2c - 3(c + 5)$
c $2c + 3(c - 5)$
d $2c - 3(c - 5)$
e $4(c - 3) + 2(c + 7)$
f $4(c - 3) - 2(c + 7)$
g $5(2c - 3) - 3(2c + 1)$
h $5(2c - 3) - 3(2c - 1)$
i $3(2c - d + 3e) + (5c + 3d - 2e)$
j $3(2c - d + 3e) - (5c + 3d - 2e)$
k $y(y^2 - 7)$
l $2z(5z^2 + 4z - 8)$
m $x(x - 3) + 5(x - 3)$
n $p(p^2 + 3p - 4) - 6(p^2 + 3p - 4)$

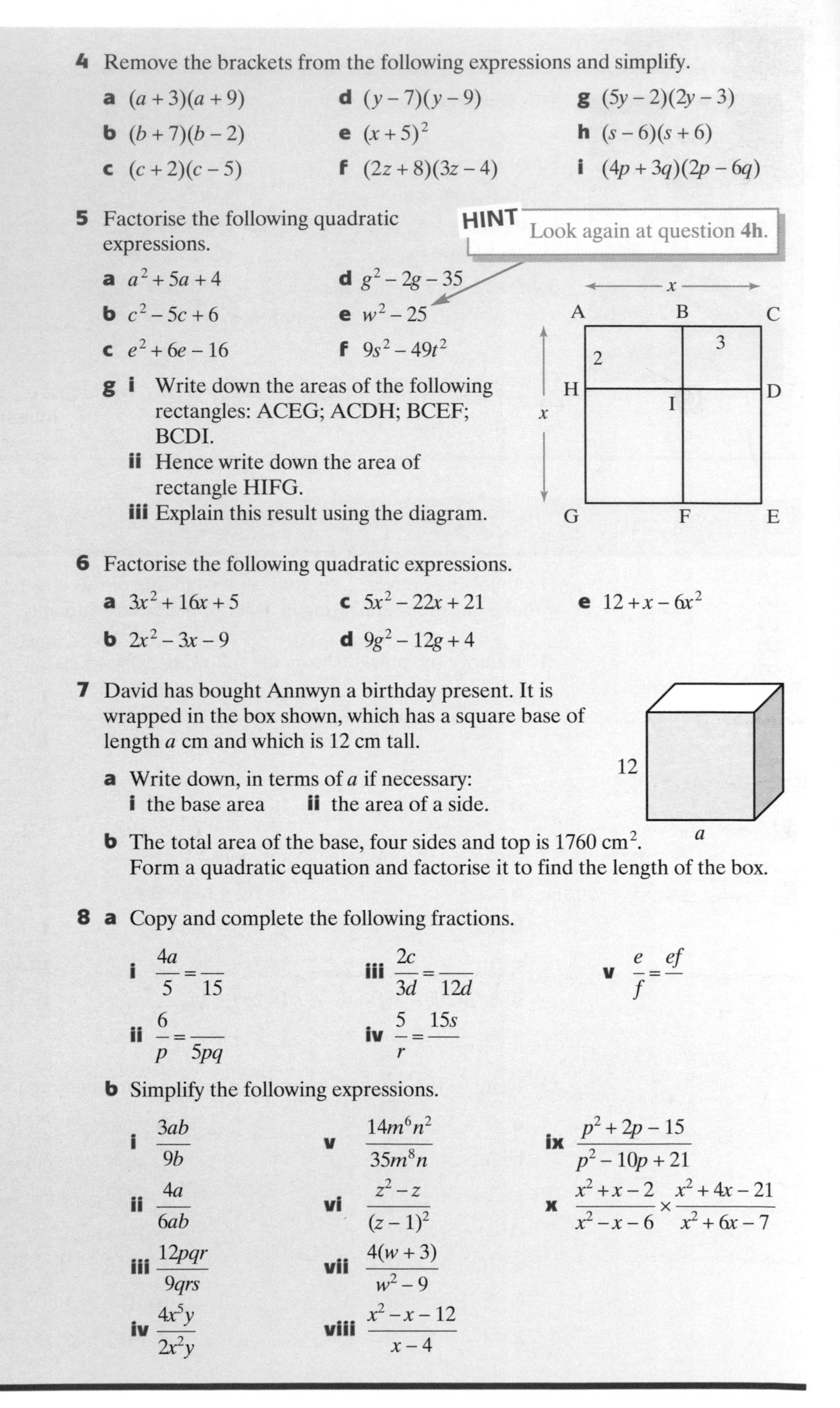

4 Remove the brackets from the following expressions and simplify.

a $(a+3)(a+9)$ **d** $(y-7)(y-9)$ **g** $(5y-2)(2y-3)$

b $(b+7)(b-2)$ **e** $(x+5)^2$ **h** $(s-6)(s+6)$

c $(c+2)(c-5)$ **f** $(2z+8)(3z-4)$ **i** $(4p+3q)(2p-6q)$

5 Factorise the following quadratic expressions.

HINT Look again at question **4h**.

a a^2+5a+4 **d** $g^2-2g-35$

b c^2-5c+6 **e** w^2-25

c $e^2+6e-16$ **f** $9s^2-49t^2$

g **i** Write down the areas of the following rectangles: ACEG; ACDH; BCEF; BCDI.

ii Hence write down the area of rectangle HIFG.

iii Explain this result using the diagram.

6 Factorise the following quadratic expressions.

a $3x^2+16x+5$ **c** $5x^2-22x+21$ **e** $12+x-6x^2$

b $2x^2-3x-9$ **d** $9g^2-12g+4$

7 David has bought Annwyn a birthday present. It is wrapped in the box shown, which has a square base of length a cm and which is 12 cm tall.

a Write down, in terms of a if necessary:
i the base area **ii** the area of a side.

b The total area of the base, four sides and top is 1760 cm^2. Form a quadratic equation and factorise it to find the length of the box.

8 **a** Copy and complete the following fractions.

i $\frac{4a}{5}=\frac{}{15}$ **iii** $\frac{2c}{3d}=\frac{}{12d}$ **v** $\frac{e}{f}=\frac{ef}{}$

ii $\frac{6}{p}=\frac{}{5pq}$ **iv** $\frac{5}{r}=\frac{15s}{}$

b Simplify the following expressions.

i $\frac{3ab}{9b}$ **v** $\frac{14m^6n^2}{35m^8n}$ **ix** $\frac{p^2+2p-15}{p^2-10p+21}$

ii $\frac{4a}{6ab}$ **vi** $\frac{z^2-z}{(z-1)^2}$ **x** $\frac{x^2+x-2}{x^2-x-6}\times\frac{x^2+4x-21}{x^2+6x-7}$

iii $\frac{12pqr}{9qrs}$ **vii** $\frac{4(w+3)}{w^2-9}$

iv $\frac{4x^5y}{2x^2y}$ **viii** $\frac{x^2-x-12}{x-4}$

5 Decimals

OBJECTIVES

D **Examiners would normally expect students who get a D grade to be able to:**

Multiply two decimals such as 2.4×0.7

Convert decimals to fractions and fractions to decimals

C **Examiners would normally expect students who get a C grade also to be able to:**

Divide a number by a decimal such as $1 \div 0.2$ and $2.8 \div 0.07$

B **Examiners would normally expect students who get a B grade also to be able to:**

Identify recurring and terminating decimals

Convert recurring decimals to fractions and fractions to recurring decimals

What you should already know ...

- Add, subtract, multiply and divide whole numbers
- Add and subtract decimals
- Estimate answers to questions involving decimals

VOCABULARY

Digit – any of the numerals from 0 to 9

Decimal – a number in which a decimal point separates the whole number part from the decimal part, for example, 24.8

Numerator – the number on the top of a fraction

Numerator → $\frac{3}{8}$ ← Denominator

Denominator – the number on the bottom of a fraction

Terminating decimal – a decimal that ends, for example, 0.3, 0.33 or 0.3333

Recurring decimal – a decimal with a repeating digit or group of digits, for example, 0.33333333333 ... (written as $0.\dot{3}$) or 0.25678678678678 ... (written as $0.25\dot{6}7\dot{8}$)

Learn 1 Multiplying decimals

Example:

Calculate 0.78×5.2

First remove decimal points: 78×52

Then multiply in your usual way (The grid method is shown here, but use your usual method.)

×	70	8
50	3500	400
2	140	16

$$\begin{array}{r} 3500 \\ 400 \\ 140 \\ +\ \ 16 \\ \hline 4056 \end{array}$$

Finally, put the decimal point back in the answer.

Estimate that 0.78×5.2 is about $1 \times 5 = 5$.

So $0.78 \times 5.2 = 4.056$

Alternatively, count up the number of decimal places in the question.

There are three decimal places in the question: 0.78×5.2

So you need three decimal places in the answer: 4.056

So $0.78 \times 5.2 = 4.056$

Apply 1

1 Use the multiplication $23 \times 52 = 1196$ to help you to complete the questions.

- **a** 2.3×52
- **b** 2.3×520
- **c** 0.23×5.2
- **d** 0.23×52
- **e** 0.23×0.52
- **f** $23\,000 \times 0.052$
- **g** $0.023 \times 520\,000$

2 Calculate:

- **a** 0.13×22
- **b** 1.5×2.3
- **c** 0.7×1.3
- **d** 1.1×4.5
- **e** 1.7×0.22
- **f** 3.2×13
- **g** 5.1×2.3
- **h** 2.7×0.13
- **i** 8.7×2.51
- **j** 8.93×162
- **k** $73.1 \times 12\,400$
- **l** $14.3 \times 223\,000$

3 Using your answers to question **2**, write down the answers to these.

- **a** 1.3×22
- **b** 1.5×0.23
- **c** 0.07×1.3
- **d** 0.11×0.45
- **e** 17×2.2
- **f** 0.032×13
- **g** 0.0051×0.23
- **h** 27×1.3
- **i** 0.087×0.0251
- **j** 0.00893×0.0162
- **k** 7.31×1.24
- **l** 1430×22.3

4 A can of Fizzicola contains 0.3 litres of drink. A box holds 36 cans. How many litres of Fizzicola are there in a box?

5 Here are two multiplagons. On each straight line, the numbers in the circles multiply together to make the number in the rectangle.

Your job is to copy and complete the multiplagons by filling in the missing numbers.

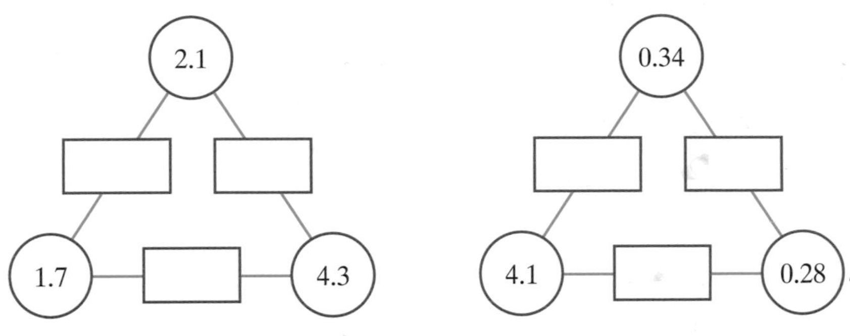

6 Get Real!

Rachel is making some curtains. She buys 4.2 metres of fabric. The fabric costs £3.85 per metre.

a How much does Rachel have to pay?

b The fabric is 1.2 metres wide. What area of fabric has Rachel bought?

7 $87 \times 132 = 11\,484$.
Use this fact to write down the answers to these multiplications.

a 8.7×1.32

b 87×0.132

c 0.87×13.2

d $870 \times 13\,200$

e 1.32×870

f $114.84 \div 132$

g $11.484 \div 8.7$

8 Toby says $0.4 \times 0.2 = 0.8$
Austin says it isn't, because $4 \times 0.2 = 0.8$
Austin says $0.4 \times 0.2 = 0.08$
Toby says it isn't because that's less than you started with.
Who is right, Toby or Austin? Give a reason for your answer.

9 **a** You know $3 \times 2 = 6$.
So what is 0.3×0.2?

b What other multiplications have the same answer as 0.3×0.2?

c Write down five multiplications with an answer of 0.12

10 Work out the area of this shape.

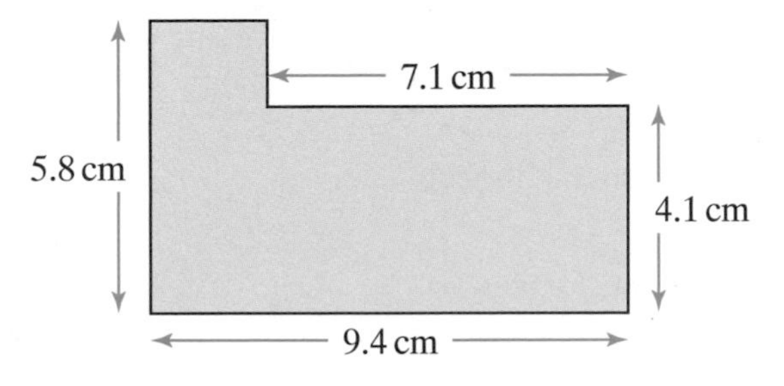

11 Think of a number.
Write it down.
Divide it by 2.
Divide the answer by 2.
Write down your answer.

Go back to your starting number.
Multiply it by 0.25
Write down your answer.
Can you explain why your answers are the same?

Explore

- Add together 1.125 and 9
- Now multiply 1.125 by 9
- You should get the same answer to both questions

Can you find other pairs of numbers with this characteristic?
Can you find a pair where the product is twice the sum?

Investigate further

Learn 2 Dividing decimals

Examples:

a Calculate $31.2 \div 0.4$

$31.2 \div 0.4 = \frac{31.2}{0.4}$

$= \frac{31.2}{0.4} = \frac{312}{4}$ ($\times 10$ numerator and denominator) — Make the fraction an equivalent fraction by multiplying the numerator and denominator by 10

$4\overline{)312}$ = 78 — Now do the division

So $31.2 \div 0.4 = 78$

b Calculate $3.8 \div 0.05$

$3.8 \div 0.05 = \frac{3.8}{0.05}$

$= \frac{3.8}{0.05} = \frac{38}{0.5} = \frac{380}{5}$ ($\times 10$, $\times 10$) — Make the fraction an equivalent fraction by multiplying the numerator and denominator by 10 and 10 again

$5\overline{)380}$ = 76 — Now do the division

So $3.8 \div 0.05 = 76$

Apply 2

1 Write these calculations as equivalent fractions and work them out.

a $3.2 \div 0.4$
b $25.4 \div 0.2$
c $2.85 \div 0.5$
d $42.2 \div 0.02$
e $53.1 \div 0.3$
f $1.74 \div 0.6$
g $0.4 \div 0.08$
h $32 \div 0.8$
i $0.056 \div 0.7$
j $13.2 \div 400$
k $0.028 \div 700$

2 Write these calculations as equivalent fractions and work them out.

a $4.07 \div 1.1$
b $22.8 \div 1.2$
c $2.73 \div 0.13$
d $0.264 \div 1.1$
e $16.8 \div 0.12$
f $25.3 \div 0.11$
g $7.392 \div 0.11$
h $0.474 \div 0.12$
i $0.0552 \div 0.012$
j $0.945 \div 1.4$
k $222.89 \div 3.1$
l $83.16 \div 2200$
m $12.45 \div 15\,000$
n $56.2 \div 250$

3 Get Real!
Malcolm the plumber has a 6 metre length of copper pipe.
He needs to cut it into 0.4 metre lengths.
How many pieces will he get?

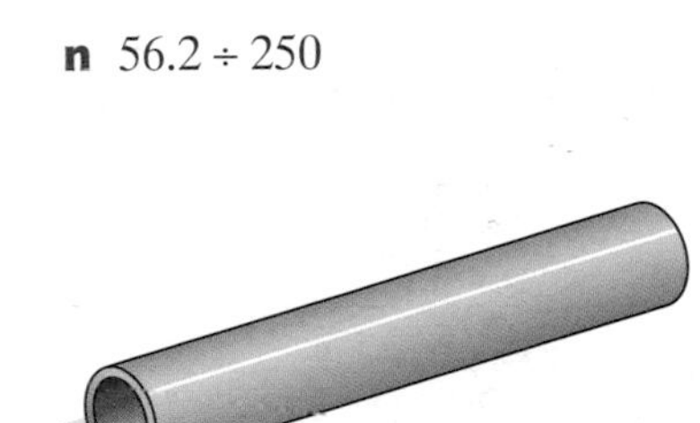

4 Get Real!
On her birthday, Bridget is given a big box of small sweets called Little Diamonds.
She wants to find out how many sweets are in the box, but it would take too long to count them.
A label on the box tells her that the total weight is 500 g.
She weighs 10 sweets. The weight of the 10 sweets is 0.4 g.

a How much does one sweet weigh?
b How many sweets are there in the box?

5 Hazel says that $48 \div 2 = 24$, so $48 \div 0.2 = 2.4$
Darren says $48 \div 2 = 24$, so $4.8 \div 0.2 = 2.4$
Harry says $48 \div 2 = 24$, so $4.8 \div 2 = 2.4$
Who is right? Give a reason for your answer.

6 €1 is worth £0.60. How many euro would you get for £7.50?

7 Here are two multiplagons.
On each straight line, the numbers in the circles multiply together to make the number in the rectangle.
Your job is to copy and complete them by filling in the missing numbers.

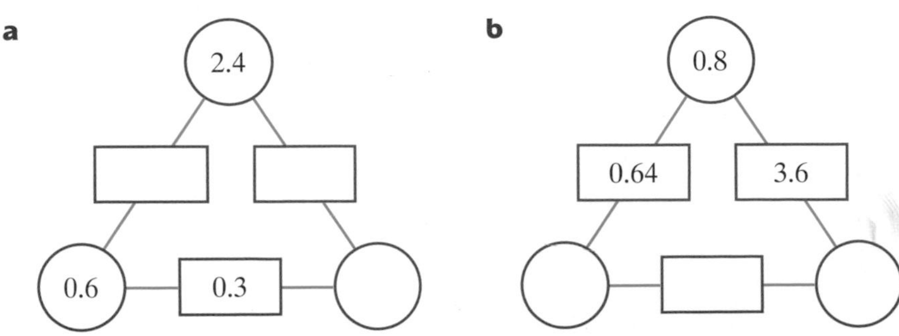

8 Carrie knows that $3.4 \div 0.4 = 8.5$
Use this to copy and fill in the gaps in these questions.

a $34 \div 0.4 = \square$
b $3.4 \div 4 = \square$
c $340 \div \square = 8.5$
d $\square \div 4 = 0.85$
e $\square \div 0.04 = 8.5$
f $0.34 \div 8.5 = \square$

9 Start with a number less than 1.
Take it away from 1.
Divide the first number by the second.
Divide this number by one more than itself.
You end up with the starting number!

Example:

		0.8
$1 - 0.8$	=	0.2
$0.8 \div 0.2 = 8 \div 2$	=	4
$4 \div 5$	=	0.8

Try this yourself, starting with

a 0.5 **b** 0.9 **c** 0.75

Check it with any number you like – although you will probably need a calculator for more difficult examples.

10 Use $43 \times 28 = 1204$ to write down the answers to these divisions.

a $1204 \div 43$
b $12.04 \div 43$
c $1.204 \div 2.8$
d $120.4 \div 2.8$
e $120\,400 \div 0.43$
f $0.1204 \div 280$
g $1.204 \div 430$
h $12.04 \div 8.6$

Explore

- Draw a grid like this one:

$\square \, . \, \square \div 0 \, . \, \square$

- Roll a dice
- Write the score in one of the empty boxes in your grid
- Roll the dice twice more, writing the score in a box after each roll
- Work out the answer to the division
- Now try again – your aim is to get the highest possible answer

Investigate further

Learn 3 Fractions and decimals

Examples:

a Write 0.72 as a fraction.

To change a decimal to a fraction, just remember the place values.

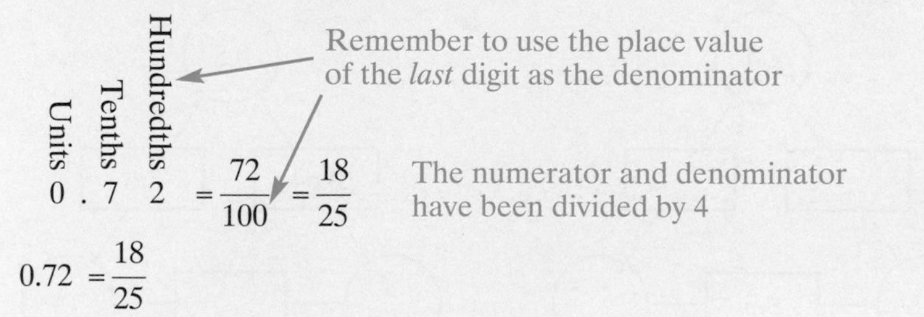

$0.72 = \frac{18}{25}$

b Write $\frac{7}{8}$ as a decimal.

$\frac{7}{8}$ means $7 \div 8$.

$$8 \overline{)7.{}^{7}0{}^{6}0{}^{4}0} = 0.875$$

You can check your answers with a calculator

$\frac{7}{8} = 0.875$

Apply 3

1 Write these decimals as fractions, giving your answers in their simplest form:

a 0.6 **b** 0.32 **c** 0.125 **d** 0.995

2 a Write these fractions as decimals.

i $\frac{2}{5}$ **ii** $\frac{3}{8}$ **iii** $\frac{7}{20}$

b Use your answers to part **a** to write the fractions in order of size, starting with the smallest.

3 Which of these fractions is closest to 0.67?

a $\frac{3}{4}$ **b** $\frac{5}{8}$ **c** $\frac{3}{5}$ **d** $\frac{13}{20}$

HINT Write the fractions as decimals.

4 a What is 2.65 as a fraction?

b What is $3\frac{7}{20}$ as a decimal?

5 Write down five fractions that are equal to 0.4

6 Get Real!
At a school fête, some children decided to raise money with a 'Guess the weight of the cake' stall.
Amy guessed 3300 g, Tariq guessed 3.28 kg and Caroline guessed $3\frac{1}{5}$ kg.
The real weight was 3.237 kg. Who won?

7 Josh divided one number by another, and 2.375 was the answer. Both numbers were less than 20. What were the two numbers?

8 Hilary says that $\frac{1}{8} = 1.8$
Nick says $\frac{3}{8} = 0.38$
Eleanor says $\frac{1}{10} = 0.10$
Jeff says $\frac{1}{20} = 0.20$
Who is correct? Correct the errors of the others.

9 Dan knows that $\frac{1}{8} = 0.125$
Use this answer to change these fractions to decimals.

a $\frac{3}{8}$ **b** $1\frac{1}{8}$ **c** $\frac{5}{8}$ **d** $\frac{1}{16}$

10 Write these fractions as decimals. Be careful – they never end! They are called recurring decimals. Stop when you have reached a repeating digit or pattern of digits.

a $\frac{2}{3}$ **b** $\frac{4}{11}$ **c** $\frac{3}{7}$

11 Find three fractions that fit all these rules:

a All three fractions must have different denominators.

b Each denominator must be less than 10.

c The fraction must be greater than 0.2

d The fraction must be less than 0.3

12 Use a calculator to change these fractions to decimals.

a $\frac{3}{16}$ **b** $\frac{7}{32}$ **c** $\frac{5}{80}$ **d** $\frac{17}{8}$ **e** $2\frac{11}{40}$ **f** $3\frac{9}{64}$

Explore

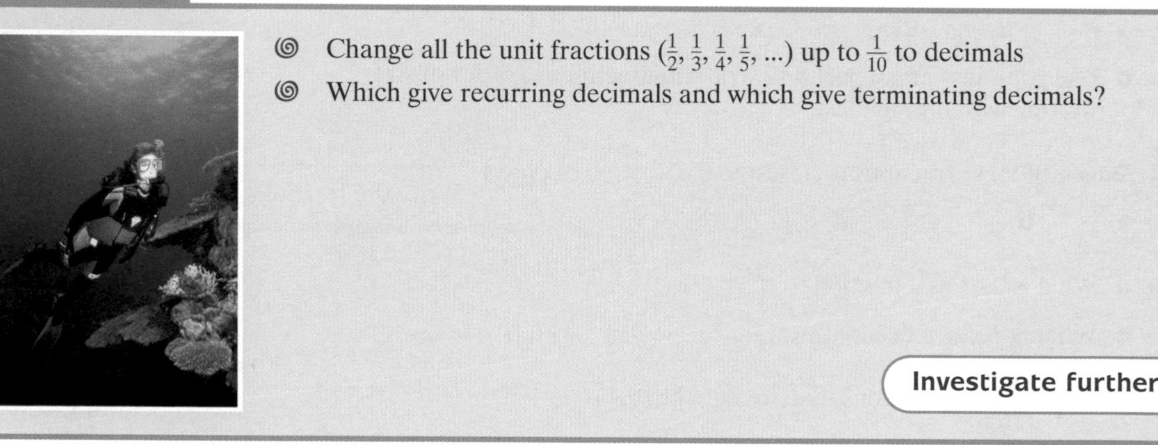

- Change all the unit fractions ($\frac{1}{2}, \frac{1}{3}, \frac{1}{4}, \frac{1}{5}, \ldots$) up to $\frac{1}{10}$ to decimals
- Which give recurring decimals and which give terminating decimals?

Investigate further

Learn 4 Recurring decimals and fractions

Example:

Write $\frac{7}{11}$ as a decimal.

$\frac{7}{11}$ means $7 \div 11$.

$$\begin{array}{r} 0.\,6\,3\,6\,3\ldots \\ 11\overline{)7.{}^{7}0{}^{4}0{}^{7}0{}^{4}0\ldots} \end{array}$$

You can check your answers with a calculator

The recurring decimal 0.6363636363 ... is written as $0.\dot{6}\dot{3}$

Similarly:
The recurring decimal 0.3333333333 ... is written as $0.\dot{3}$
The recurring decimal 0.54789789789 ... is written as $0.54\dot{7}8\dot{9}$

Apply 4

1 Which of these fractions will give recurring decimals?

a $\frac{4}{15}$ **b** $\frac{7}{18}$ **c** $\frac{5}{16}$ **d** $\frac{5}{6}$ **e** $\frac{3}{20}$ **f** $\frac{5}{9}$ **g** $\frac{9}{60}$ **h** $\frac{13}{125}$

2 Write these fractions as recurring decimals.

a $\frac{2}{3}$ **b** $\frac{7}{11}$ **c** $\frac{4}{7}$ **d** $\frac{5}{13}$

3 Get Real!

According to the rules of football, the circumference of the ball must be between 68 cm and 70 cm.

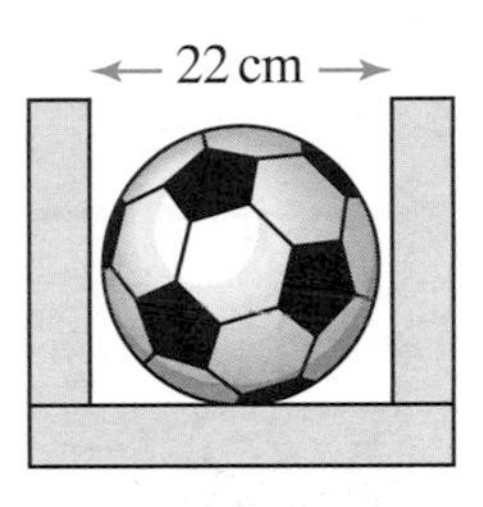

To make it easier to check, a club chairman makes up a frame as shown.
He says that a ball with a circumference of 69 cm will just fit inside the frame.

a Divide the circumference of 69 cm by the diameter, 22 cm.

b Explain why this answer is about the right size.

4 Suppose we want to change $0.\dot{1}6\dot{2}$ to a fraction.
Follow the reasoning below.

Suppose $x = 0.\dot{1}6\dot{2}$ (A)
Then $1000x = 162.\dot{1}6\dot{2}$ (B)
Subtracting (B) – (A): $999x = 162$
$x = \frac{162}{999} = \frac{18}{111} = \frac{6}{37}$

Use the same approach to turn these recurring decimals into fractions in their smallest form.

a $0.\dot{1}\dot{8}$ **b** $0.\dot{4}2\dot{3}$ **c** $0.\dot{1}38\dot{6}$

5 Can you write a simple rule for changing any recurring decimal to a fraction?

6 Charlotte knows that $\frac{4}{11} = 0.\dot{3}\dot{6}$, but she mistakenly writes it as $0.3\dot{6}$
Write $0.3\dot{6}$ as a fraction in its lowest terms.

7 What is $0.4\dot{5} - 0.\dot{4}\dot{5}$? Give your answer as a fraction or a decimal.

8 Change all the sevenths ($\frac{1}{7}, \frac{2}{7}$, ... up to $\frac{6}{7}$) to decimals.
What do you notice about the answers?
Try the same for the elevenths and the thirteenths. What do you notice?

9 The circumference of a circle divided by its diameter is π.
$\pi = 3.14159$ ...
People occasionally use $3\frac{1}{7}$ as an approximate value.

a What is $3\frac{1}{7}$ as a decimal?

b Can you find a fraction that is a better approximation?

10 Unit fractions are fractions with a numerator of 1.

a How many unit fractions are there with denominators of less than 100?

b How many of these will give recurring decimals?

Explore

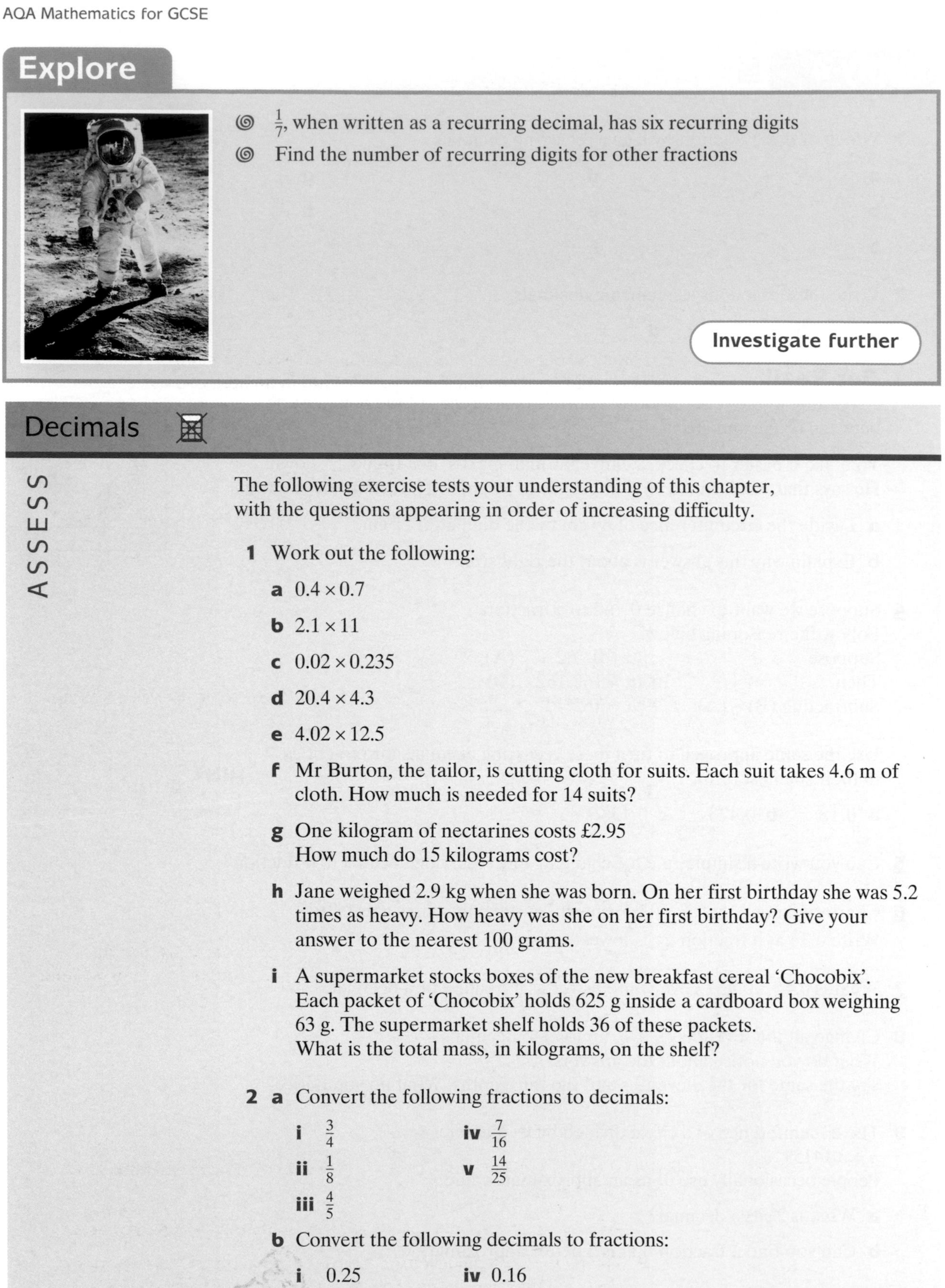

- $\frac{1}{7}$, when written as a recurring decimal, has six recurring digits
- Find the number of recurring digits for other fractions

Investigate further

Decimals

ASSESS

The following exercise tests your understanding of this chapter, with the questions appearing in order of increasing difficulty.

1 Work out the following:

a 0.4×0.7

b 2.1×11

c 0.02×0.235

d 20.4×4.3

e 4.02×12.5

f Mr Burton, the tailor, is cutting cloth for suits. Each suit takes 4.6 m of cloth. How much is needed for 14 suits?

g One kilogram of nectarines costs £2.95
How much do 15 kilograms cost?

h Jane weighed 2.9 kg when she was born. On her first birthday she was 5.2 times as heavy. How heavy was she on her first birthday? Give your answer to the nearest 100 grams.

i A supermarket stocks boxes of the new breakfast cereal 'Chocobix'. Each packet of 'Chocobix' holds 625 g inside a cardboard box weighing 63 g. The supermarket shelf holds 36 of these packets.
What is the total mass, in kilograms, on the shelf?

2 **a** Convert the following fractions to decimals:

i $\frac{3}{4}$ **iv** $\frac{7}{16}$

ii $\frac{1}{8}$ **v** $\frac{14}{25}$

iii $\frac{4}{5}$

b Convert the following decimals to fractions:

i 0.25 **iv** 0.16

ii 0.375 **v** 0.6875

iii 0.45

3 a Work out the following:

i $1.68 \div 0.4$

ii $220 \div 0.05$

iii $16.9 \div 1.3$

iv $6.25 \div 0.25$

v $49.2 \div 1.2$

b Road Runner travels 3.64 m in 0.7 seconds.
How fast is this in metres per second?

c Naomi sees 54 suspect cells under her microscope in an area of 0.06 cm^2.
How many cells would she expect to find in an area of 1 cm^2?

d A bag of sweets weighing 95 g includes wrappings of 0.5 g.
Each sweet weighs 4.5 g.
How many sweets are in the bag?

4 a Convert the following fractions to decimals without using a calculator:

i $\frac{1}{3}$

ii $\frac{2}{7}$

iii $\frac{7}{9}$

iv $\frac{6}{11}$

v $\frac{7}{15}$

vi $\frac{5}{24}$

vii $3\frac{7}{12}$

b Use the method shown in Apply **4** question **4** to convert the following decimals to fractions:

i $0.\dot{7}$

ii $0.\dot{2}\dot{7}$

iii $0.\dot{1}0\dot{1}$

iv $0.\dot{1}2\dot{3}$

v $0.5\dot{3}$ (beware: this is harder!)

5 a A rectangular plot, 10.4 m by 7.5 m, is to be sown with grass seed.
The gardener needs 30 g of grass seed for each square metre of ground.
Grass seed costs £4.80 per kilogram.
Find:

i the area of the plot

ii the mass of grass seed needed

iii the cost of the seed.

b A beer glass, full of beer, has a mass of 1.43 kg. The glass alone has a mass of 810 g.

i What is the total mass when the glass is half full of the same beer?

ii The glass is now filled with a 'heavier' beer. The total weight of the glass and 'heavier' beer is now 1.74 kg. How many times heavier is the new beer compared to the old?

c Paul travels from his home in Eastbourne to meet Andrew in Manchester. He travels at 64 miles each hour on average until he stops in a service area. It is 275 miles from Eastbourne to Manchester.

i How far has he travelled after 2 h and 45 min?

ii How far has he still left to travel?

iii Paul now travels at 45 miles each hour on average.
How much longer will his journey take?
(Give your answer in hours and minutes.)

6 a Express $0.\dot{5}\dot{1}$ as a fraction in its simplest form.

b Express $0.4\dot{5}\dot{1}$ as a fraction in its simplest form.

6 Area and volume

OBJECTIVES

D **Examiners would normally expect students who get a D grade to be able to:**

Find the area of a triangle, parallelogram, kite and trapezium

Find the area and perimeter of compound shapes

Convert between measures of area

Calculate the circumference of a circle to an appropriate degree of accuracy

Calculate the area of a circle to an appropriate degree of accuracy

C **Examiners would normally expect students who get a C grade also to be able to:**

Calculate the surface areas of prisms and cylinders

Calculate volumes of triangular prisms, parallelogram-based prisms and cylinders

Calculate the perimeter and area of a semicircle

Convert between measures of volume

A **Examiners would normally expect students who get an A grade also to be able to:**

Find the volume and surface area of pyramids, cones and spheres

Find the volume of the top cone of a truncated cone

Find the length of a major arc of a circle

Find the area of a major sector of a circle

Find the area of a segment of a circle

A* **Examiners would normally expect students who get an A* grade also to be able to:**

Find the volume of the frustum of a truncated cone

What you should already know ...

- Find the area of a rectangle, triangle, parallelogram and trapezium
- Find the volume of a solid by counting cubes, and state the units
- Find the volume of water in a container with dimensions such as 30 cm by 30 cm by 20 cm
- Find the height of a cuboid, given its volume, length and breadth
- The vocabulary specific to circles, such as circumference, radius, diameter, arc, chord, sector and segment

VOCABULARY

Shape – an enclosed space

Polygon – a closed two-dimensional shape made from straight lines

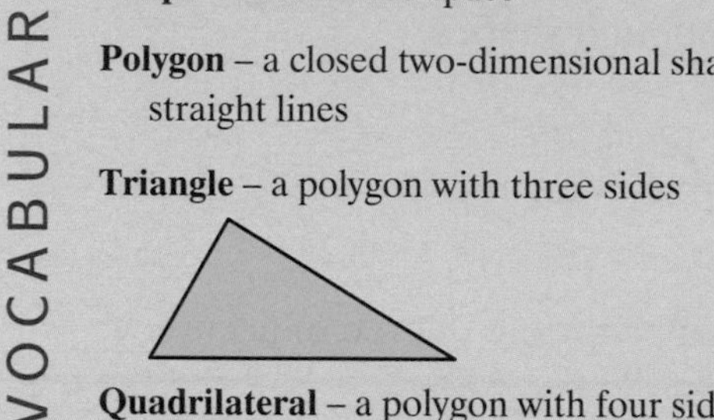

Triangle – a polygon with three sides

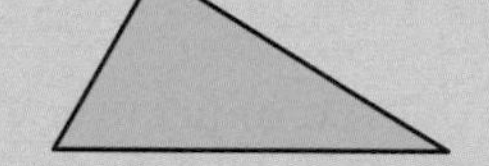

Quadrilateral – a polygon with four sides

Square – a quadrilateral with four equal sides and four right angles

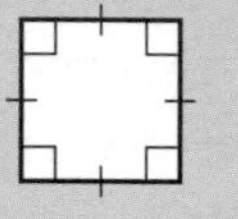

Rectangle – a quadrilateral with four right angles, and opposite sides equal in length

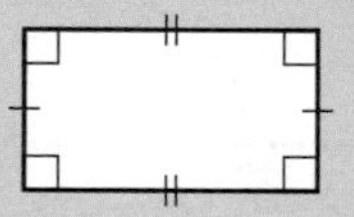

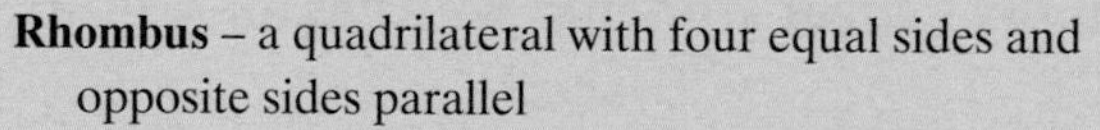

Rhombus – a quadrilateral with four equal sides and opposite sides parallel

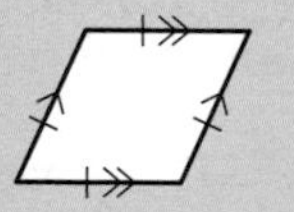

Parallelogram – a quadrilateral with opposite sides equal and parallel

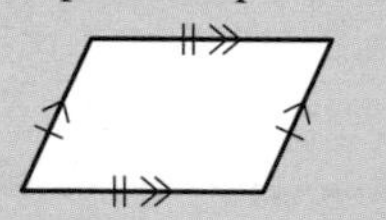

Trapezium (pl. **trapezia**) – a quadrilateral with one pair of parallel sides

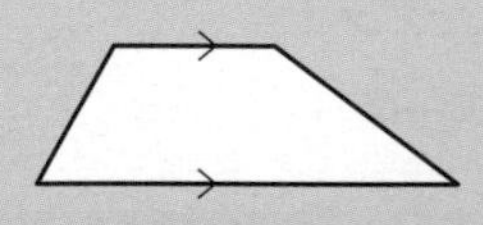

Kite – a quadrilateral with two pairs of equal adjacent sides

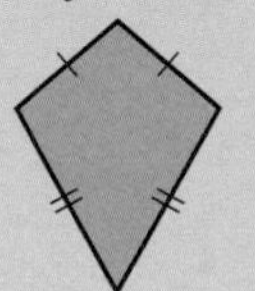

Pentagon – a polygon with five sides

Hexagon – a polygon with six sides

Octagon – a polygon with eight sides

Area – the amount of enclosed space inside a shape

Perimeter – the distance around an enclosed shape

Solid – a three-dimensional shape

Face – one of the flat surfaces of a solid

Cube – a solid with six identical square faces

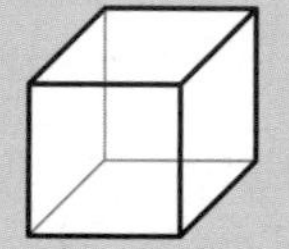

Cuboid – a solid with six rectangular faces (two or four of the faces can be squares)

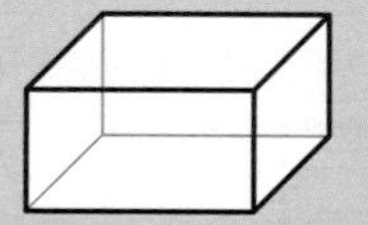

Vertex (pl. **vertices**) – the point where two or more edges meet

Edge – a line segment that joins two vertices of a solid

Cross-section – a cut at right angles to a face and usually at right angles to the length of a prism

Prism – a three-dimensional solid with two cross-sectional faces that are identical polygons, parallel to each other; all other faces are either parallelograms or rectangles

Prisms are named according to the cross-sectional face; for example,

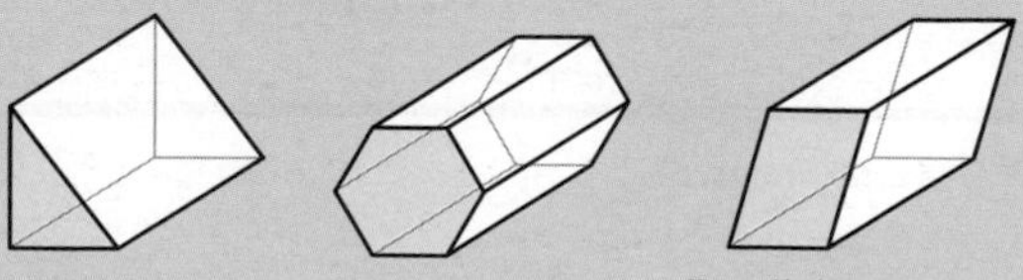

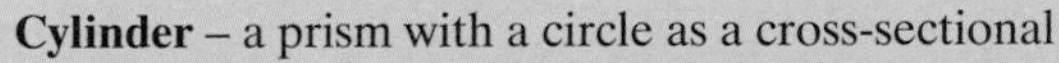

Cylinder – a prism with a circle as a cross-sectional face

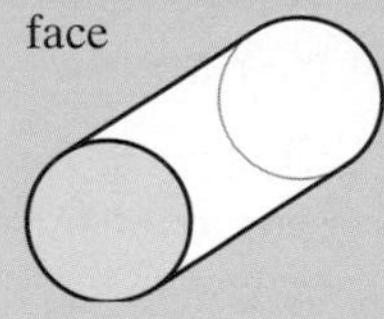

Volume – a measure of how much space fills a solid, commonly measured in cubic centimetres (cm^3) or cubic metres (m^3)

Capacity – the amount of liquid a hollow container can hold, commonly measured in litres (1 litre = 1000 cm^3)

Net – a two-dimensional shape made of polygons that can be folded to make a three-dimensional solid, for example,

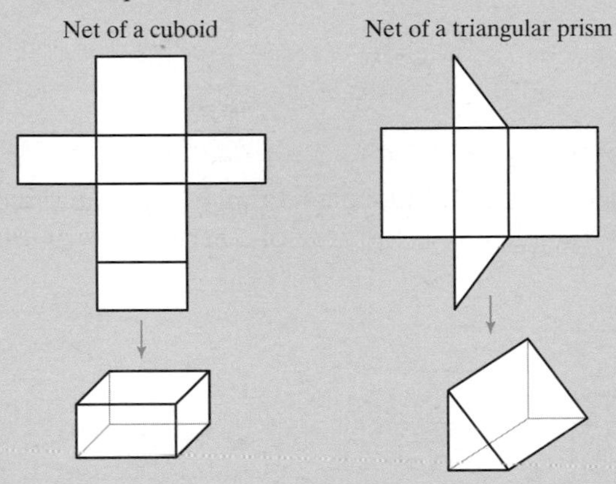

Sphere – a solid in which all the points on the surface are the same distance from the centre

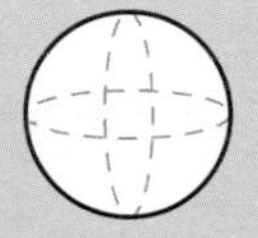

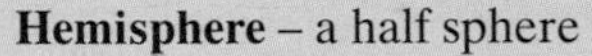

Hemisphere – a half sphere

Pyramid – a solid with a polygon as the base and one other vertex; all the vertices of the base are joined to this vertex forming triangular faces. Pyramids are named according to their base, for example,

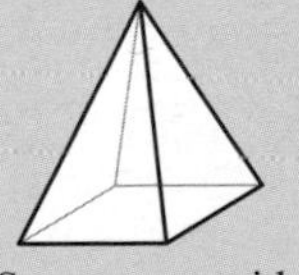

Square pyramid

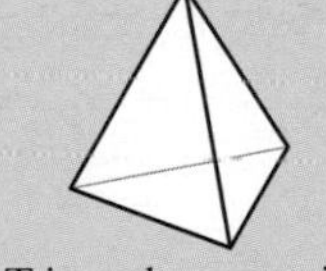

Triangular pyramid

Cone – a pyramid with a circular base and a curved surface rising to a vertex

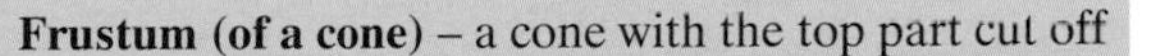

Frustum (of a cone) – a cone with the top part cut off

Circle – a shape formed by a set of points that are all the same distance from a fixed point (the centre of the circle)

Semicircle – one half of a circle

Quadrant (of a circle) – one quarter of a circle

Radius – the distance from the centre of a circle to any point on the circumference

Chord – a straight line joining two points on the circumference of a circle

Diameter – a chord passing through the centre of a circle; the diameter is twice the length of the radius

Circumference – the perimeter of a circle

Arc (of a circle) – part of the circumference of a circle; a minor arc is less than half the circumference and a major arc is greater than half the circumference

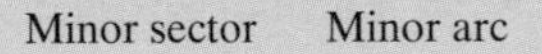

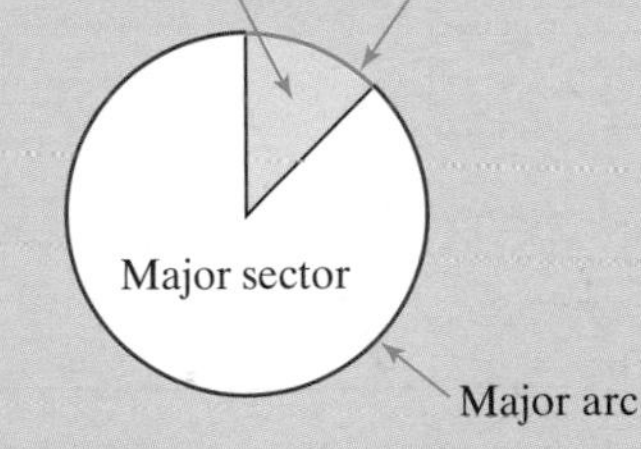

Sector (of a circle) – a region in a circle bounded by two radii and an arc

Segment – the region bounded by an arc and a chord

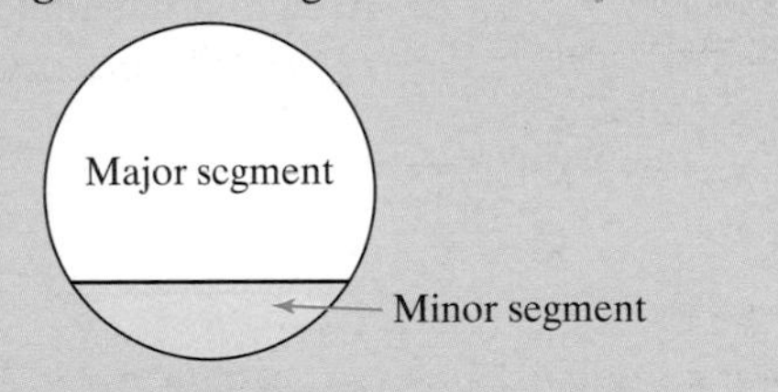

Tangent (to a circle) – a straight line that touches the circle at only one point

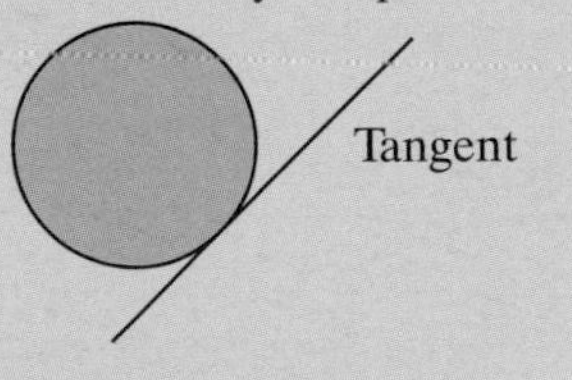

Learn 1 Perimeters and areas of triangles and parallelograms

Examples: Find the perimeter and area of the following shapes.

a Triangle

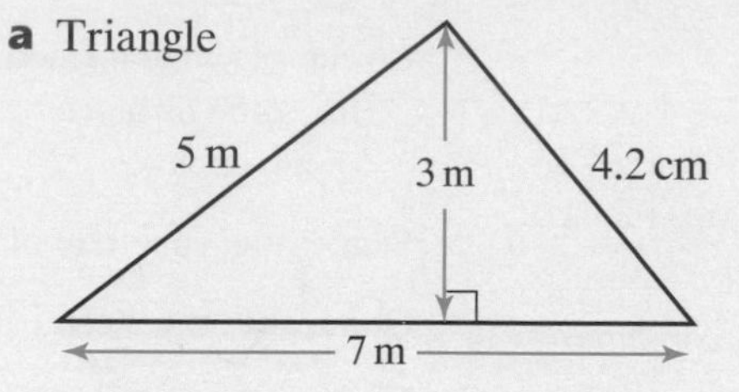

Not drawn accurately

Perimeter = 7 + 5 + 4.2 = 16.2 m
Area = $\frac{1}{2} \times 7 \times 3 = 10.5 \text{ m}^2$

Area of a triangle = $\frac{1}{2}$ × base × perpendicular height

Area = $\frac{1}{2} \times b \times h$

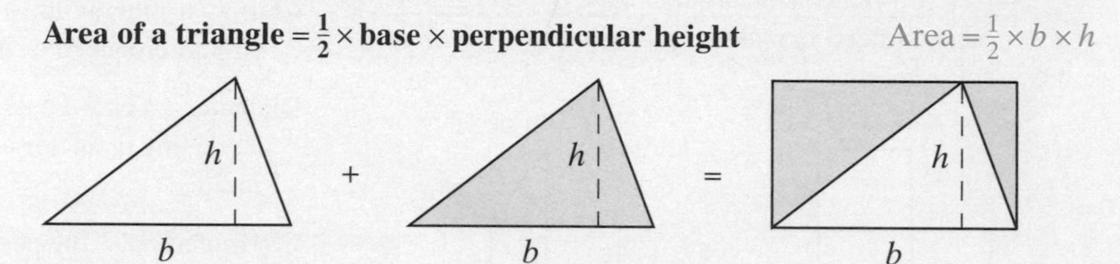

Two triangles can be joined together to make a rectangle with the same base and height. The area of a triangle is half the area of a rectangle with the same base and height.

b Parallelogram

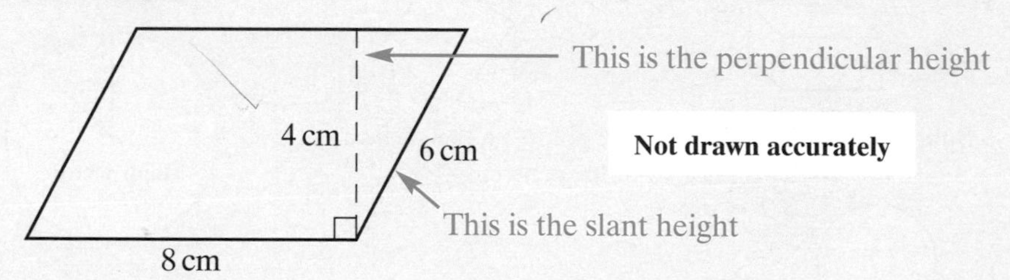

Not drawn accurately

Perimeter = 8 + 6 + 8 + 6 = 28 cm
Area = $8 \times 4 = 32 \text{ cm}^2$

Area of a parallelogram = base × perpendicular height

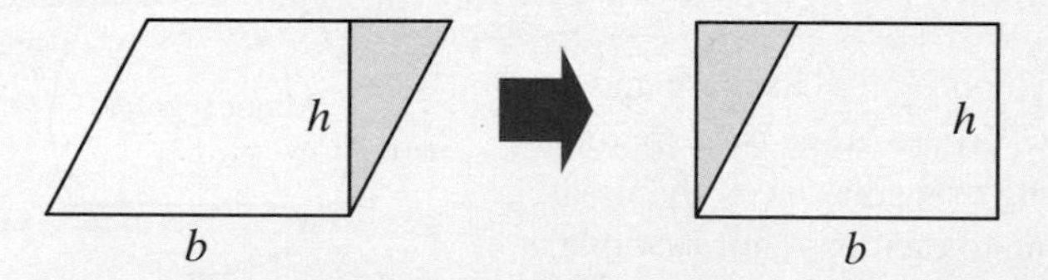

Area = $b \times h$

A parallelogram can be transformed into a rectangle as shown above. They both have the same area.

Apply 1

1 a Copy and complete the table for the following shapes made from 1 cm squares.

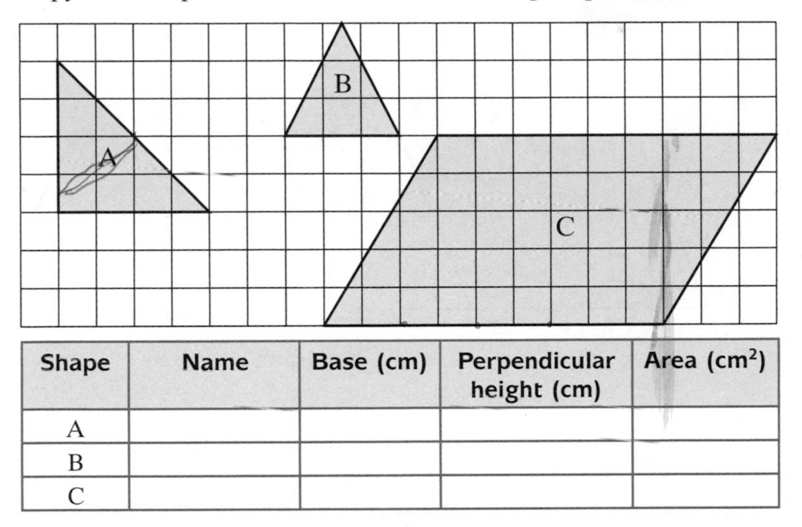

Shape	Name	Base (cm)	Perpendicular height (cm)	Area (cm^2)
A				
B				
C				

b Copy and complete the statements:

The area of a triangle = ... × ... × ...

The area of a parallelogram = ... × ...

2 Find the area of each of the following shapes:

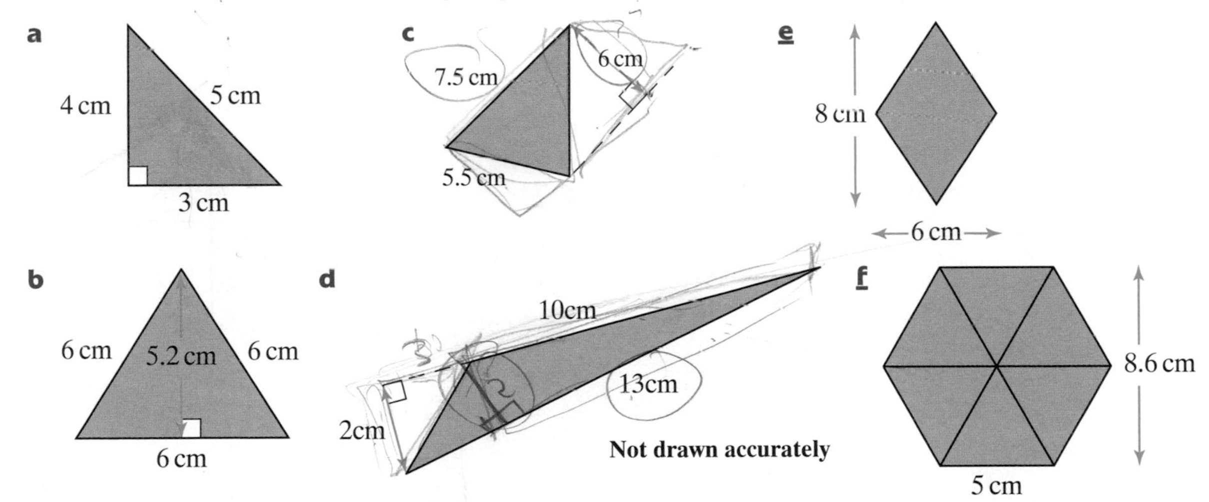

3 Five students are trying to find the area of the following triangle:

- Sameera thinks the answer is 48 cm^2 because $6 \times 8 = 48$
- Bruce thinks the answer is 30 cm^2 because $\frac{1}{2} \times 6 \times 10 = 30$
- Cassie thinks the answer is 24 cm^2 because $6 + 8 + 10 = 24$
- Des thinks the answer is because 40 cm^2 because $\frac{1}{2} \times 8 \times 10 = 40$
- Elliot thinks the answer is 24 cm^2 because $\frac{1}{2} \times 6 \times 8 = 24$

Who is correct? What mistakes have the other students made?

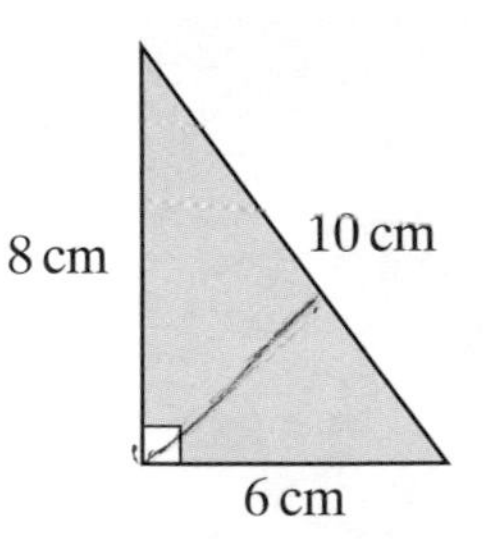

4 a The area of a parallelogram is 60 cm^2. Sketch five parallelograms with that area, showing the dimensions in each case.

b The area of a triangle is 30 cm^2. Sketch five triangles with that area, showing the dimensions in each case.

5 Find the area of each of the following parallelograms:

a 5 cm, 6 cm

b 4.5 cm, 4 cm, 6 cm

c 3 cm, 15 cm, 6 cm

Not drawn accurately

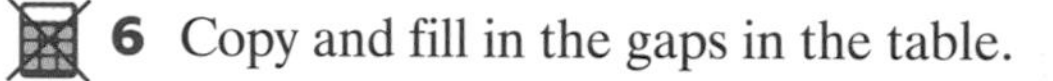

6 Copy and fill in the gaps in the table.

	Shape (Parallelogram/Triangle)	Base	Perpendicular height	Area
a	Parallelogram	5 cm	4 cm	
b	Triangle	5 cm	4 cm	
c		10 cm	2 cm	10 cm^2
d	Parallelogram	4 cm		8 cm^2
e	Triangle	4 cm		8 cm^2
f	Parallelogram	0.5 m	20 cm	

7 Get Real!

Reece wants to make a kite. The yellow silk costs £5 per square metre and the green silk costs £7 per square metre. Will he be able to buy enough silk with £3?

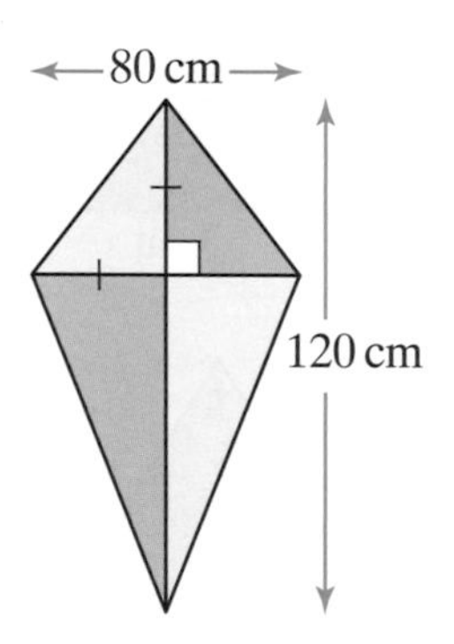

Explore

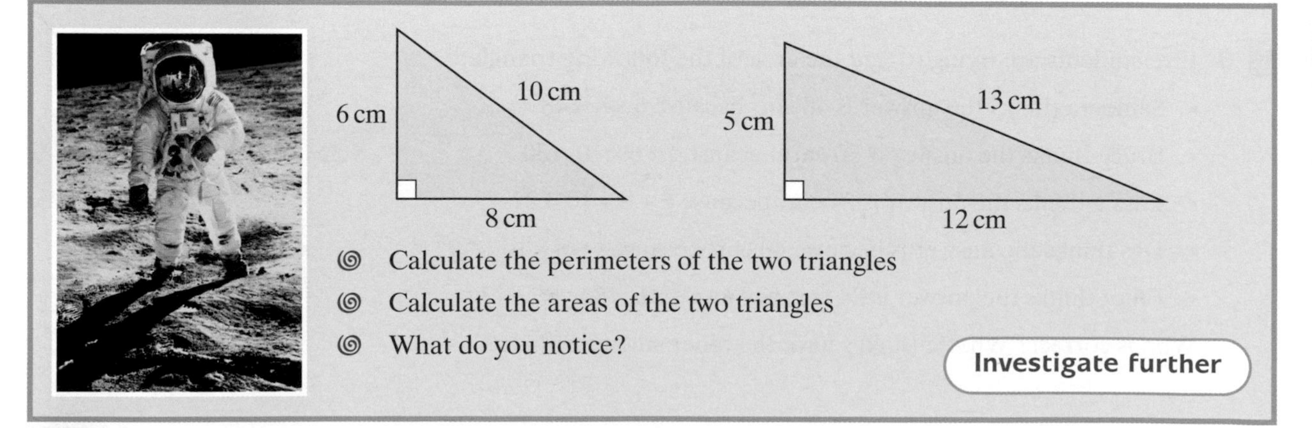

- Calculate the perimeters of the two triangles
- Calculate the areas of the two triangles
- What do you notice?

Investigate further

Explore

- Calculate the area of the parallelogram
- Find the dimensions of parallelograms with an area that is one half of the area of the parallelogram shown
- Find the dimensions of parallelograms with an area that is one quarter of the area of the parallelogram shown

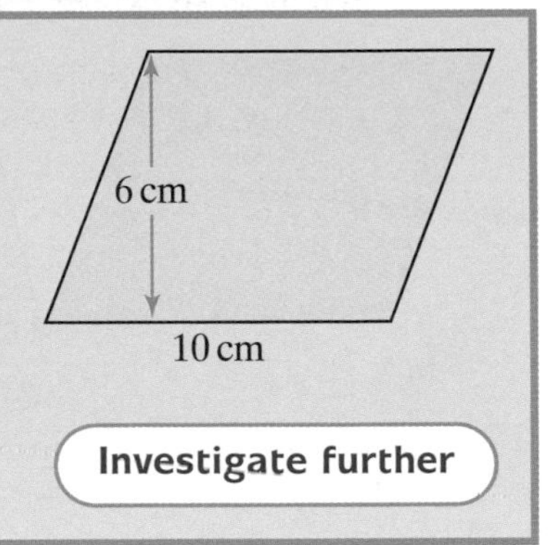

Learn 2 Areas of compound shapes

Examples: Find the area of the following shapes:

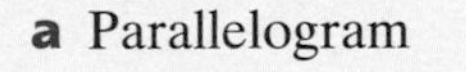

a Parallelogram

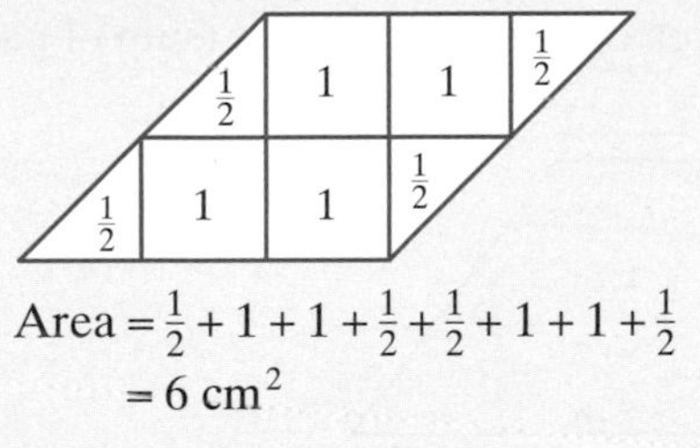

Area $= \frac{1}{2} + 1 + 1 + \frac{1}{2} + \frac{1}{2} + 1 + 1 + \frac{1}{2}$

$= 6 \text{ cm}^2$

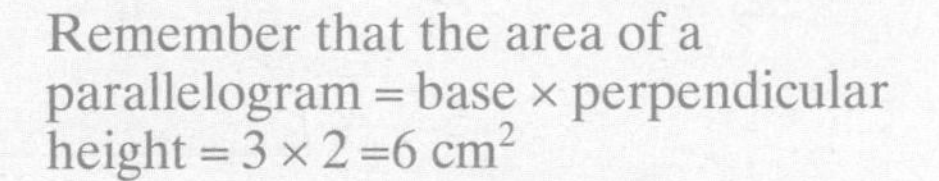

Remember that the area of a parallelogram = base × perpendicular height = $3 \times 2 = 6 \text{ cm}^2$

b Kite

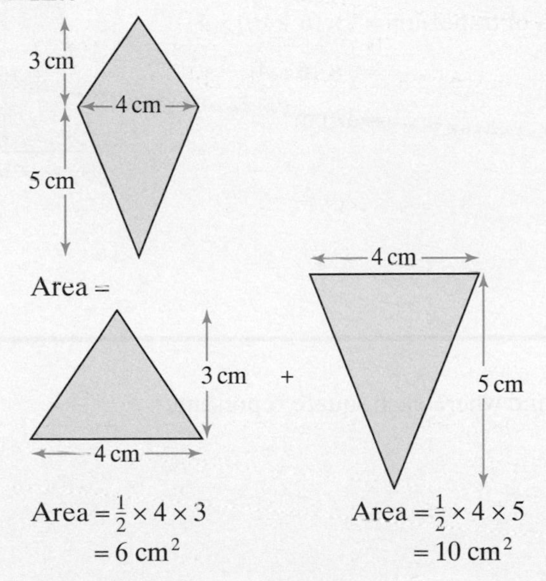

Area $= \frac{1}{2} \times 4 \times 3$
$= 6 \text{ cm}^2$

Area $= \frac{1}{2} \times 4 \times 5$
$= 10 \text{ cm}^2$

Total area $= 6 + 10 = 16 \text{ cm}^2$

In general the area of a kite $= \frac{1}{2}$ height × width

$= \frac{1}{2} \times (3 + 5) \times 4$

$= \frac{1}{2} \times 8 \times 4$

$= 16 \text{ cm}^2$

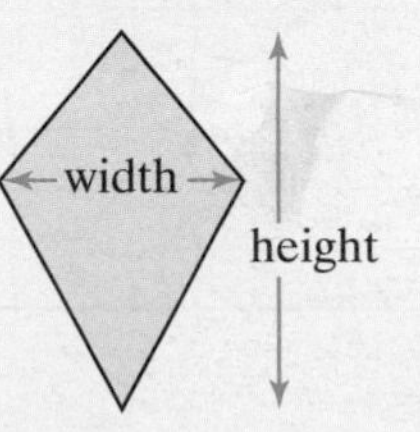

c Trapezium

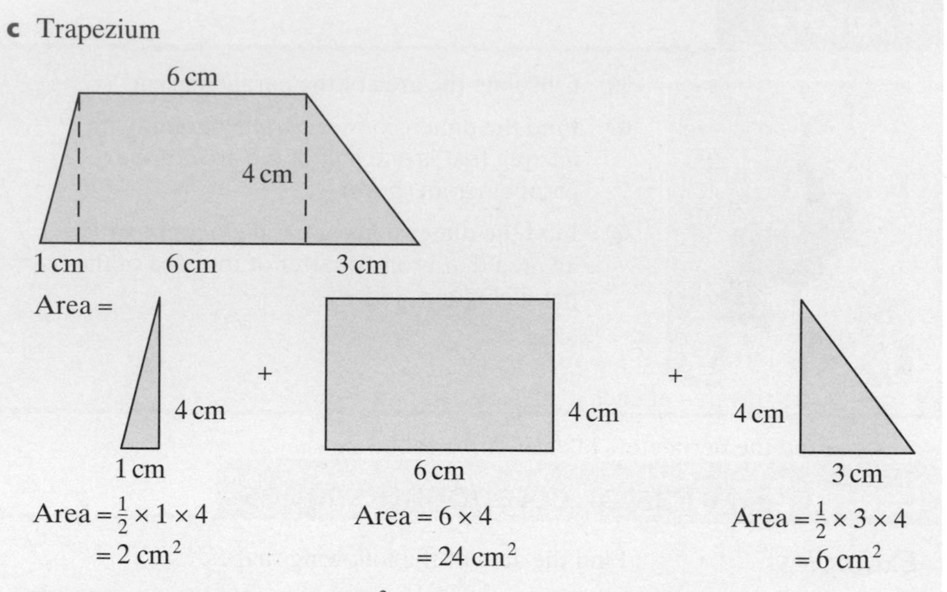

Area $= \frac{1}{2} \times 1 \times 4$
$= 2\ \text{cm}^2$

Area $= 6 \times 4$
$= 24\ \text{cm}^2$

Area $= \frac{1}{2} \times 3 \times 4$
$= 6\ \text{cm}^2$

Total area $= 2 + 24 + 6 = 32\ \text{cm}^2$

In general the area of a trapezium $= \frac{1}{2} \times$ (sum of parallel sides) $\times h$

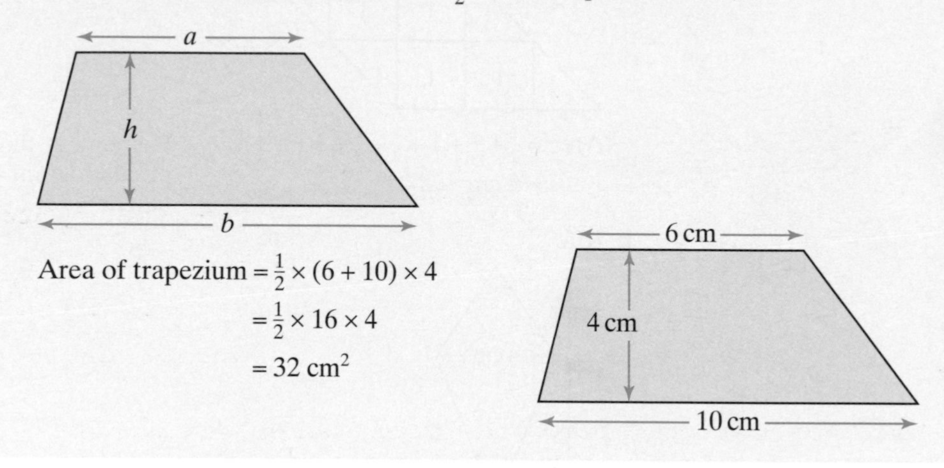

Area of trapezium $= \frac{1}{2} \times (6 + 10) \times 4$
$= \frac{1}{2} \times 16 \times 4$
$= 32\ \text{cm}^2$

Apply 2

1 Estimate the area of the island where each square represents one square mile.

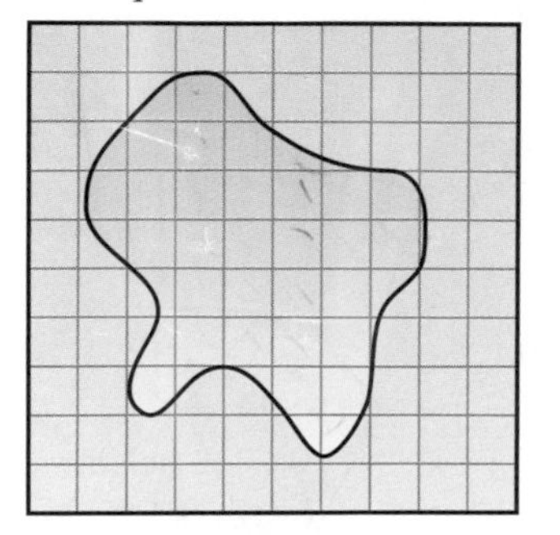

2

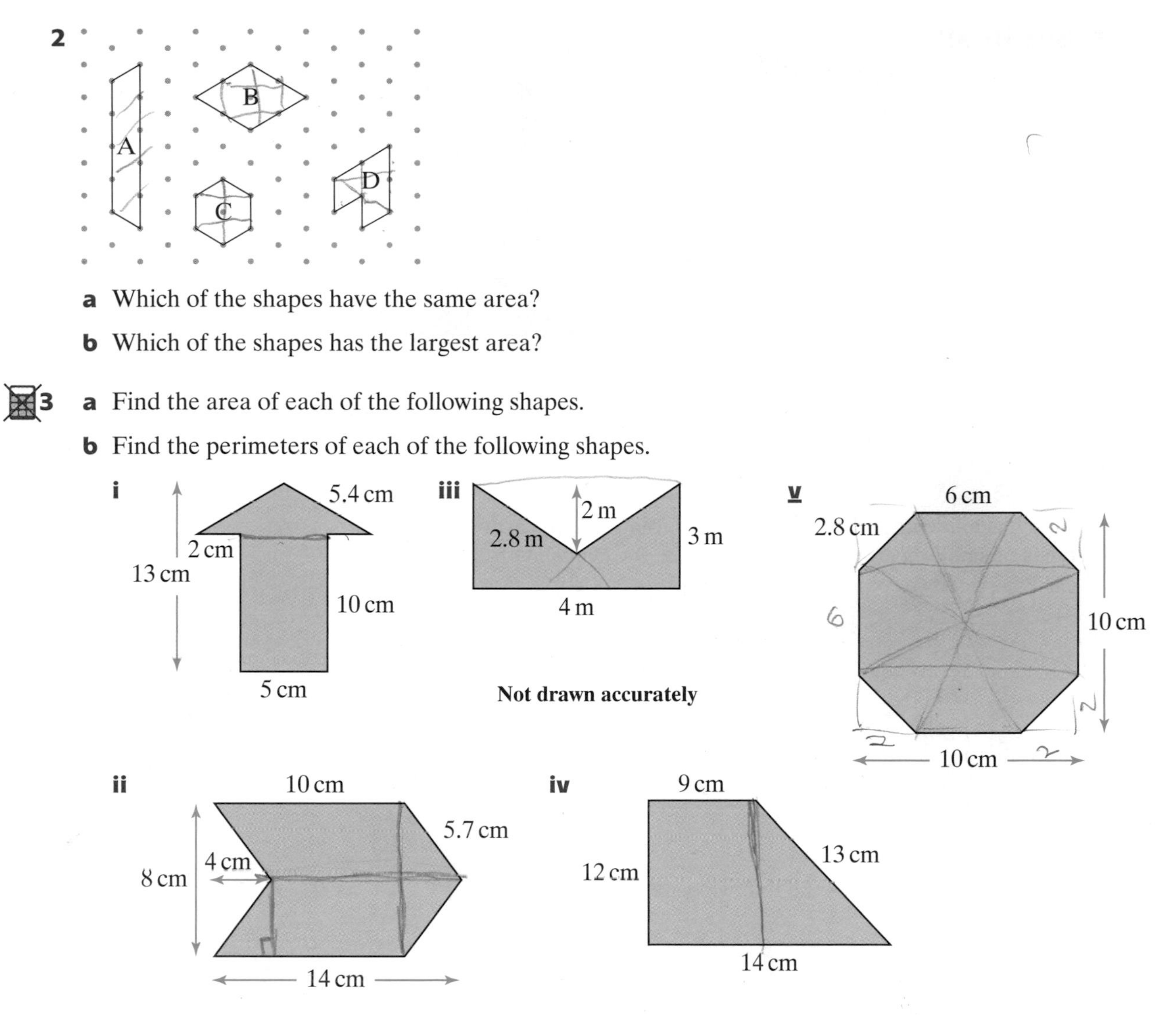

a Which of the shapes have the same area?

b Which of the shapes has the largest area?

3 **a** Find the area of each of the following shapes.

b Find the perimeters of each of the following shapes.

i, ii, iii, iv, v

4 Roberta is finding the area of a hexagon. She spots that she can split it into a rectangle and two triangles.

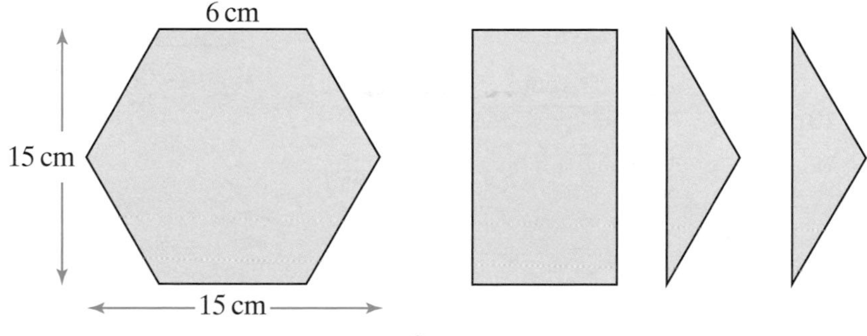

Area of rectangle $= 6 \times 15 = 90\ \text{cm}^2$
Area of triangle $= 15 \times 4.5 = 67.5\ \text{cm}^2$

Area of hexagon $= 90 + 2 \times 67.5 = 225\ \text{cm}^2$

Do you agree with Roberta? Give reasons for your answer.

5 Get Real!

George is varnishing the front of Fred's doghouse. How many 1 litre tins of varnish does he need? (The label on the tin has instructions that 1 litre of varnish covers $0.5\ \text{m}^2$.)

6 Get Real!

Jane has made an 'EXIT' sign using a large piece of cardboard and two types of small rectangles to make the letters. Calculate the area of grey card she needs to paint.

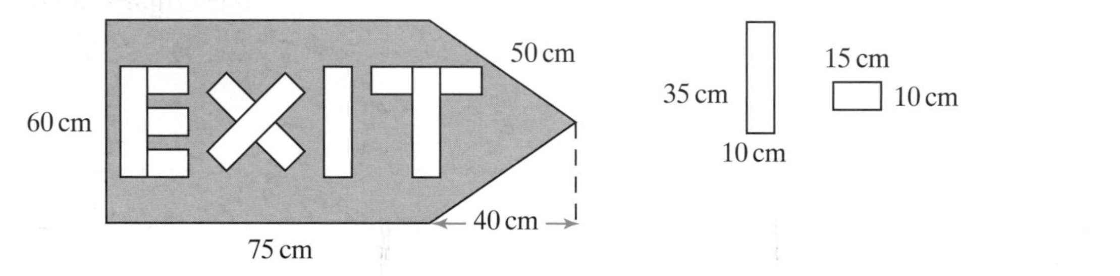

7 Find the area of each of the following shapes.

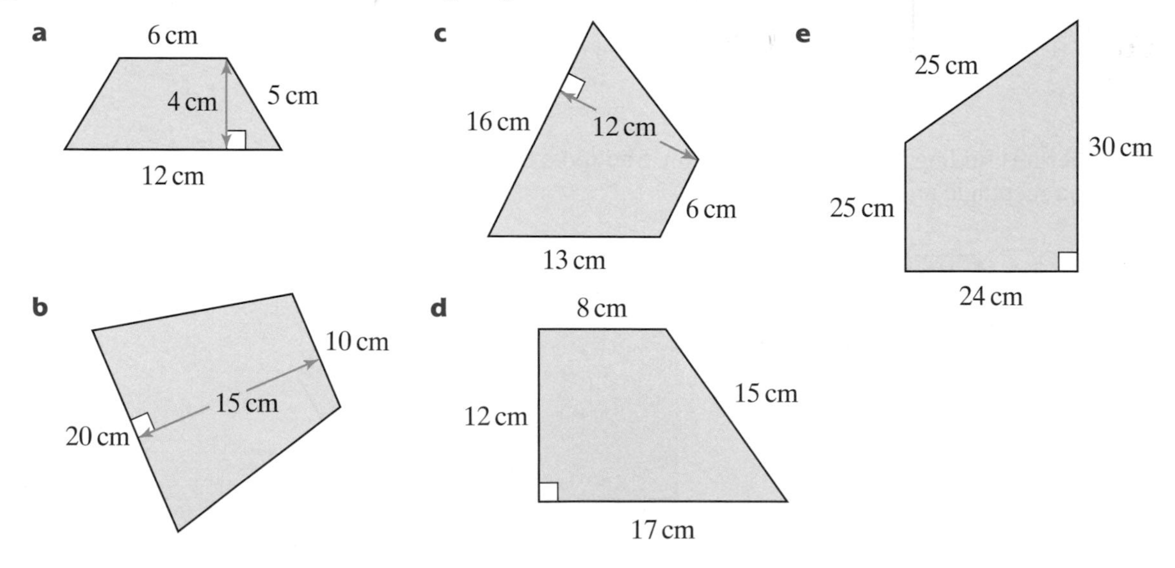

8 Sketch five trapezia with the area $25\ \text{cm}^2$, stating clearly the dimensions in each case.

Explore

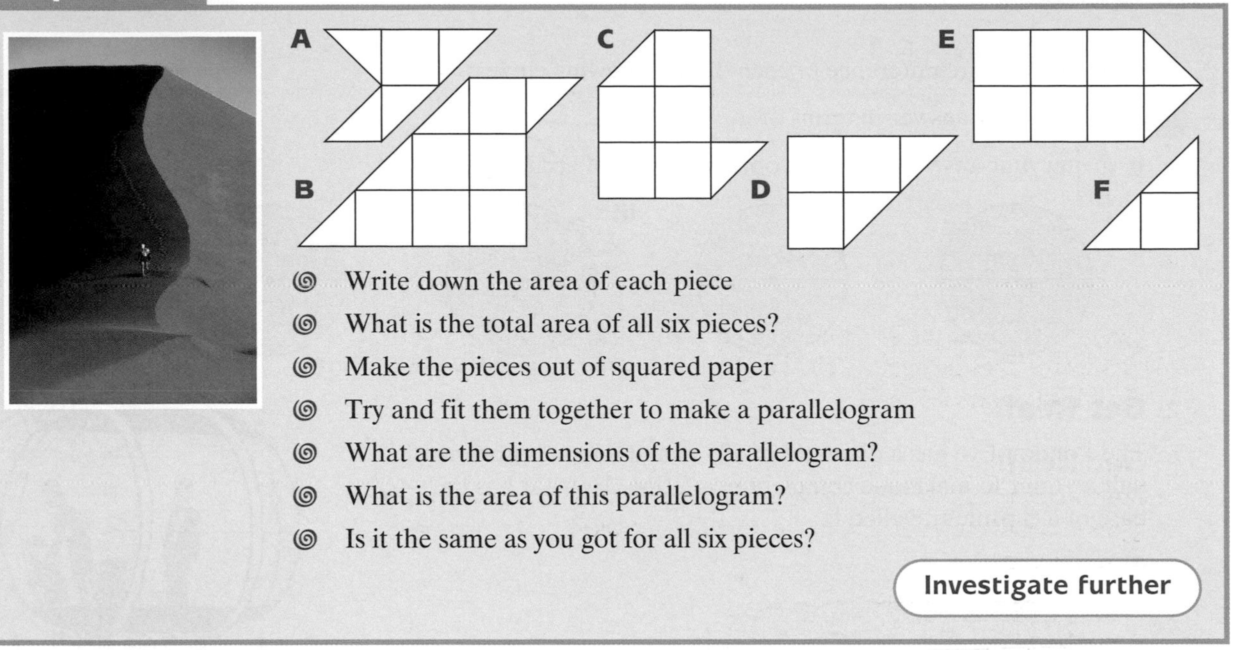

- Write down the area of each piece
- What is the total area of all six pieces?
- Make the pieces out of squared paper
- Try and fit them together to make a parallelogram
- What are the dimensions of the parallelogram?
- What is the area of this parallelogram?
- Is it the same as you got for all six pieces?

Investigate further

Explore

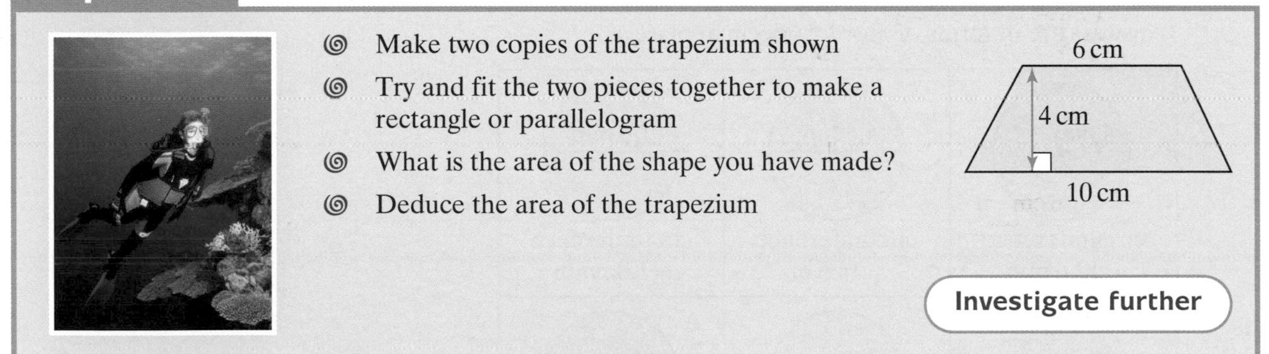

- Make two copies of the trapezium shown
- Try and fit the two pieces together to make a rectangle or parallelogram
- What is the area of the shape you have made?
- Deduce the area of the trapezium

Investigate further

Learn 3 Circumferences of circles

Examples:

Calculate the circumference of this circle:

a leaving your answer in terms of π

b giving your answer to 3 significant figures.

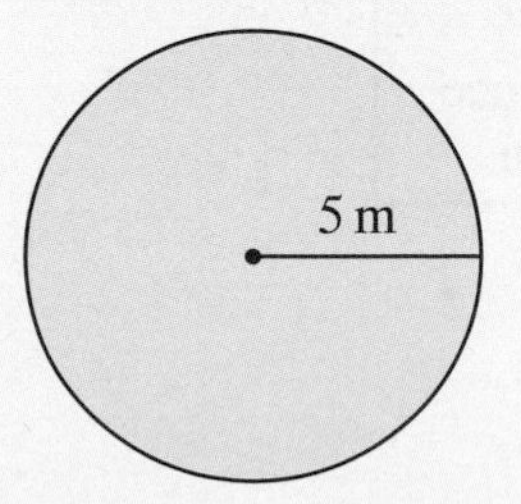

You may be asked to leave your answers in terms of π on the non-calculator paper

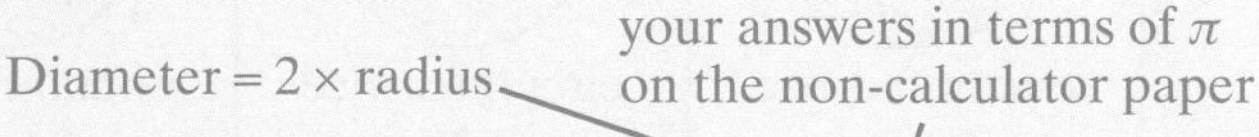

Diameter = 2 × radius

Circumference is the length around the outside of a circle

a Circumference $= \pi d = \pi \times 10 = 10\pi$ m

b Circumference $= \pi d = \pi \times 10 = 31.4$ m (3 s.f.)

You can take π as 3.14 or use your calculator

Apply 3

1 Calculate the circumference of each of the following circles:

a leaving your answer in terms of π

b giving your answer to an appropriate degree of accuracy.

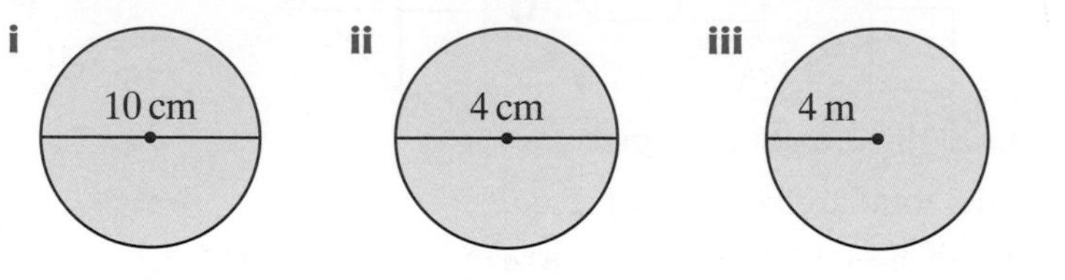

2 Get Real!

The London Eye has a diameter of 135 m and takes approximately half an hour to make one complete revolution. How far has the base of a capsule travelled in:

a 30 minutes

b 15 minutes

c 1 hour?

3 The circumference of a circle is π times d.
Can you find a correct line of three?
(Answers are in terms of π or to 1 decimal place.)

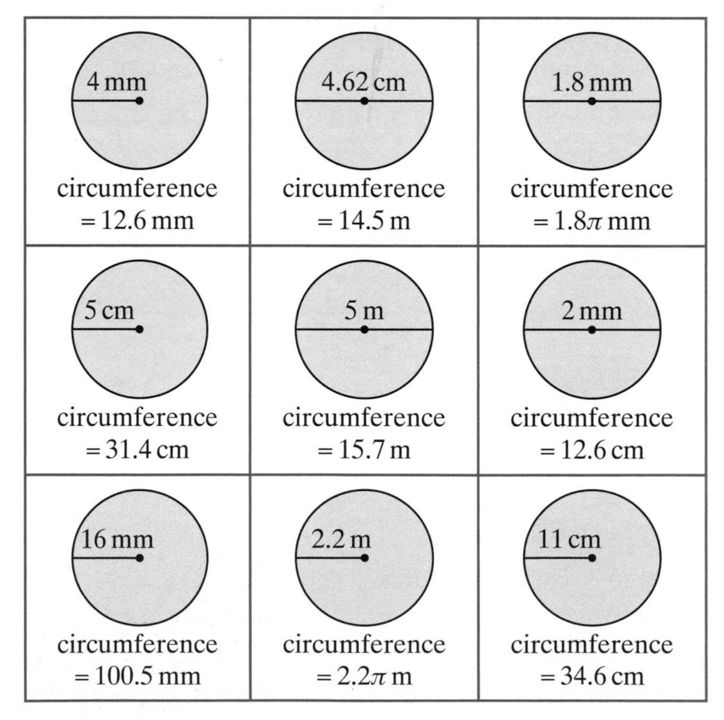

4 mm circumference = 12.6 mm	4.62 cm circumference = 14.5 m	1.8 mm circumference = 1.8π mm
5 cm circumference = 31.4 cm	5 m circumference = 15.7 m	2 mm circumference = 12.6 cm
16 mm circumference = 100.5 mm	2.2 m circumference = 2.2π m	11 cm circumference = 34.6 cm

4 Calculate the perimeter of each of the following shapes:

a leaving your answer in terms of π

b giving your answers to an appropriate degree of accuracy.

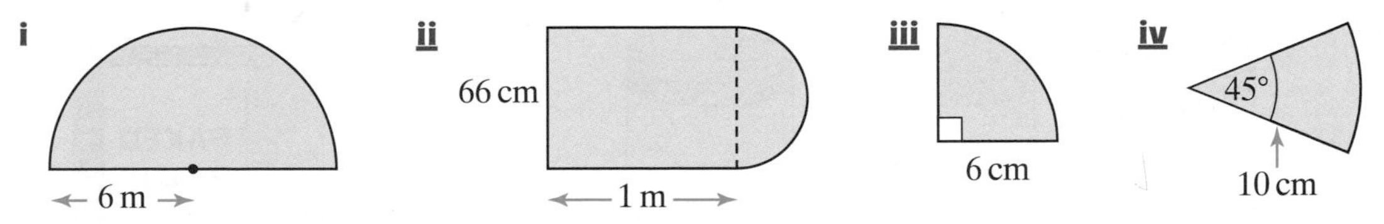

5 Get Real!

Ahmed lives 1.5 km from school. The diameter of his bike's wheel is 80 cm. How many complete revolutions does the wheel turn during Ahmed's journey to school?

6 Get Real!

Jack and Susan have a race around the track shown in the diagram. Jack is in lane 1 and Susan in lane 8.

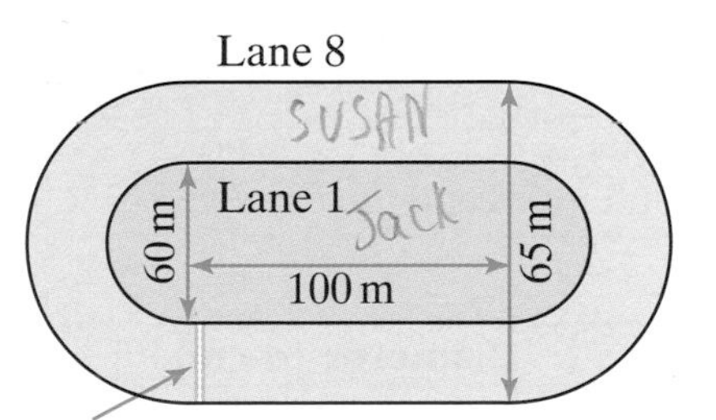

a How far does Jack run?

b How far does Susan run?

c How can you make the race fair?

7 Copy and complete the following table, giving your answers to an appropriate degree of accuracy.

Radius	Diameter	Circumference
	5 cm	
4 m		
		10 mm
		15π cm

8 Get Real!

Ahmed's CDs have a circumference of 40 cm.
He wants to make square covers for them.
Find, correct to 1 d.p., the dimensions of the smallest square into which the CDs will fit.

9 Get Real!

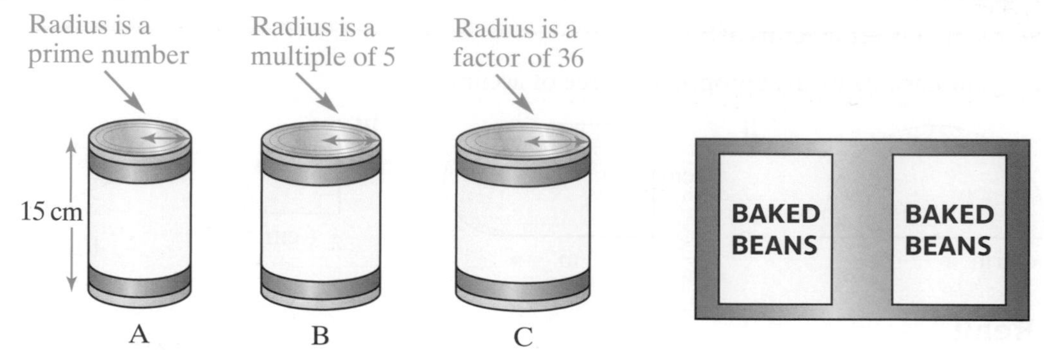

The labels have fallen off three tins.
The area of the 'Baked Beans' label is 942.5 cm^2 (1 d.p.).

a Which tin does it fit – A, B or C?

b Is the area of the label approximately 1 square metre?
Give a reason for your answer.

Explore

- Find five circular objects
- For each object, measure the circumference and the diameter
- Copy and complete the table

Object	Circumference (c cm)	Diameter (d cm)	$c \div d$

Investigate further

Learn 4 Areas of circles

Examples:

Calculate the area of this circle:

a leaving your answer in terms of π

b giving your answer to 3 significant figures.

Use the fact that $A = \pi \times r^2$ where A is the area and r is the radius

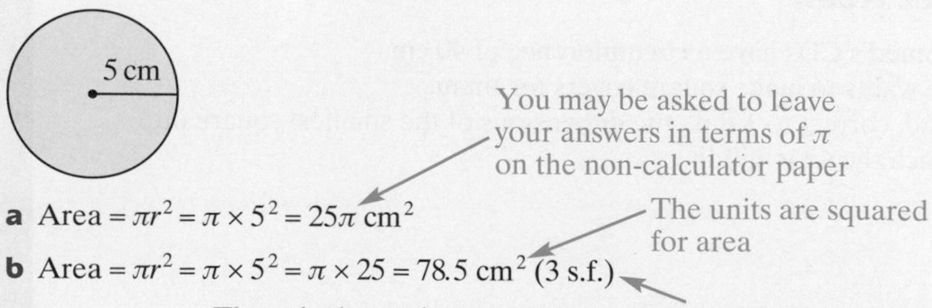

You may be asked to leave your answers in terms of π on the non-calculator paper

a Area $= \pi r^2 = \pi \times 5^2 = 25\pi\text{ cm}^2$

The units are squared for area

b Area $= \pi r^2 = \pi \times 5^2 = \pi \times 25 = 78.5\text{ cm}^2$ (3 s.f.)

The calculator gives more decimal places but you need to round to an appropriate degree of accuracy (usually 3 s.f. or 2 d.p.)

Apply 4

1 Calculate the area of each of the following circles:

a leaving your answer in terms of π

b giving your answer to 2 decimal places.

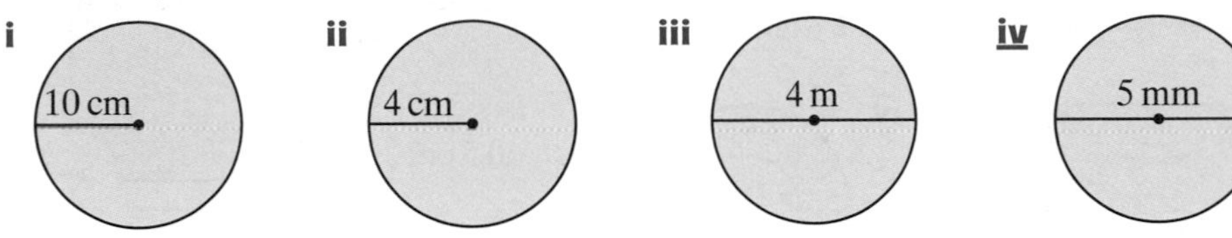

2 Joy and Jan are finding the area of a CD.

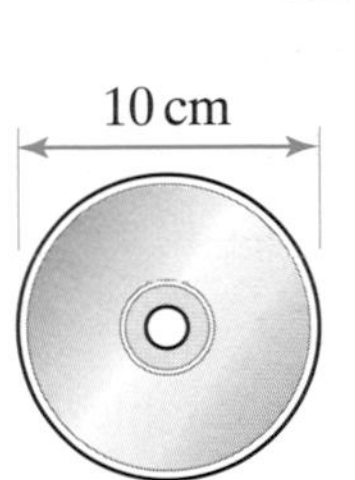

a This is Joy's method:
Area $= \pi \times d = \pi \times 10 = 31.42\ \text{cm}^2$ (to 2 d.p.)
Do you agree with Joy's answer? Justify your answer.

b This is Jan's method:
Area of the circle $= \pi r^2 = \pi \times 5^2 = 246.74\ \text{cm}^2$ (to 2 d.p.)
Calculator buttons used: [π] [×] [5] [=] [x^2] [=]

Do you agree with Jan's answer? Justify your answer.

3 Copy this and match each circle with the correct area.
Fill in the missing areas and the missing radius.

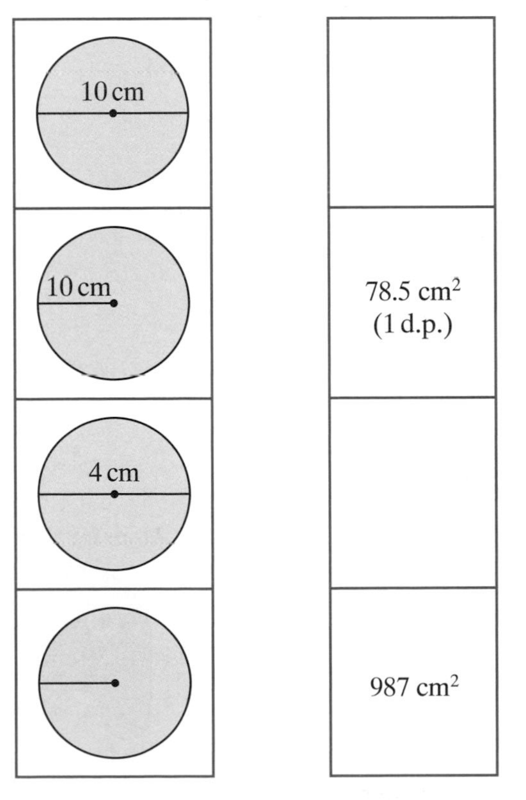

4 Calculate the total area of each of the following shapes:

a leaving your answer in terms of π

b giving your answer to an appropriate degree of accuracy.

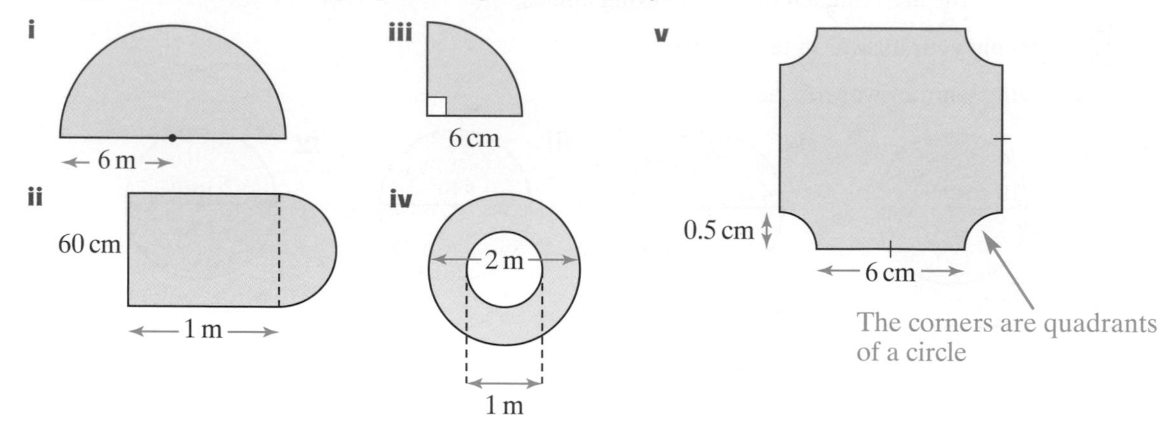

5 Copy and complete this table, giving your answers to an appropriate degree of accuracy.

Radius	Diameter	Area
	5 cm	
4 m		
		$10\ mm^2$
		$16\pi\ cm^2$

6 **Get Real!**

Steve wants to paint a tribute to his favourite computer game on his bedroom wall. His parents aren't so keen! They will only allow him to do it if the black paint covers no more than two thirds of the wall.

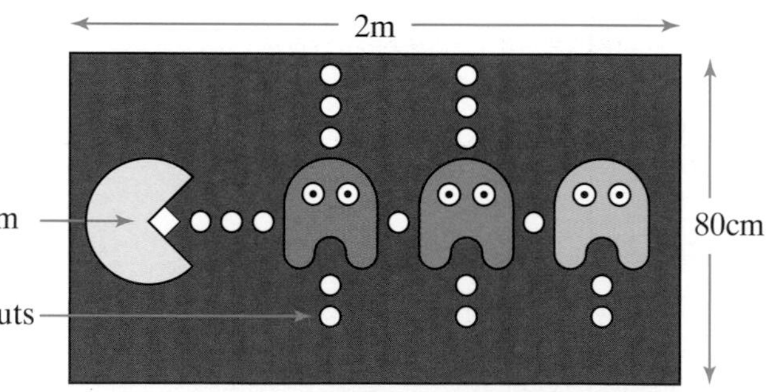

a Calculate the area of one monster (including the eyes).

b Calculate the total area of the doughnuts if each one has a radius of 5 cm.

c Calculate the area covered by the yellow doughnut eater.

HINT Note the right angle.

d Calculate the total area covered by the monsters, doughnuts and doughnut eater.

e What percentage of the wall will be black?

f Will Steve's parents allow him to paint his wall?

Monster details

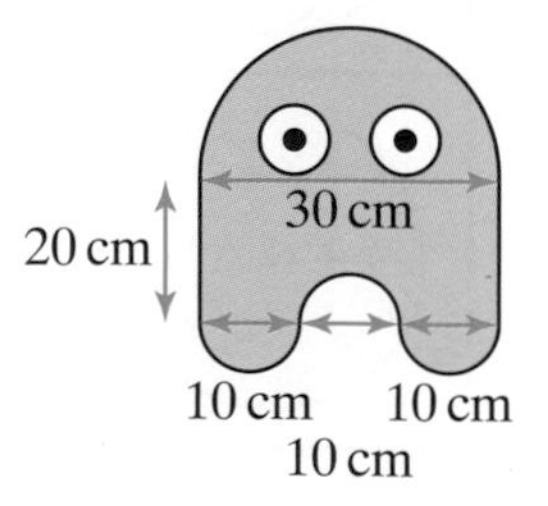

Explore

- Divide a circle into six equal pieces and cut out the pieces
- Try to make the following shape:

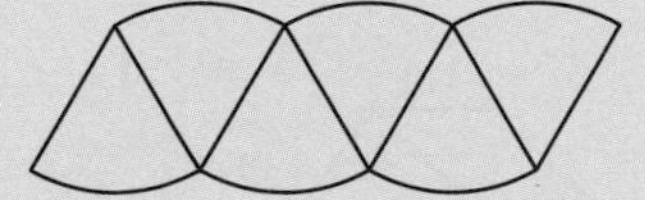

- Measure the length and height of your shape
- Work out the approximate area of the shape
- Repeat with the same size circle but this time divided into 10 pieces

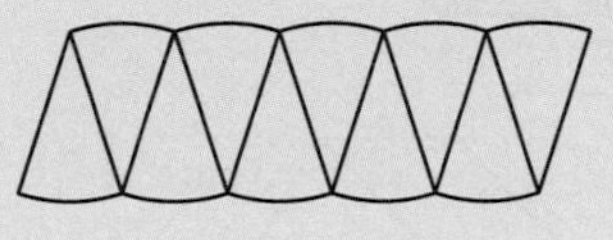

Investigate further

Learn 5 Volumes of cubes, cuboids, prisms and cylinders

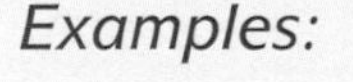

Examples:

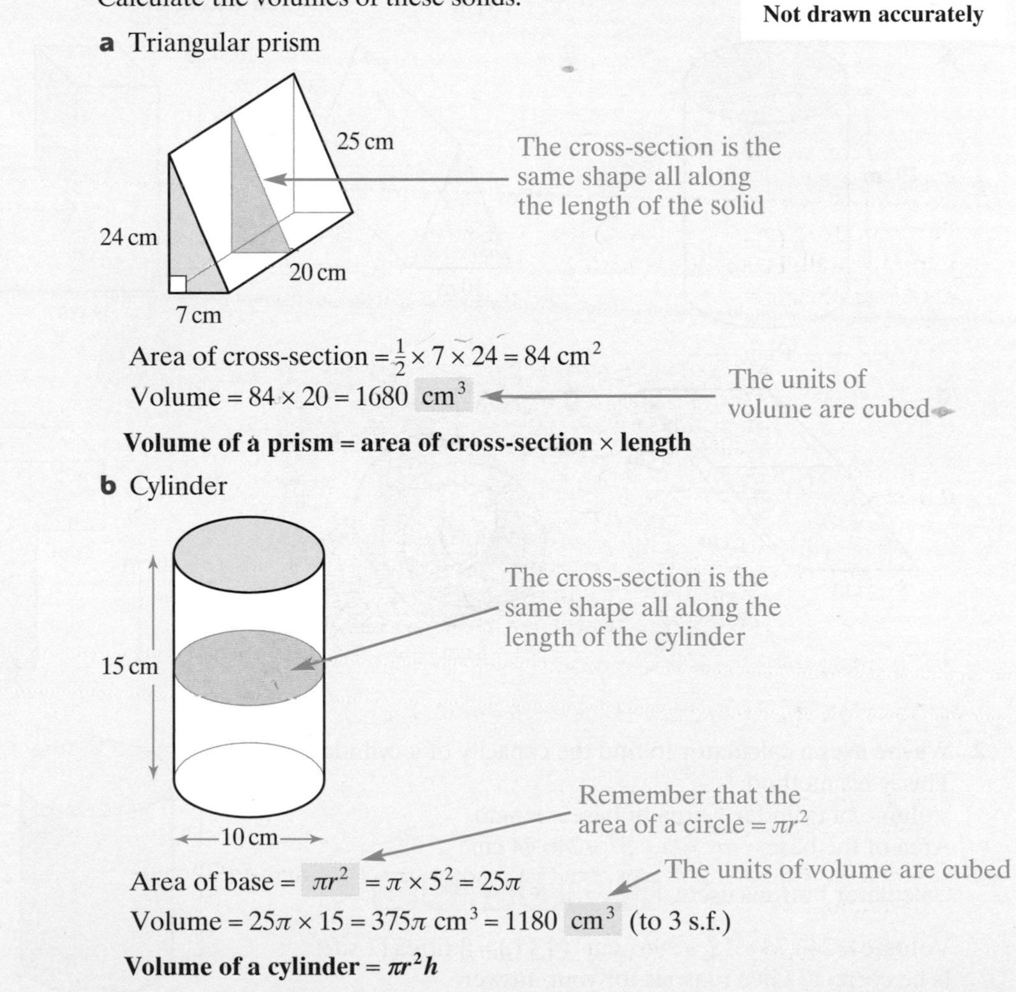

Calculate the volumes of these solids.

Not drawn accurately

a Triangular prism

Area of cross-section $= \frac{1}{2} \times 7 \times 24 = 84\text{ cm}^2$

Volume $= 84 \times 20 = 1680\text{ cm}^3$ — The units of volume are cubed

Volume of a prism = area of cross-section × length

b Cylinder

Area of base $= \pi r^2 = \pi \times 5^2 = 25\pi$ — Remember that the area of a circle $= \pi r^2$

Volume $= 25\pi \times 15 = 375\pi\text{ cm}^3 = 1180\text{ cm}^3$ (to 3 s.f.) — The units of volume are cubed

Volume of a cylinder $= \pi r^2 h$

You may be asked to leave your answer in terms of π on a non-calculator paper

Apply 5

1 Calculate the volumes of these solids leaving your answer as multiples of π where appropriate.

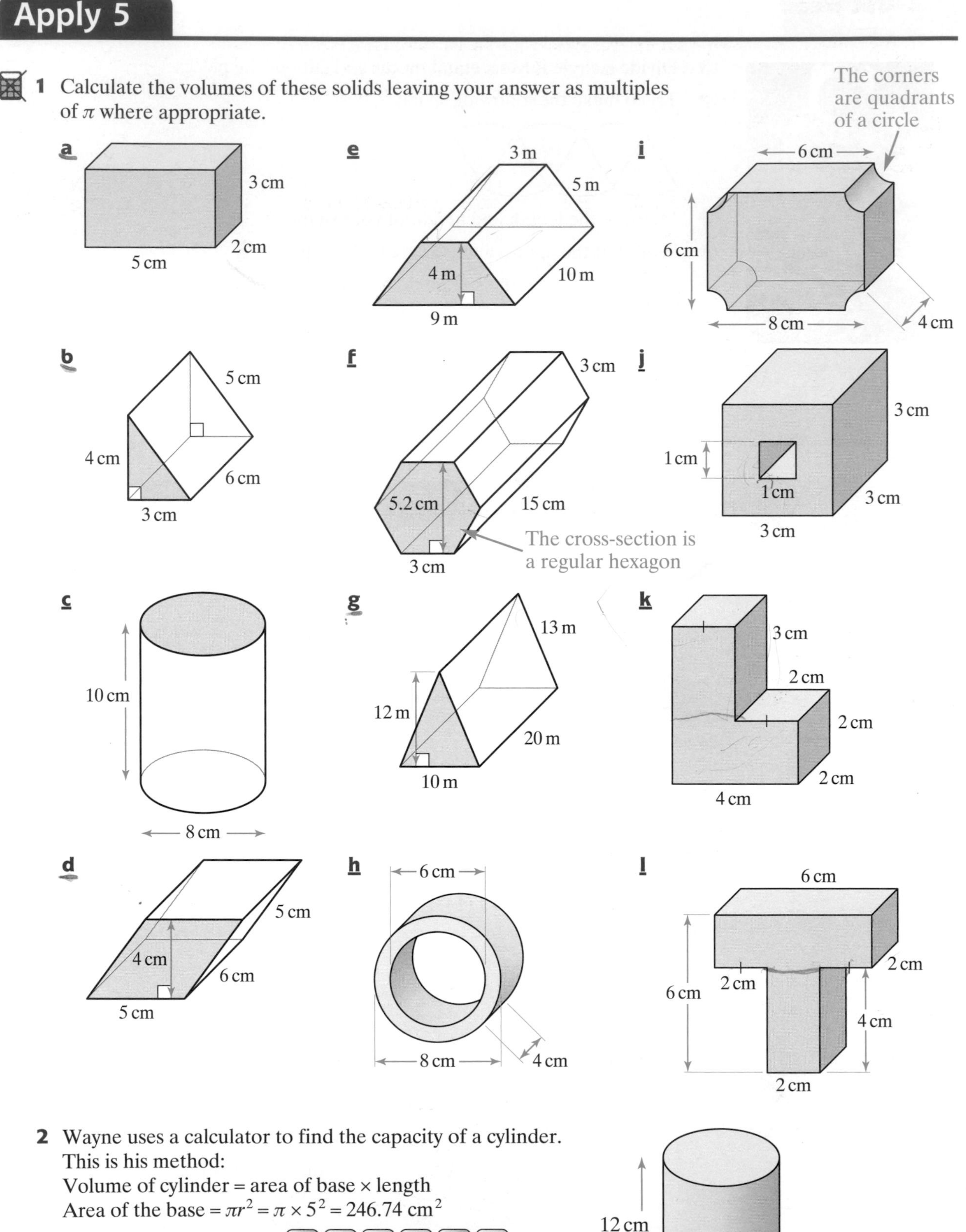

2 Wayne uses a calculator to find the capacity of a cylinder.
This is his method:
Volume of cylinder = area of base × length
Area of the base = $\pi r^2 = \pi \times 5^2 = 246.74\ \text{cm}^2$

Calculator buttons used: π × 5 = ² =

Volume = $246.74 \times 12 = 2961\ \text{cm}^3$ (4 s.f.) = 3 litres (1 s.f.)
Is he correct? Give reasons for your answer.

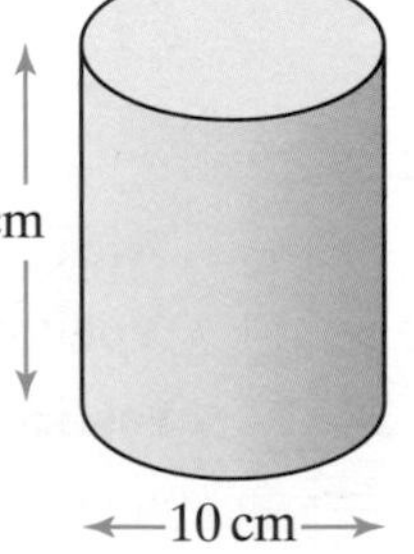

3 Get Real!

Steve has made a swing set for his children in the garden. The swing set has 4 legs. He must cement the legs of the frame into the ground. Each hole must measure 30 cm × 30 cm × 30 cm. How many litres of cement does Steve need?

4 Get Real!

Calculate the capacity of Farmer Vines' barn, to 3 significant figures.

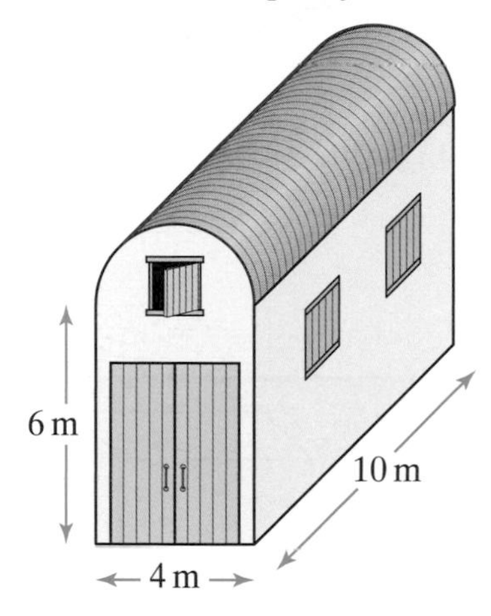

The cross-section of the barn is:

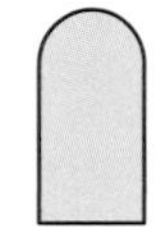

5 Get Real!

a Beth is having a paddling pool party. The party starts in 30 minutes. Beth's dad is filling the paddling pool with his hosepipe at a rate of 150 litres/min. Will it be ready in time?

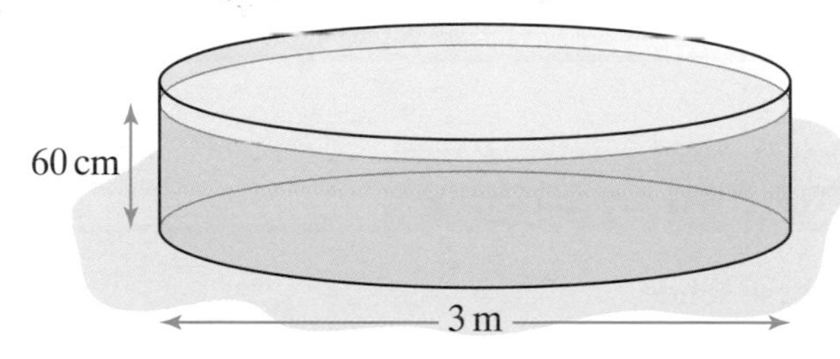

b Given that 1 cm^3 of water has a mass of 1 g, how heavy is the water in the paddling pool?

6 The volume of a solid is 60 cm^3 (to 1 significant figure). Find the dimensions of five different solids with this volume.

7 a Calculate the volumes of these cubes.

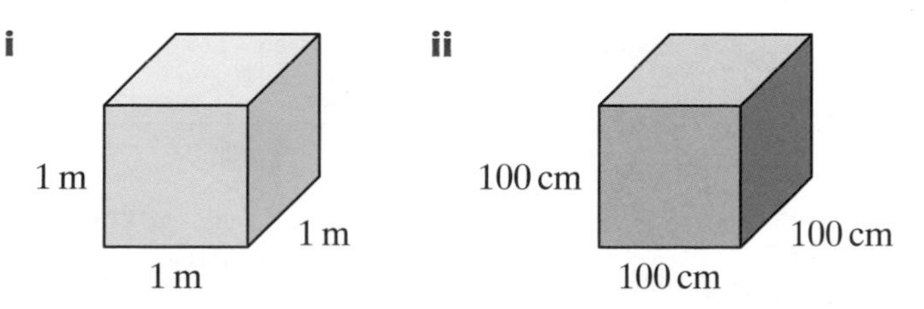

b What do you notice about the two cubes?

c Look at your answers to part **a** and complete the statement:
1 m^3 = ... cm^3.

8 Get Real!

Freda is relaxing in a jacuzzi. Her husband joins her and she notices that the water level rises by 3 cm. She thinks her husband weighs $16\frac{1}{2}$ stone. Given that the jacuzzi is an octagonal prism and 1 cm^3 of water has a mass of 1 g, do you agree with Freda's estimate of her husband's weight?

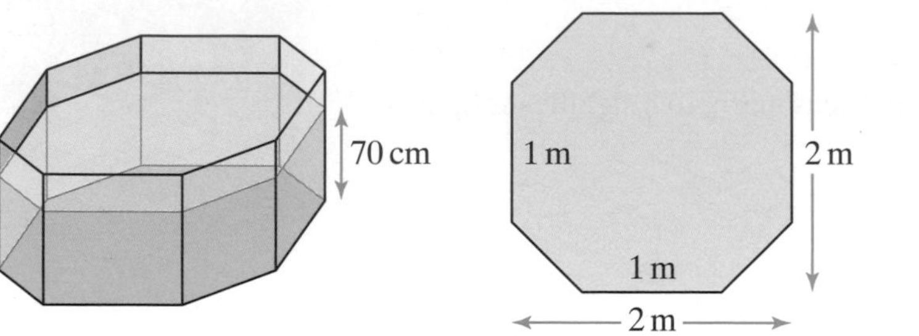

1 kg = 2.2 pounds
14 pounds = 1 stone

Explore

- Calculate the volume of this cuboid
- Find the dimensions of some cuboids with a volume that is one half of the volume of this cuboid
- Find the dimensions of some cuboids with a volume that is one quarter of the volume of this cuboid

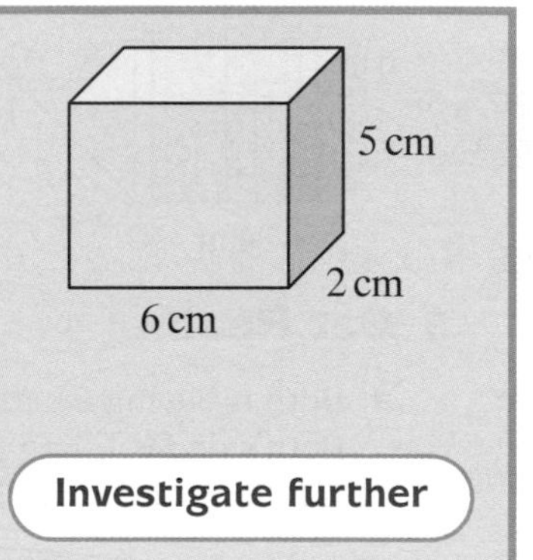

Investigate further

Learn 6 Surface areas of prisms and cylinders

Examples:

Find the surface areas of these solids.

Not drawn accurately

a Triangular prism

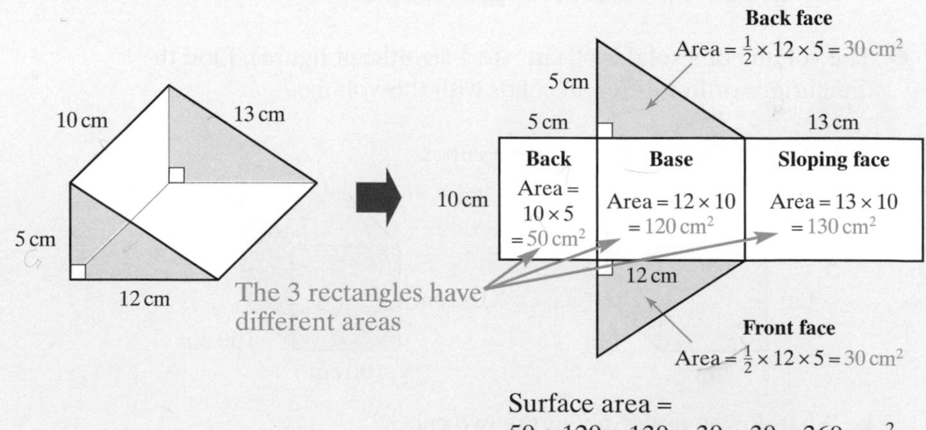

Surface area =
$50 + 120 + 130 + 30 + 30 = 360$ cm^2

The front face is the same as the back face

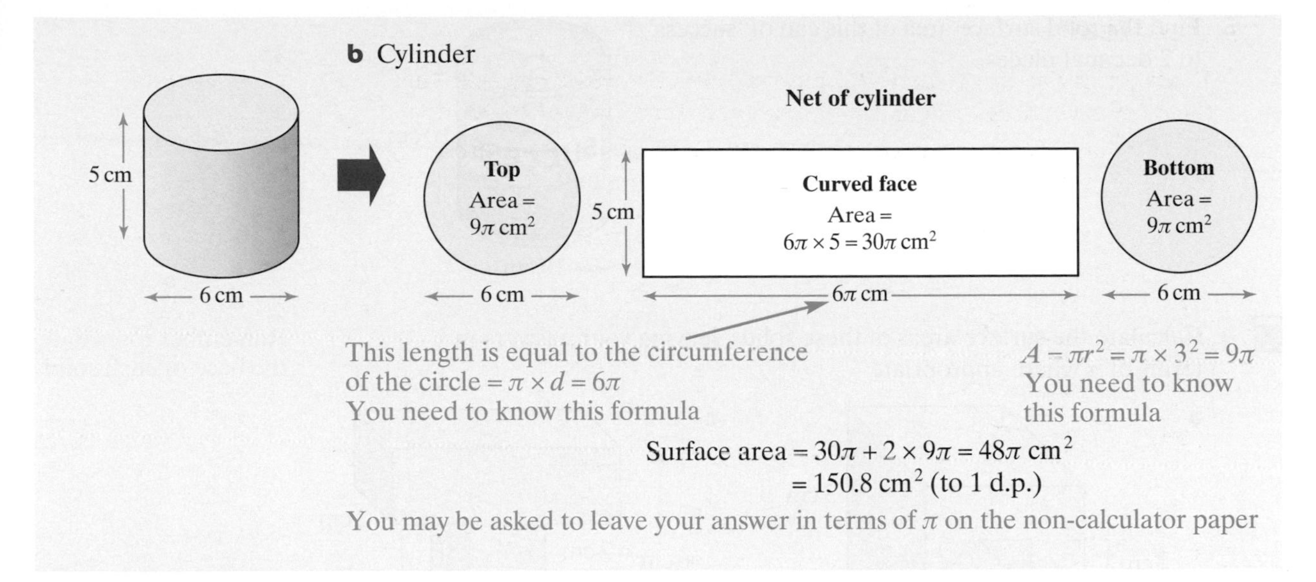

Apply 6

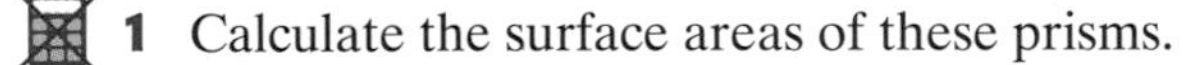

1 Calculate the surface areas of these prisms.

Not drawn accurately

a 3 cm, 2 cm, 5 cm

b 10 cm, 8 cm, 5 cm, 6 cm

c 13 m, 12 m, 20 m, 10 m

d 3 m, 5 m, 4 m, 10 m, 9 m

2 Albert finds the surface area of this triangular prism as follows:

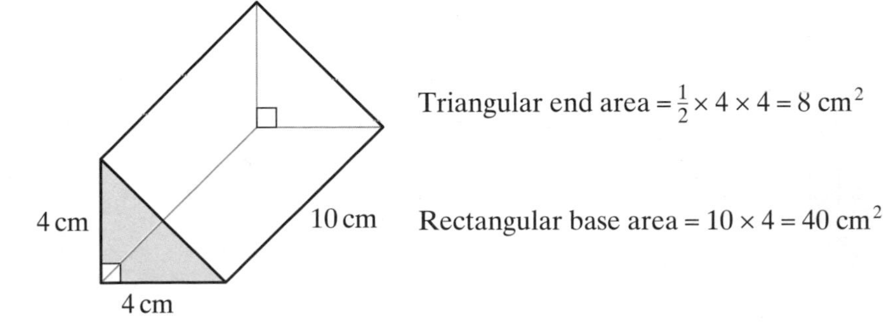

Albert notices that the triangular prism is made up of two triangles and three rectangles so:
Surface area = 2 ends + 3 sides = $(2 \times 8) + (3 \times 40) = 16 + 120 = 136$ cm^2
Is Albert correct?
Explain your answer.

3 Find the total surface area of this can of 'success' to 2 decimal places.

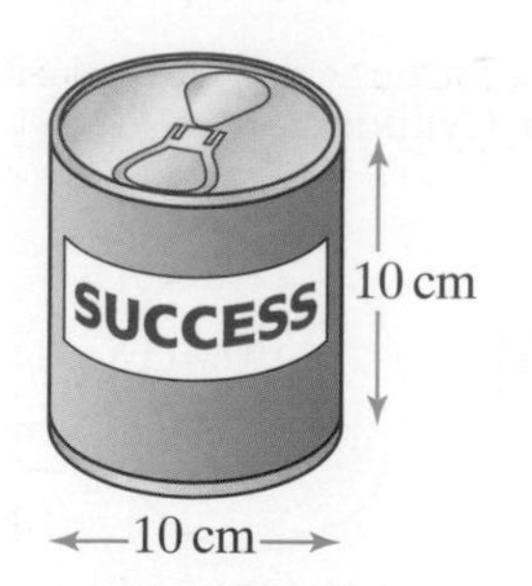

4 Calculate the surface areas of these solids, leaving your answers in terms of π where appropriate.

Remember to include the base of each solid

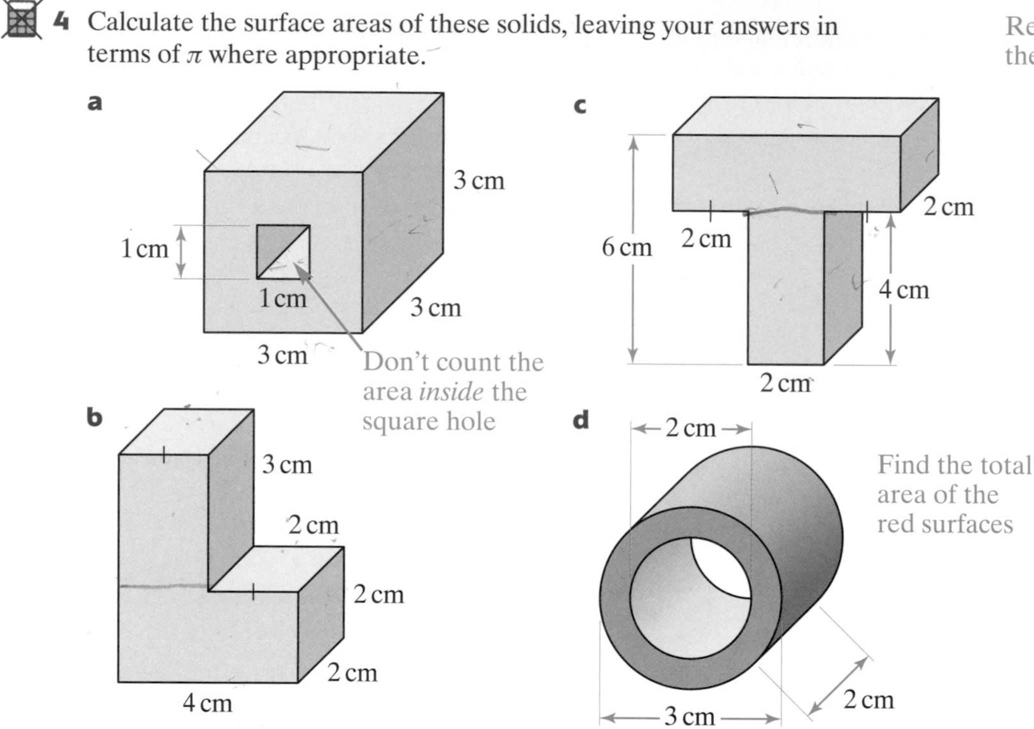

5 The surface area of a solid is between 60 cm^2 and 100 cm^2.
Find the dimensions of five different solids with this surface area.

6 Get Real!

Trevor wants to paint the outside walls and roof of his play house.

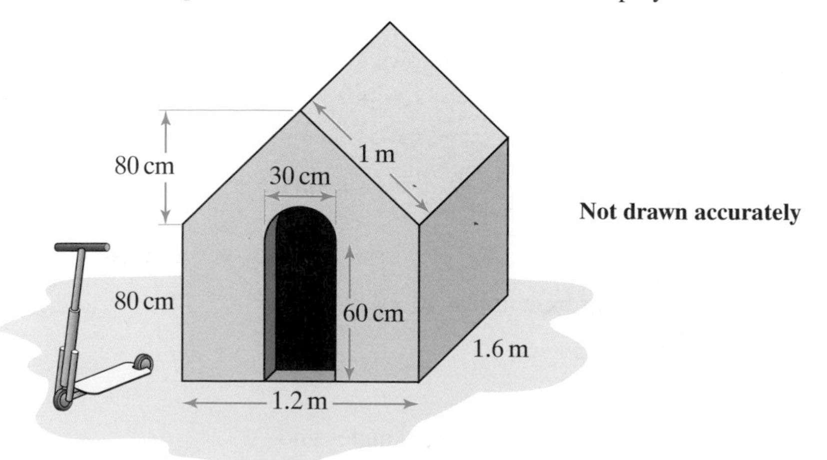

The label on the tin says that 1 litre of paint covers 4 m^2.
How many tins of paint does he need?

7 **a** Calculate the surface areas of these cubes.

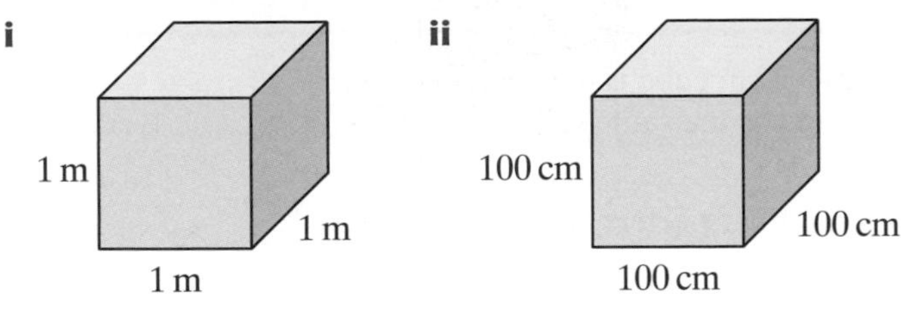

b What do you notice about the two cubes?

c Looking at your answers to part **a**, copy and complete the statement: $1\ \text{m}^2 = \ldots\ldots\ \text{cm}^2$.

d Using your answer from part **c**, convert your answers to question **1** parts **c** and **d** to cm^2.

8 Show that the surface area of this cylinder can be calculated using the formula:

Surface area of a cylinder $= \dfrac{\pi d}{2}(d + 2h)$

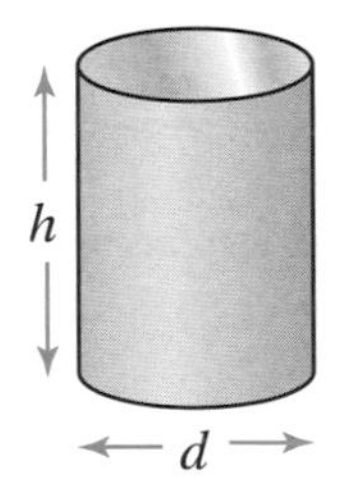

Explore

Some restaurants now serve chunky chips as well as french fries

Does the shape of a chip make it a healthier option?

Suppose the chips are cuboids with a volume of $24\ \text{cm}^3$, for example, $1\ \text{cm} \times 1\ \text{cm} \times 24\ \text{cm}$ makes a very long french fry whereas $3\ \text{cm} \times 3\ \text{cm} \times 4\ \text{cm}$ makes a chunky chip

- Find five different size chips that have a volume of $24\ \text{cm}^3$
- Calculate the surface area of each chip
- Healthier chips have a smaller surface area for the fat globules to attach themselves to
  Which of your chips has the smallest surface area?

Investigate further

Explore

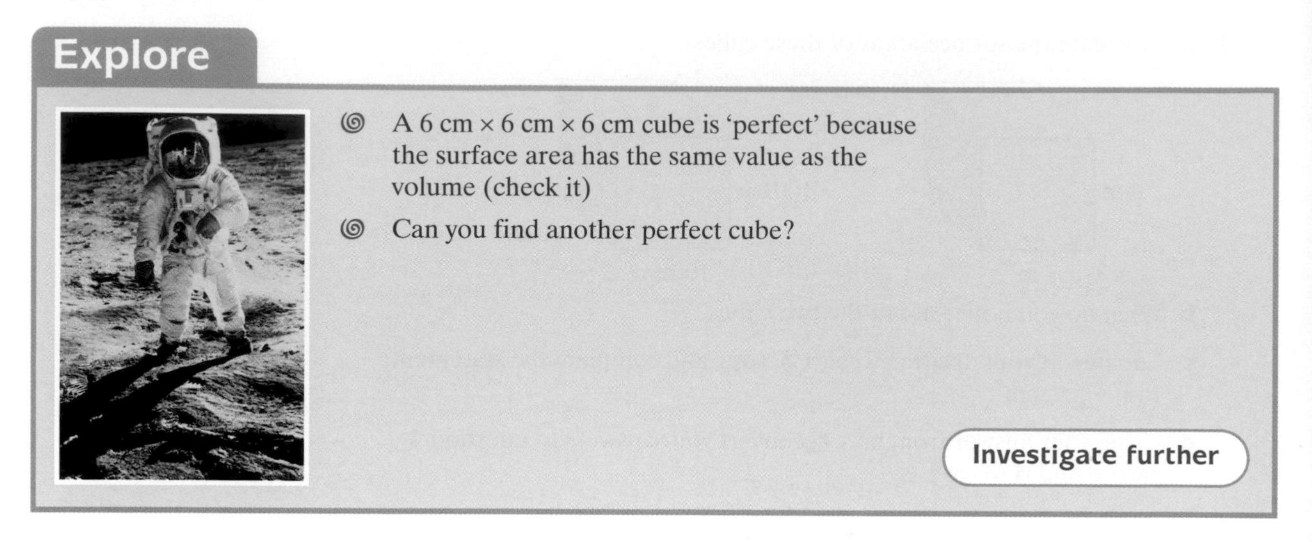

- A 6 cm × 6 cm × 6 cm cube is 'perfect' because the surface area has the same value as the volume (check it)
- Can you find another perfect cube?

Investigate further

Learn 7 Volumes and surface areas of pyramids, cones and spheres

Examples:

Find **i** the volume and **ii** the surface area of the following solids.

Not drawn accurately

a Sphere

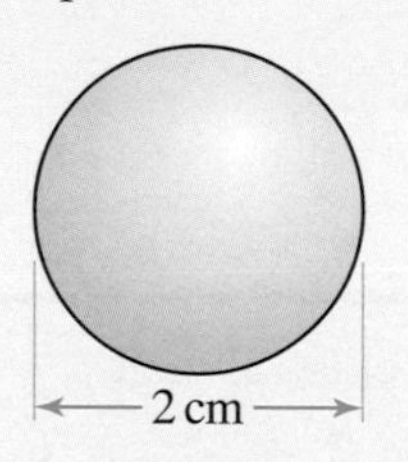

i Volume of sphere $= \frac{4}{3}\pi r^3 = \frac{4}{3} \times \pi \times r^3 = \frac{4}{3} \times \pi \times 2^3 = \frac{32\pi}{3}$ cm^3 = 33.5 cm^3 (to 3 s.f.)

ii Surface area of sphere $= 4\pi r^2 = 4 \times \pi \times 2^2 = 16\pi$ cm^2 = 50.3 cm^2 (to 3 s.f.)

You may be asked to leave your answer in terms of π on a non-calculator paper

b Frustum (a truncated cone)

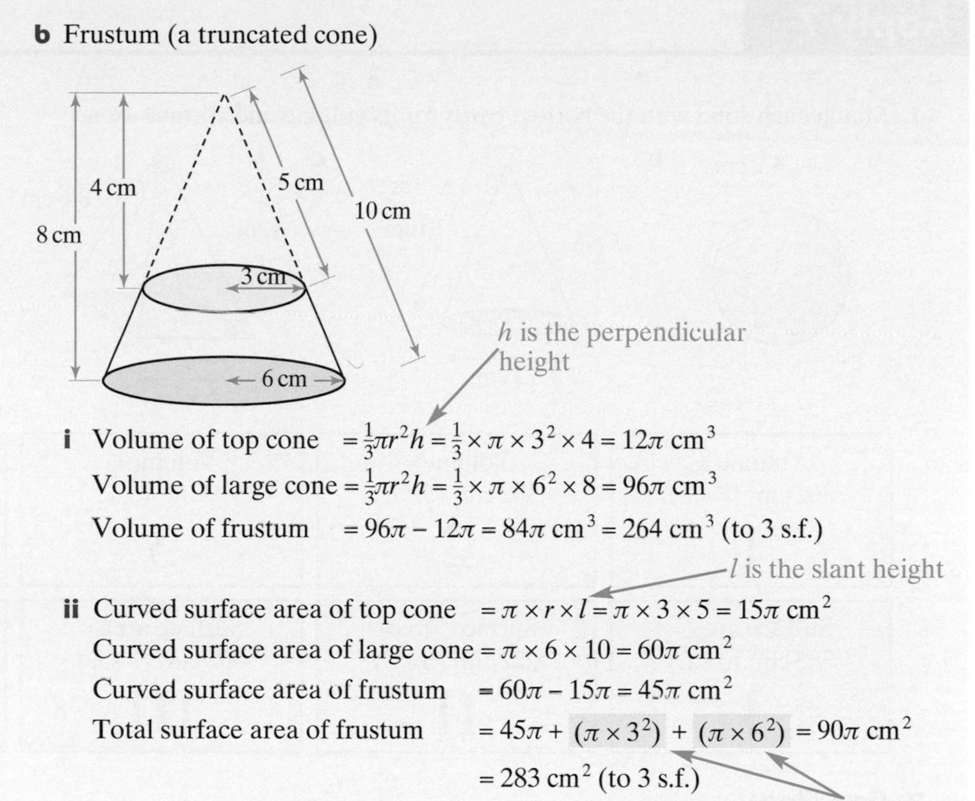

i Volume of top cone $= \frac{1}{3}\pi r^2 h = \frac{1}{3} \times \pi \times 3^2 \times 4 = 12\pi \text{ cm}^3$

Volume of large cone $= \frac{1}{3}\pi r^2 h = \frac{1}{3} \times \pi \times 6^2 \times 8 = 96\pi \text{ cm}^3$

Volume of frustum $= 96\pi - 12\pi = 84\pi \text{ cm}^3 = 264 \text{ cm}^3$ (to 3 s.f.)

l is the slant height

ii Curved surface area of top cone $= \pi \times r \times l = \pi \times 3 \times 5 = 15\pi \text{ cm}^2$

Curved surface area of large cone $= \pi \times 6 \times 10 = 60\pi \text{ cm}^2$

Curved surface area of frustum $= 60\pi - 15\pi = 45\pi \text{ cm}^2$

Total surface area of frustum $= 45\pi + (\pi \times 3^2) + (\pi \times 6^2) = 90\pi \text{ cm}^2$

$= 283 \text{ cm}^2$ (to 3 s.f.)

Don't forget to add on the areas of the two circles

You need to be able to use these formulae.

	Volume	Surface area
Cone (h, l, r)	$V = \frac{1}{3}\pi r^2 h$	Curved surface area $= \pi r l$
Pyramid	$V = \frac{1}{3} \times$ base area $\times$ perpendicular height	Depends on the base
Sphere (r)	$V = \frac{4}{3}\pi r^3$	Surface area $= 4\pi r^2$

Apply 7

1 Match each solid with the correct cards for its volume and surface area.

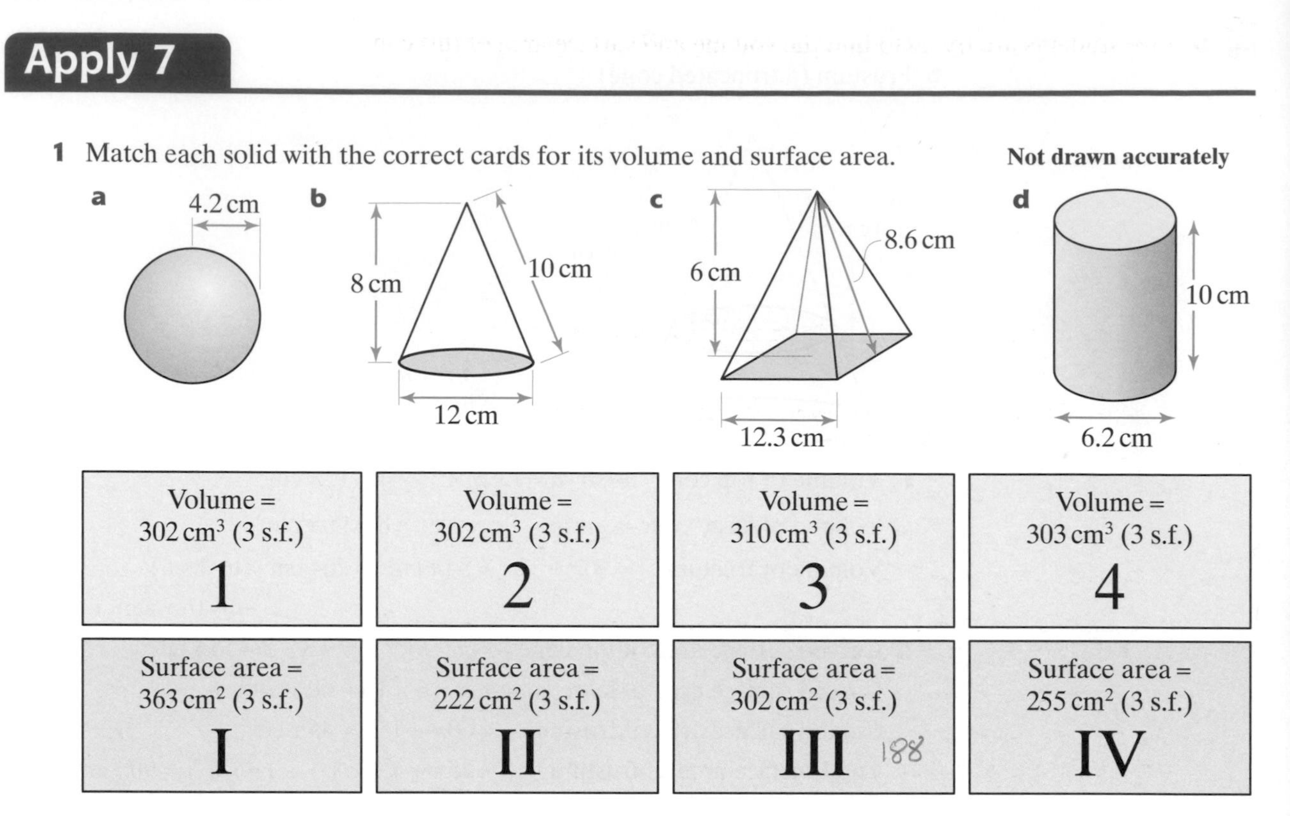

2 Get Real!
Anna is making ice balls for the drinks for her party. If the moulds in the tray have a diameter of 4 cm, how many ice balls can Anna make using 1 litre of water? ($1000\text{ cm}^3 = 1$ litre)

3 The capacity of a container is 1 litre (to 1 significant figure).
Give the dimensions of a sphere, cone and a pyramid with this capacity.

4 Get Real!
The diagram shows a popular toy whose top is a cone and whose base is a hemisphere.

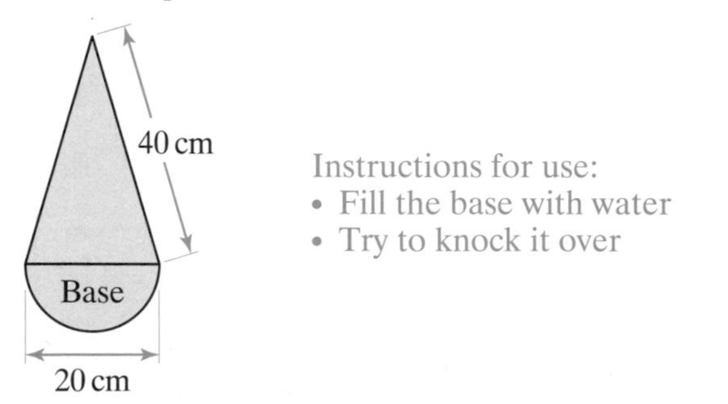

a How many litres of water are needed to fill the base?

b Calculate the surface area of the toy (to 4 significant figures).

5 Five students are trying to find the volume and surface area of this cone.

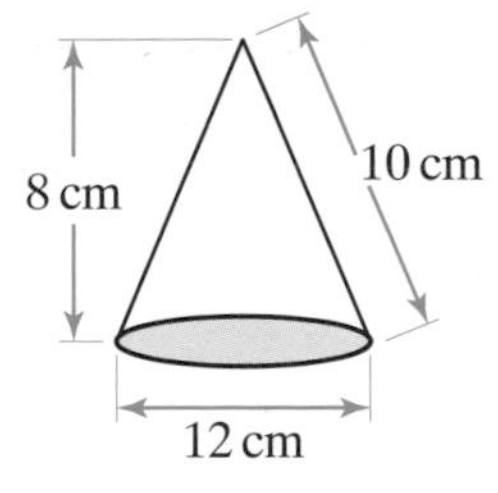

a Anna thinks the volume is 1206 cm^3 (3 s.f.) because $\frac{1}{3} \times \pi \times 12^2 \times 8 = 1206.371579$ cm^3.

b Bruce thinks the surface area is 188 cm^2 (3 s.f.) because $\pi \times 6 \times 10 = 188.4955592$ cm^2.

c Cassie thinks the volume is 377 cm^3 because $\frac{1}{3} \times \pi \times 6^2 \times 10 = 376.9911184$ cm^3.

d Derek finds the volume of the equivalent cylinder (same base) and halves the answer.

e Edna finds the surface area by finding the area of the base ($\pi \times 6^2$) and adding on the area of two triangles, each with an area of $\frac{1}{2} \times 12 \times 8 = 48$ cm^2.

Who is correct? Who is wrong?
Give reasons for your answers.

6 **Get Real!**

The Tropical Biome can be modelled as a hemisphere with diameter 110 m.

a Estimate the amount of tropical air inside the Biome.

b Estimate the surface area of the Biome to 2 s.f.

c The Biomes are situated at the bottom of an old pit. The pit has the shape of an inverted cone of approximate radius 220 m and depth 50 m. Calculate the volume of earth that would have to be excavated to create a similar space.

7 The surface area of a sphere is 320 cm^2. Calculate the radius of the sphere.

8 Find expressions for the volume and surface area of a sphere of diameter d.

9 If a cone with perpendicular height $6h$ and radius $2h$ has the same volume as a sphere of radius r, show that $r = \sqrt[3]{6}h$.

10 **Get Real!**

The diagram shows part of one wall of the Great Pyramid in Egypt. When it was built the pyramid was 146.7 m high with a square base of side 230 m.

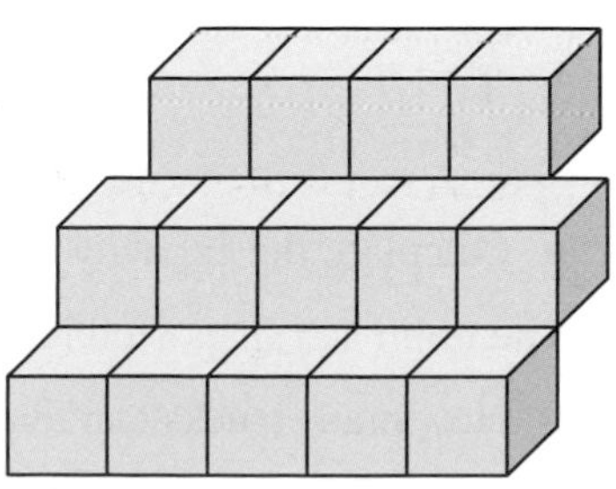

a Calculate the exact volume of the Great Pyramid.

b Taking the blocks as cubes, estimate the number of blocks of side 1.5 m used to make the Great Pyramid.

11 Calculate the volumes of these frustums to 4 significant figures.

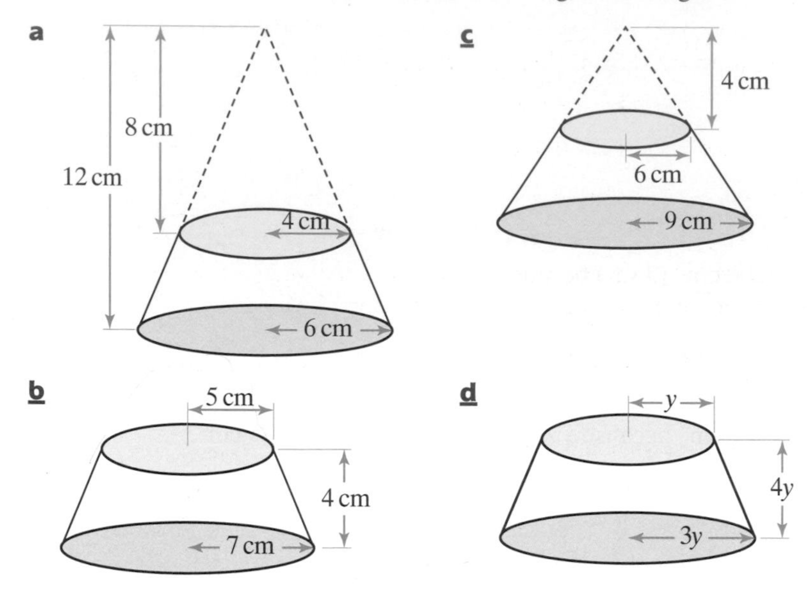

12 Get Real!

A DIY store advertises a '10 litre bucket' as shown.
Will the bucket hold 10 litres?

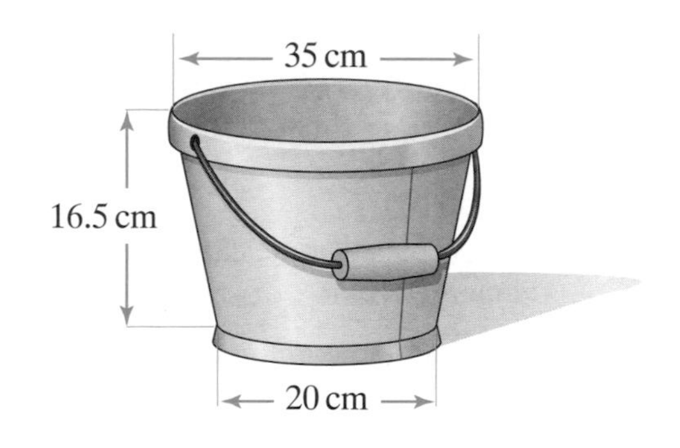

13 Get Real!

Farmer Gadd stores her grain in this hopper.

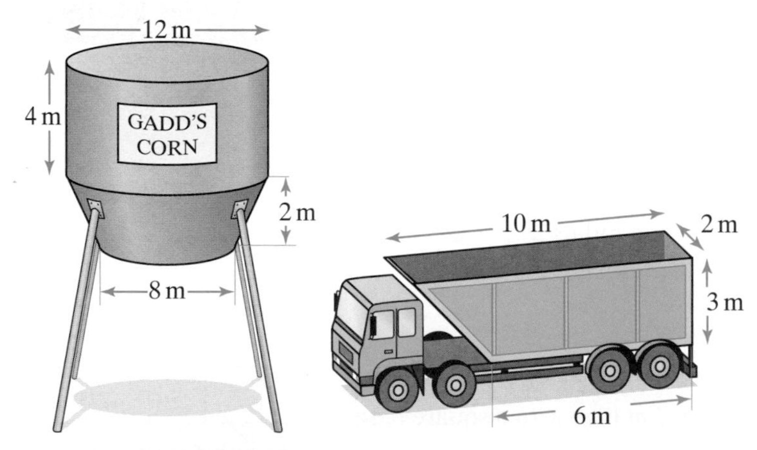

a Calculate the capacity of the hopper.

The grain is transported in trucks.

b How many trucks can the hopper fill when working at maximum capacity?

Explore

Bob makes a pyramid using 2 cm multilink cubes

- How many cubes has he used?
- What is the volume of one cube?
- What is the total volume of all the cubes used?
- What is the volume of a cuboid with the same base?
- What fraction do you think Bob states as the link between a pyramid and a prism of the same base?

Roberta takes a much smoother approach and makes a pyramid out of 2 cm diameter marbles

- How many marbles has she used?
- What is the volume of one marble?
- What is the total volume of all the marbles used?
- What is the volume of a cuboid with the same base?
- What fraction do you think Roberta states as the link between a pyramid and a prism of the same base?

Investigate further

Explore

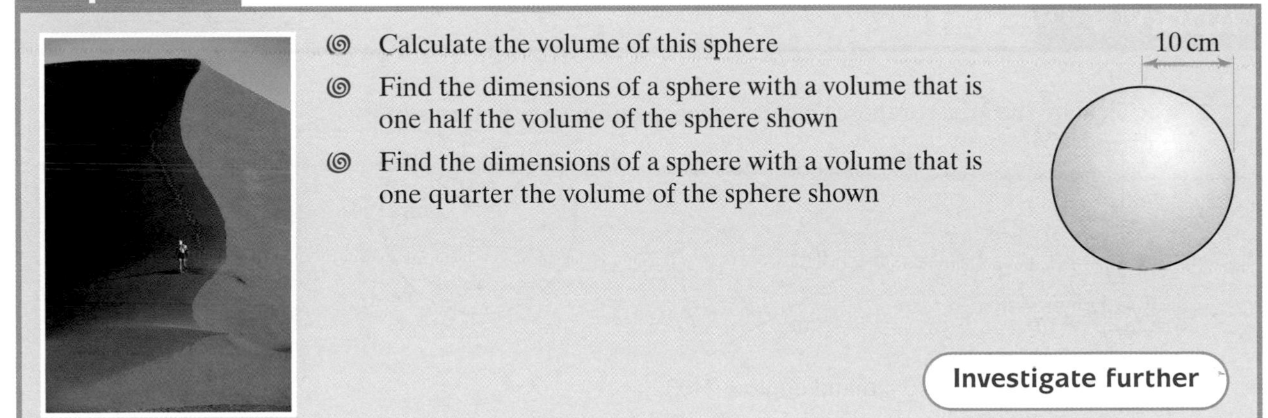

- Calculate the volume of this sphere
- Find the dimensions of a sphere with a volume that is one half the volume of the sphere shown
- Find the dimensions of a sphere with a volume that is one quarter the volume of the sphere shown

Investigate further

Learn 8 Lengths of arcs and areas of sectors and segments

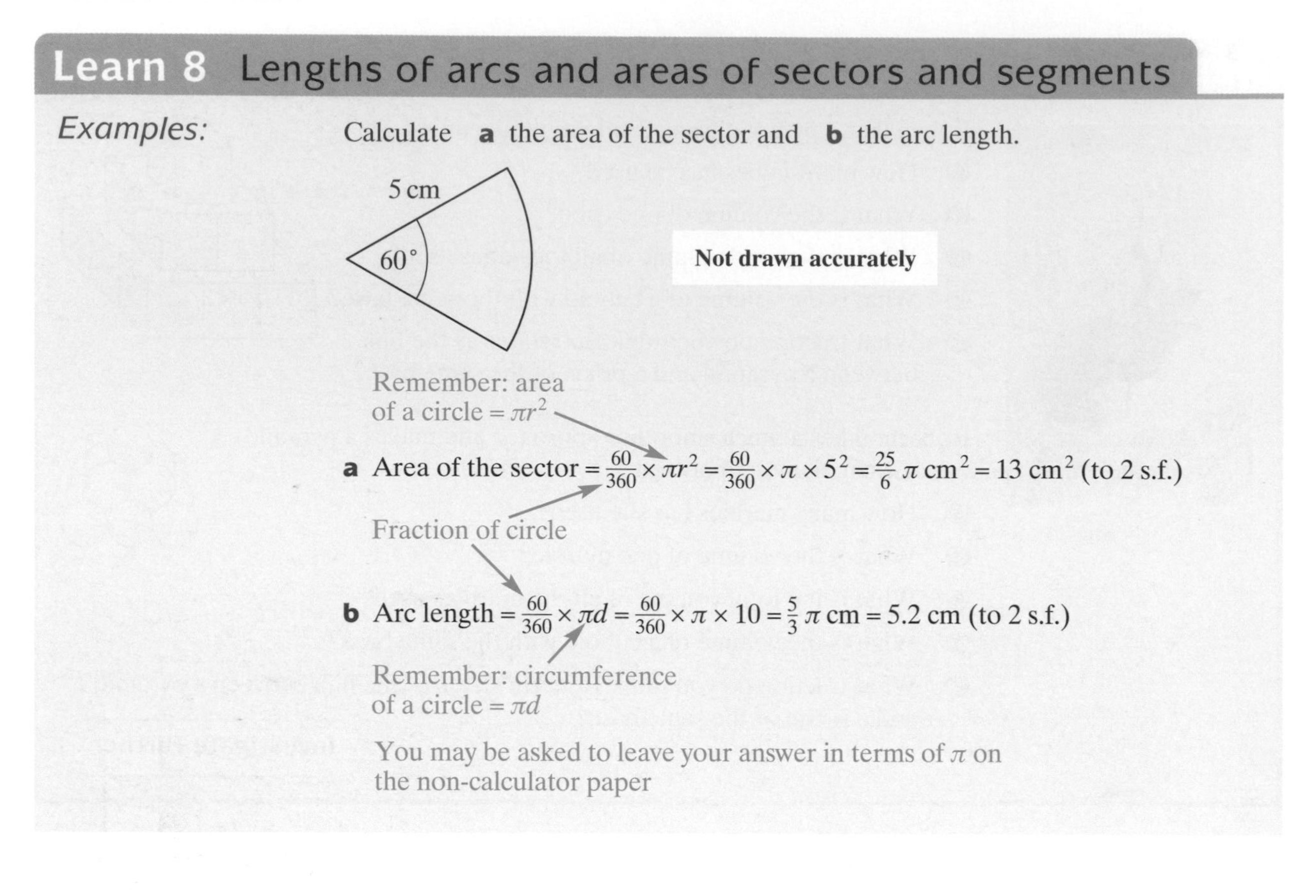

Examples: Calculate **a** the area of the sector and **b** the arc length.

Remember: area of a circle $= \pi r^2$

a Area of the sector $= \frac{60}{360} \times \pi r^2 = \frac{60}{360} \times \pi \times 5^2 = \frac{25}{6}\pi \text{ cm}^2 = 13 \text{ cm}^2$ (to 2 s.f.)

Fraction of circle

b Arc length $= \frac{60}{360} \times \pi d = \frac{60}{360} \times \pi \times 10 = \frac{5}{3}\pi \text{ cm} = 5.2 \text{ cm}$ (to 2 s.f.)

Remember: circumference of a circle $= \pi d$

You may be asked to leave your answer in terms of π on the non-calculator paper

Apply 8

1 a Calculate the areas of these shapes, leaving your answer in terms of π.

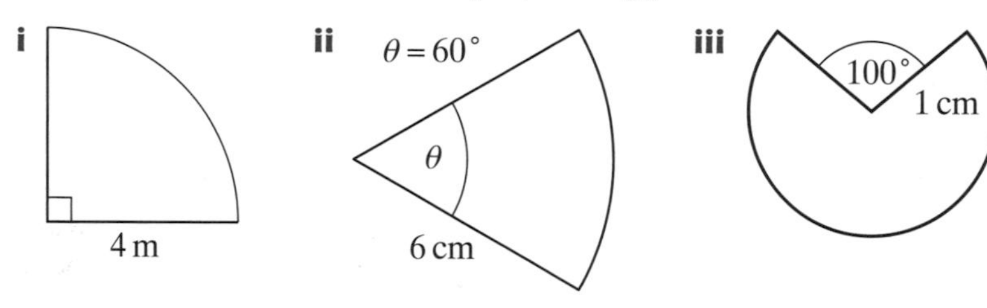

iv Major sector: $r = 3$ cm and angle $= 240°$

b Calculate the perimeter of each shape in part **a**, leaving your answers in terms of π.

2 a Copy and complete this table.

	Radius (cm)	Angle	Minor arc (cm)	Major arc (cm)	Area of minor sector (cm²)	Area of major sector (cm²)
i	5	40°				
ii		90°	10			
iii	20			78		
iv		45°			50	
v	10					220

b Calculate the area of the minor segment of the circle in part **ii**.

3 Get Real!

The London Eye has a diameter of 135 m and takes approximately 30 minutes to make one complete revolution. How far has the base of a capsule travelled after:

a 5 minutes

b 11 minutes

c 23 minutes

d 4 hours 15 minutes?

Give all answers to 3 s.f.

4 The length of an arc is 25 cm (to the nearest cm).
Find the sector angle and radius of five different arcs with this length.

5

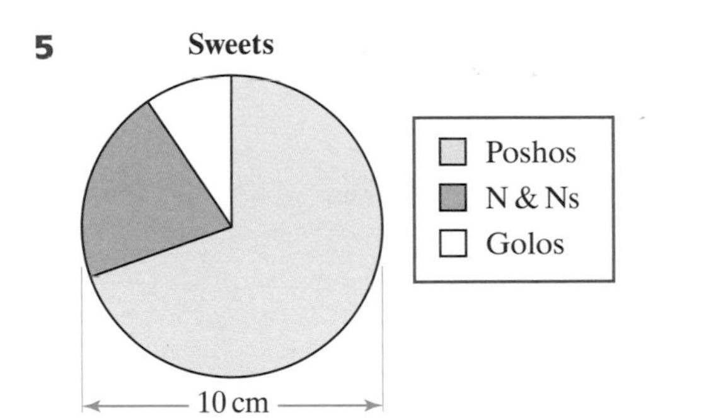

Rashid estimates that 70% of the class buy Poshos from the vending machine. He calculates the Poshos area of the pie chart as follows:

$$\frac{70}{360} \times \pi \times 5^2 = 15.3 \text{ cm}^2$$

Do you agree? Give a reason for your answer.

6 Calculate the area of the shaded region in each diagram, giving your answer to an appropriate degree of accuracy.

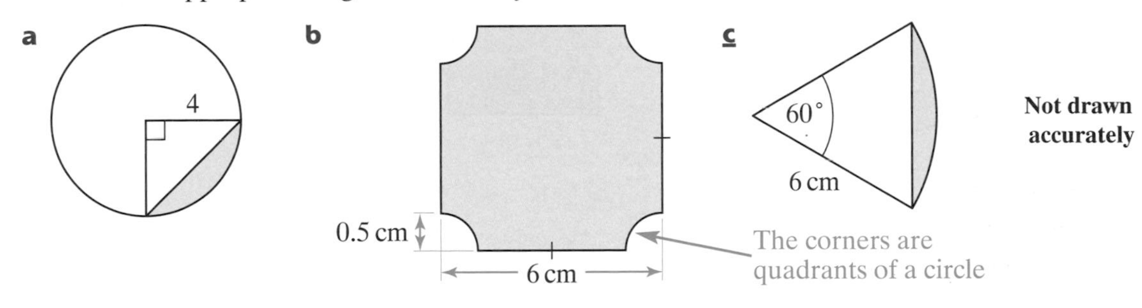

7 Get Real!

Samson's legendary guitar plectrums are made by pouring molten liquid into moulds as shown here.

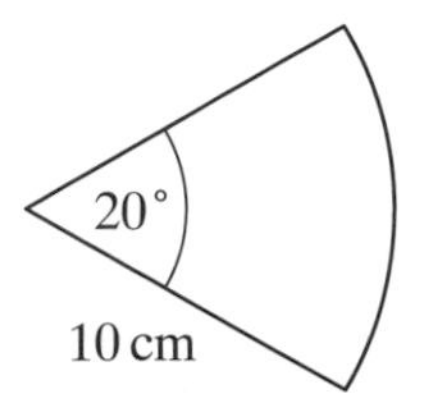

If the plectrums are 3 mm thick, how many can be made from 1 litre of molten liquid?

8 Calculate the area of the segment in the diagram.
Leave your answer in terms of π.

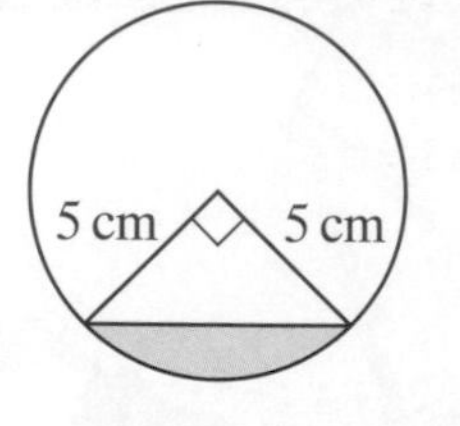

9 **Get Real!**

Mrs Johnson's lily pads are out of control. The goldfish must have at least 70% of clear surface to live and feed safely. Are the goldfish in danger?
The large lilypads have a radius of 15 cm and the small ones 8 cm.

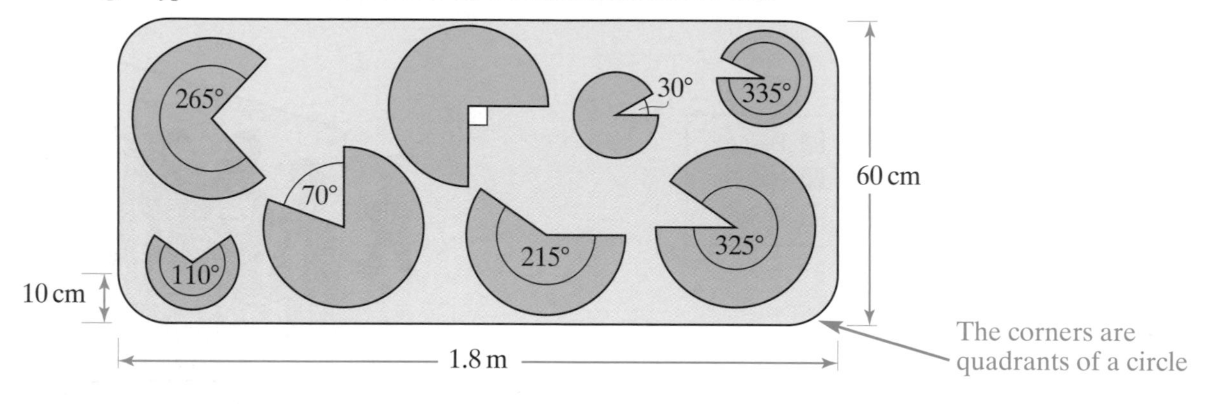

10 How many crackers can be spread with one cheese 'triangle'?
Assume that the cheese is 2 cm thick and the spread is 2 mm thick.

Cheese 'triangle'
(A prism with a sector of a circle as the cross-section)

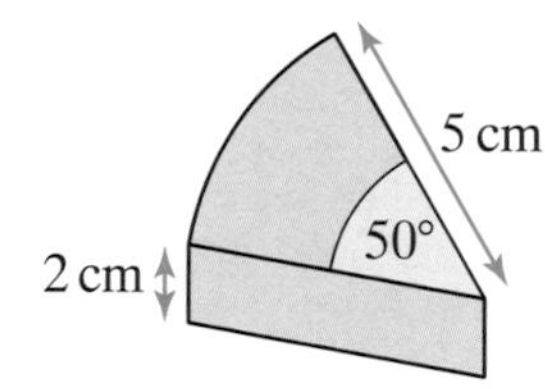

Cracker

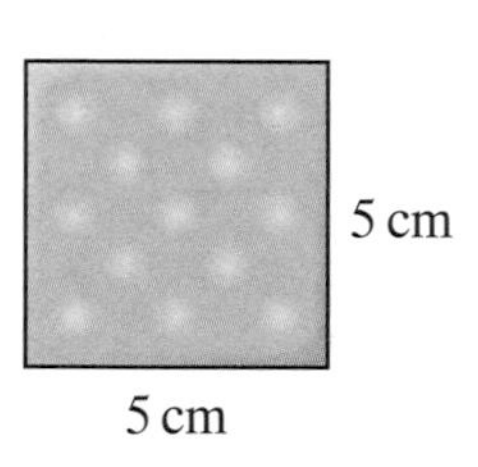

Explore

- Draw a circle of radius 5 cm
- Mark a starting point on the circumference with X
- Put a second cross on the circumference where you think the length of the arc is equal to the radius of the circle
- Measure the angle at the centre of the circle
- Compare your angle with your neighbour's

Investigate further

Area and volume

ASSESS

The following exercise tests your understanding of this chapter, with the questions appearing in order of increasing difficulty.

1 Find the areas of the following shapes.

a 3 mm, 10 mm, 2 mm, 18 mm

b 8.4 mm, 9.9 mm, 2.7 mm, 2.7 mm, 3.7 mm, 2.7 mm

c 5 in, 12 in, 9 in

d 3.7 cm, 6.1 cm

e 6.4 ft, 14.6 ft

Not drawn accurately

f 7 cm, 8 cm, 19 cm

2 **a** Find the circumference of a circle of diameter 12.6 cm. (Take $\pi = 3.14$)

b A garden reel contains 30 m of hosepipe.
The reel has a diameter of 20 cm.
Calculate the number of times the reel rotates when the complete length of the hosepipe is unwound.
(You may ignore the thickness of the hosepipe.)

3 A cylindrical waste paper basket has a base radius of 5 in and a height of 16 in. Calculate its volume.

4 A door wedge is in the shape of a triangular prism 3 cm wide.
The triangular cross-section has a base of 6.4 cm and a height of 4.2 cm.
Find its volume.

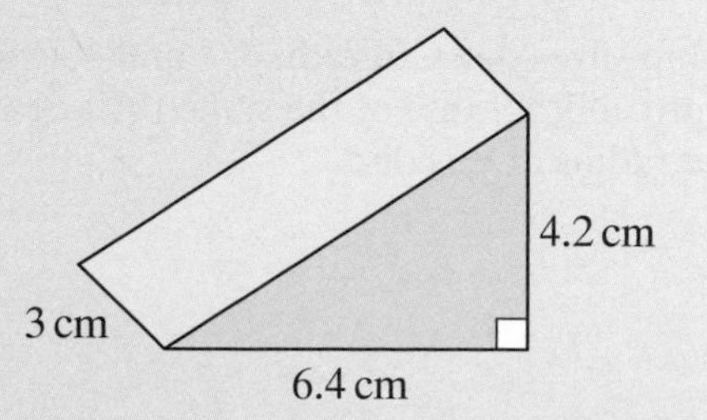

5 An unsharpened wooden pencil is a hexagonal prism.
Each side of the hexagon is 0.15 in and the perpendicular distance across the hexagon is 0.26 in.
The pencil is 6 in long.

Calculate:

a the area of the hexagonal cross-section

b the volume of the pencil.

6 The British kitemark can be drawn as two semicircles on top of an equilateral triangle of side 4 cm, as shown below.

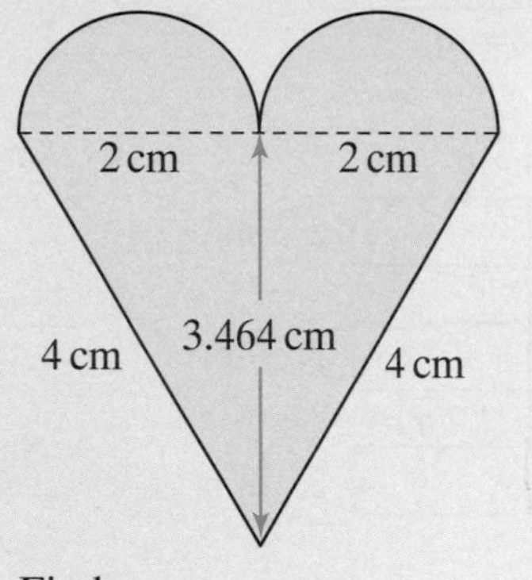

Find:

a its perimeter

b its area.

7 **a** A washer has an outer radius of 3.6 cm and a hole of radius 0.4 cm.

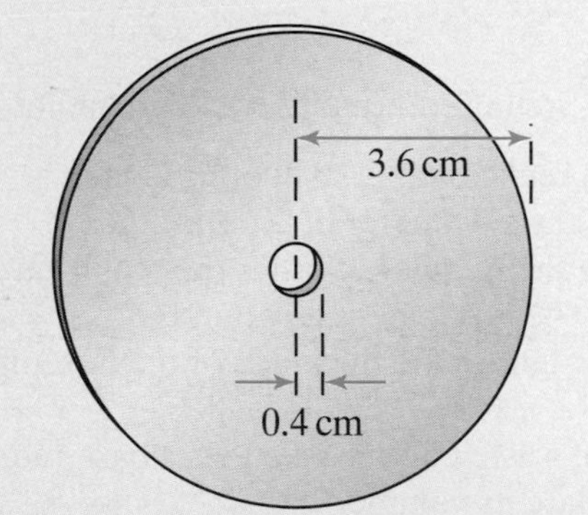

Calculate the area of the face of the washer, giving your answer:

i in terms of π

ii to 3 significant figures.

b Three thin silver discs, of radii 4, 7 and 10 cm, are melted down and recast into another disc of the same thickness.
Find the radius of this disc.

8

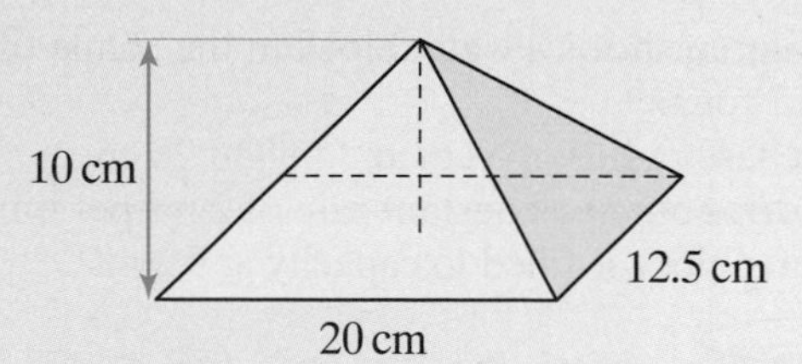

A lead pyramid with a rectangular base, 20 cm by 12.5 cm and height 10 cm, is melted down and recast into a number of lead cubes of side 2 cm.
How many **complete** cubes can be made?

9 An iron pipe is 50 cm long. It has an outer radius of 1 cm and a central hole of radius 0.5 cm.

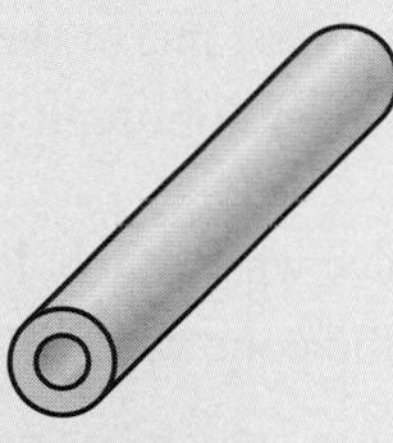

a Find the volume of iron in the pipe.

b Find the mass of the pipe if the density of iron is 7.86 g/cm^3.

10 Assume that the Earth is a sphere of radius 3960 miles with mean density 5.35 lb/in^3.

HINT lb/in^3 means pounds per cubic inch.

a Calculate the approximate volume of the Earth.

b Calculate the approximate mass of the Earth.
(1 mile = 1760 yards, 1 yard = 36 inches)

The surface area of a sphere is given by the formula $A = 4\pi r^2$.

c Calculate the approximate surface area of the Earth.

A rampant virus kills every living thing in its tracks.
It infects 50 000 square miles of the Earth every day.
Scientists are desperately searching for an effective vaccine.

d How long do they have before life ceases to survive?

11 The diagram shows a water clock in the shape of an inverted cone.
It has a base radius of 6 cm and height 20 cm.
Water drips out at a constant rate of 2 mℓ per minute.
The water clock is filled to capacity at 9 a.m.

Calculate:

a the original volume, to the nearest mℓ, of the full cone

b the length of time, to the nearest minute, it takes to empty the water clock

c the radius of the water surface when the height of water is 10 cm

d the volume of water left in the cone when the height is 10 cm

e the time, to the nearest minute, which has passed since the water started dripping out

f the actual time when the height of water is 10 cm.

7 Fractions

OBJECTIVES

D **Examiners would normally expect students who get a D grade to be able to:**

Do calculations with simple fractions involving subtraction

C **Examiners would normally expect students who get a C grade also to be able to:**

Do calculations with simple fractions involving division

Do calculations with mixed numbers

What you should already know ...

- Understand fractions including equivalent fractions
- Simplify fractions and arrange them in order

- Calculate fractions of quantities
- Work out one number as a fraction of another number

VOCABULARY

Fraction or **simple fraction** or **common fraction** or **vulgar fraction** – a number written as one whole number over another, for example, $\frac{3}{8}$ (three eighths), which has the same value as $3 \div 8$

Numerator – the number on the top of a fraction

Numerator → $\frac{3}{8}$ ← Denominator

Denominator – the number on the bottom of a fraction

Unit fraction – a fraction with a numerator of 1, for example, $\frac{1}{5}$

Proper fraction – a fraction in which the numerator is smaller than the denominator, for example, $\frac{5}{13}$

Improper fraction or **top-heavy fraction** – a fraction in which the numerator is bigger than the denominator, for example, $\frac{13}{5}$, which is equal to the mixed number $2\frac{3}{5}$

Mixed number or **mixed fraction** – a number made up of a whole number and a fraction, for example, $2\frac{3}{5}$, which is equal to the improper fraction $\frac{13}{5}$

Decimal fraction – a fraction consisting of tenths, hundredths, thousandths, and so on, expressed in a decimal form, for example, 0.65 (6 tenths and 5 hundredths)

Equivalent fraction – a fraction that has the same value as another, for example, $\frac{3}{5}$ is equivalent to $\frac{30}{50}, \frac{6}{10}, \frac{60}{100}, \frac{15}{25}, \frac{1.5}{2.5}, \ldots$

Simplify a fraction or **express a fraction in its simplest form** – to change a fraction to the simplest equivalent fraction; to do this divide the numerator and the denominator by a common factor (this process is called cancelling or reducing or simplifying the fraction)

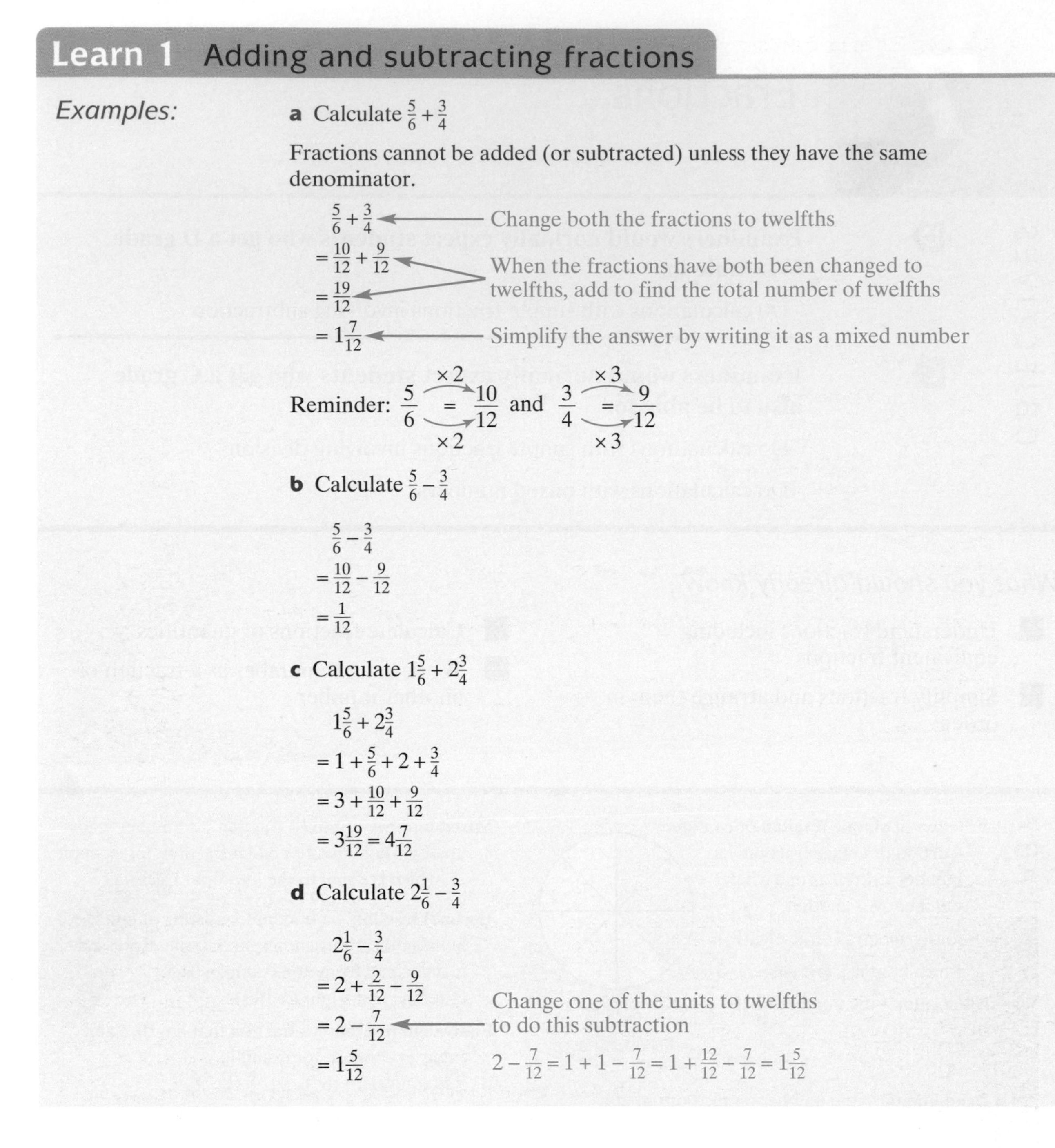

Learn 1 Adding and subtracting fractions

Examples:

a Calculate $\frac{5}{6}+\frac{3}{4}$

Fractions cannot be added (or subtracted) unless they have the same denominator.

$\frac{5}{6}+\frac{3}{4}$ ← Change both the fractions to twelfths

$=\frac{10}{12}+\frac{9}{12}$ ← When the fractions have both been changed to twelfths, add to find the total number of twelfths

$=\frac{19}{12}$

$=1\frac{7}{12}$ ← Simplify the answer by writing it as a mixed number

Reminder: $\frac{5}{6} = \frac{10}{12}$ (×2 top and bottom) and $\frac{3}{4} = \frac{9}{12}$ (×3 top and bottom)

b Calculate $\frac{5}{6}-\frac{3}{4}$

$\frac{5}{6}-\frac{3}{4}$

$=\frac{10}{12}-\frac{9}{12}$

$=\frac{1}{12}$

c Calculate $1\frac{5}{6}+2\frac{3}{4}$

$1\frac{5}{6}+2\frac{3}{4}$

$=1+\frac{5}{6}+2+\frac{3}{4}$

$=3+\frac{10}{12}+\frac{9}{12}$

$=3\frac{19}{12}=4\frac{7}{12}$

d Calculate $2\frac{1}{6}-\frac{3}{4}$

$2\frac{1}{6}-\frac{3}{4}$

$=2+\frac{2}{12}-\frac{9}{12}$

$=2-\frac{7}{12}$ ← Change one of the units to twelfths to do this subtraction

$=1\frac{5}{12}$

$2-\frac{7}{12}=1+1-\frac{7}{12}=1+\frac{12}{12}-\frac{7}{12}=1\frac{5}{12}$

Apply 1

Questions like this will be on the non-calculator paper so make sure you can do them without using your calculator.

1 Work out: **a** $\frac{3}{4}+\frac{2}{3}$ **b** $\frac{3}{4}-\frac{2}{3}$ **c** $\frac{5}{6}+\frac{2}{5}$ **d** $\frac{5}{6}-\frac{2}{5}$

2 Work out: **a** $3\frac{3}{4}+1\frac{2}{3}$ **b** $3\frac{3}{4}-1\frac{2}{3}$ **c** $2\frac{5}{6}+1\frac{2}{5}$ **d** $2\frac{5}{6}-1\frac{2}{5}$

3 Work out: **a** $3\frac{2}{3}+1\frac{3}{4}$ **b** $3\frac{2}{3}-1\frac{3}{4}$ **c** $2\frac{2}{5}+1\frac{5}{6}$ **d** $2\frac{2}{5}-1\frac{5}{6}$

4 Sue says, 'I add up fractions like this: $\frac{5}{6} + \frac{2}{5} = \frac{5+2}{6+5} = \frac{7}{11}$'
Is Sue right? Explain your answer.

5 Find two fractions with:

a a sum of $1\frac{1}{4}$

b a difference of $1\frac{1}{4}$

c a sum of $3\frac{1}{3}$

d a difference of $3\frac{1}{3}$

6 Get Real!

Anne is making custard, which needs $\frac{1}{3}$ of a cup of sugar.

Then she makes biscuits, which need $\frac{3}{4}$ of a cup of sugar.

Anne only has 1 cup of sugar.
Does she have enough to make the custard and the biscuits?
Show how you got your answer.

7 Get Real!

In America, lengths of fabric for making clothes are measured in yards and fractions of yards.
A tailor is making a suit for a customer. The jacket needs $2\frac{1}{4}$ yards of fabric and the trousers need $1\frac{1}{3}$ yards.
The tailor has 4 yards of fabric.
How much will be left over when he has made the jacket and the trousers?

Explore

A man was riding a camel across a desert, when he came across three young men arguing. Their father had died, leaving seventeen camels as his sons' inheritance. The eldest son was to receive half of the camels; the second son, one-third of the camels and the youngest son, one-ninth of the camels. The sons asked him how they could divide seventeen camels in this way.

The man added his camel to the 17. Then, he gave $\frac{1}{2}$ of the camels to the eldest son, $\frac{1}{3}$ of the camels to the second son and $\frac{1}{9}$ of the camels to the youngest son. Having solved the problem, the stranger mounted his own camel and rode away.

How does this work?

Investigate further

Learn 2 Multiplying and dividing fractions

Examples:

a Calculate:

i $12 \times \frac{1}{3}$ **ii** $12 \times \frac{2}{3}$ **iii** $\frac{3}{4} \times \frac{2}{3}$

i $12 \times \frac{1}{3}$ ← Multiplying by $\frac{1}{3}$ is the same as dividing by 3

$= \frac{12}{3}$

$= 4$

ii $12 \times \frac{2}{3}$ ← Multiplying by $\frac{2}{3}$ is the same as dividing by 3 and multiplying by 2

$= \frac{24}{3}$

$= 8$

iii $\frac{3}{4} \times \frac{2}{3}$

$= \frac{3 \times 2}{4 \times 3}$

$= \frac{6}{12}$

$= \frac{1}{2}$

b Calculate:

i $8 \div \frac{1}{3}$ **ii** $8 \div \frac{2}{3}$ **iii** $\frac{3}{4} \div \frac{2}{3}$

i $8 \div \frac{1}{3}$ ← Dividing 8 by a third means finding how many thirds there are in 8. There are three thirds in each whole, so there are 8×3 thirds in 8

$= 8 \times \frac{3}{1}$

$= 24$

ii $8 \div \frac{2}{3}$ ← The number of two-thirds in 8 is half the number of thirds in 8. Dividing by $\frac{2}{3}$ is the same as multiplying by $\frac{3}{2}$

$= 8 \times \frac{3}{2}$

$= 12$

iii $\frac{3}{4} \div \frac{2}{3}$

$= \frac{3}{4} \times \frac{3}{2}$ ← Dividing by a fraction is the same as multiplying by the reciprocal (upside-down) fraction

$= \frac{3 \times 3}{4 \times 2}$

$= \frac{9}{8}$

$= 1\frac{1}{8}$

Apply 2

1 Work out:

a $18 \times \frac{1}{3}$ **c** $35 \times \frac{1}{7}$ **e** $40 \times \frac{2}{5}$ **g** $24 \times \frac{5}{8}$

b $28 \times \frac{1}{4}$ **d** $40 \times \frac{1}{5}$ **f** $12 \times \frac{3}{4}$ **h** $42 \times \frac{5}{6}$

2 Work out:

a $\frac{4}{5} \times \frac{1}{2}$ **c** $\frac{5}{6} \times \frac{1}{4}$ **e** $\frac{5}{8} \times \frac{3}{5}$ **g** $\frac{11}{12} \times \frac{4}{5}$

b $\frac{3}{8} \times \frac{1}{3}$ **d** $\frac{9}{10} \times \frac{1}{6}$ **f** $\frac{8}{9} \times \frac{3}{4}$ **h** $\frac{9}{10} \times \frac{2}{3}$

3 Work out:

a $18 \div \frac{1}{3}$ **c** $6 \div \frac{1}{5}$ **e** $18 \div \frac{2}{3}$ **g** $28 \div \frac{4}{5}$

b $5 \div \frac{1}{4}$ **d** $10 \div \frac{1}{8}$ **f** $12 \div \frac{3}{4}$ **h** $35 \div \frac{5}{6}$

4 Work out:

a $\frac{7}{8} \div \frac{1}{2}$ **c** $\frac{4}{9} \div \frac{2}{5}$ **e** $\frac{1}{3} \div \frac{1}{3}$ **g** $\frac{3}{5} \div \frac{7}{10}$

b $\frac{1}{6} \div \frac{2}{3}$ **d** $\frac{2}{7} \div \frac{2}{3}$ **f** $\frac{1}{3} \div \frac{1}{5}$ **h** $\frac{11}{12} \div \frac{3}{4}$

5 Paula is working out $\frac{4}{15} \div \frac{3}{8}$
She says, 'I can cancel the 4 into the 8 and the 3 into the 15.'
Then she writes down $\frac{1}{5} \div \frac{1}{2} = \frac{1}{5} \times \frac{2}{1} = \frac{2}{5}$
Is this correct? Explain your answer.

6 Ali says, 'This is how to divide fractions: $\frac{5}{6} \div \frac{2}{5} = \frac{6}{5} \times \frac{2}{5} = \frac{12}{25}$'
Is Ali right? Explain your answer.

7 Without working out the answers, say which of these gives an answer greater than 1:
$\frac{9}{10} \times \frac{4}{5}$ or $\frac{9}{10} \div \frac{4}{5}$?
Give a reason for your answer.

8 Write down two fractions that:

a multiply to give 1 **b** divide to give 1.

9 Get Real!
Two thirds of the teachers in a school are women and three quarters of these are over 40. What fraction of the teachers in the school are women over 40?

10 Get Real!
Seven eighths of the members of the running club train on Wednesday evening and four fifths of them are male. What fraction of the members are males who train on Wednesday evenings?

Fractions

ASSESS

The following exercise tests your understanding of this chapter, with the questions appearing in order of increasing difficulty.

1 a Scrooge collects money.

$\frac{3}{10}$ of his fortune is in brass coins, $\frac{8}{15}$ is in silver and the rest is in notes.

What fraction of Scrooge's fortune is in notes?

b In the 4 × 100 m relay, the first runner took $\frac{1}{5}$ of his team's total time.

The second runner took $\frac{7}{30}$ of their total time.

The third runner took $\frac{3}{10}$ of their total time.

i What fraction of their time was taken by the fourth member of the team?

ii Which team member ran the fastest leg of the race?

iii Which team member ran the slowest leg of the race?

c Delia is cooking. She has a $1\frac{1}{2}$ kg bag of flour and needs $\frac{3}{8}$ of it in a recipe. What fraction of a kilogram does she need and what is this in grams?

d S. Crumpy has an orchard.

The orchard contains $4\frac{1}{3}$ hectares of apple trees.

Today he needs to treat $\frac{4}{5}$ of the area for disease prevention.

What area does he need to treat?

e In a football match the goalkeeper kicked the ball from the goal line for $\frac{5}{8}$ of the length of the pitch and a player then kicked it a further $\frac{5}{24}$.

The length of the pitch is 90 yards.

How far is the ball from the opposing goal line?

90 yards

Goal line

Opposing goal line

2 a Work out the following:

i $6\frac{3}{7} + 3\frac{6}{7}$	**iii** $9\frac{4}{5} + 6\frac{3}{8}$	**v** $11\frac{2}{7} - 6\frac{4}{5}$	**vii** $10 \div \frac{2}{3}$
ii $8\frac{1}{4} - 4\frac{5}{8}$	**iv** $7\frac{5}{8} - 3\frac{1}{4}$	**vi** $\frac{2}{3} \times \frac{5}{6}$	**viii** $3\frac{1}{5} \div 1\frac{3}{5}$

b i Titus Lines, the fisherman, catches one fish of mass $2\frac{1}{3}$ kg and another of mass $3\frac{1}{4}$ kg. What total mass of fish does he catch?

ii What is the perimeter of a triangle of sides $2\frac{1}{4}$, $3\frac{1}{5}$ and $4\frac{3}{10}$ inches?

iii A can holds $2\frac{8}{9}$ litres of oil. Hakim uses $1\frac{4}{15}$ litres. How much is left?

iv Deirdre drops Ken off at work after driving from home for $4\frac{5}{12}$ miles. She drives $7\frac{1}{4}$ miles altogether to her own place of work. How far is Ken's workplace from Deirdre's workplace?

v Milo is $1\frac{2}{5}$ metres tall.

He is $\frac{3}{8}$ metre taller than Fizz. How tall is Fizz?

8 Surds

OBJECTIVES

A **Examiners would normally expect students who get an A grade to be able to:**

Rationalise the denominator of a surd, such as $\frac{2}{\sqrt{5}}$

A* **Examiners would normally expect students who get an A* grade also to be able to:**

Simplify surds, such as write $(3 - \sqrt{5})^2$ in the form $a + b\sqrt{5}$

What you should already know ...

- Squares of numbers up to 15
- Multiplying fractions, and converting fractions to decimals and vice versa
- Prime factors and the expression of numbers as products of prime factors

VOCABULARY

Rational number – a number that can be expressed in the form $\frac{p}{q}$ where p and q are both integers, for example, $1(=\frac{1}{1})$, $2\frac{1}{3}(=\frac{7}{3})$, $\frac{3}{5}$, $0.\dot{1}(=\frac{1}{9})$; rational numbers, when written as decimals, are terminating decimals or recurring decimals

Irrational number – a number that is not an integer and cannot be written as a fraction, for example, $\sqrt{2}, \sqrt{3}, \sqrt{5}$ and π; irrational numbers, when expressed as decimals, are infinite, non-recurring decimals

Surd – a number containing an irrational root, for example, $\sqrt{2}$ or $3 + 2\sqrt{7}$

Learn 1 Simplifying surds

Examples:

a Are these numbers rational or irrational?

i $\sqrt{7}$ **ii** $\sqrt{16}$ **iii** $\frac{\pi}{2}$ **iv** $\frac{\pi}{2\pi}$

i $\sqrt{7}$ is irrational.

ii $\sqrt{16} = 4$; it is rational.

iii $\frac{\pi}{2}$ is irrational because π is irrational.

iv $\frac{\pi}{2\pi} = \frac{1}{2}$; it is rational. ← Do not forget to simplify if you can

b Simplify these.

i $\sqrt{6} \times \sqrt{12}$ **ii** $3\sqrt{2} + \sqrt{32}$ **iii** $\frac{\sqrt{40}}{\sqrt{5}}$

i $\sqrt{6} \times \sqrt{12}$

$= \sqrt{6 \times 12}$ ← $\sqrt{a} \times \sqrt{b} = \sqrt{ab}$

$= \sqrt{72}$

$= \sqrt{36 \times 2}$

$= \sqrt{36} \times \sqrt{2}$ ← $\sqrt{ab} = \sqrt{a} \times \sqrt{b}$

$= 6\sqrt{2}$ ← $\sqrt{36} = 6$

Remember:

$\sqrt{ab} = \sqrt{a} \times \sqrt{b}$

$\sqrt{\frac{a}{b}} = \frac{\sqrt{a}}{\sqrt{b}}$

$a\sqrt{c} + b\sqrt{c} = (a + b)\sqrt{c}$

$a\sqrt{c} - b\sqrt{c} = (a - b)\sqrt{c}$

ii $3\sqrt{2} + \sqrt{32}$

$= 3\sqrt{2} + \sqrt{16 \times 2}$

$= 3\sqrt{2} + \sqrt{16} \times \sqrt{2}$

$= 3\sqrt{2} + 4\sqrt{2}$

$= 7\sqrt{2}$ ← $a\sqrt{c} + b\sqrt{c} = (a + b)\sqrt{c}$

iii $\frac{\sqrt{40}}{\sqrt{5}}$

$= \sqrt{\frac{40}{5}}$ ← $\frac{\sqrt{a}}{\sqrt{b}} = \sqrt{\frac{a}{b}}$

$= \sqrt{8}$

$= 2\sqrt{2}$

Apply 1

1 Write these numbers in the form $\frac{a}{b}$, giving your answers in their lowest terms.

a 0.7 **b** 0.26 **c** 1.4 **d** 6 **e** $0.\dot{1}\dot{8}$

2 Which of these are rational, and which are irrational? Give a reason for your answer.

a $\sqrt{24}$ **f** $\sqrt{6}$ **k** $\sqrt{3} \times \sqrt{3}$ **p** $(\sqrt{16})^3$

b $\sqrt{25}$ **g** $\sqrt{6} \times \sqrt{6}$ **l** $\sqrt{3} \times \sqrt{12}$ **q** $\sqrt{(3^2)}$

c $\frac{4}{11}$ **h** $6 \times \sqrt{6}$ **m** $(\sqrt{5})^2$ **r** $\sqrt[3]{(8^2)}$

d $\frac{\sqrt{4}}{11}$ **i** $\sqrt{3}$ **n** $(\sqrt[3]{7})^3$

e $\frac{4}{\sqrt{11}}$ **j** $\sqrt{3} \times 3$ **o** $(\sqrt{11})^3$

3 Get Real!

A draughtsman is trying to draw a square with an area of $2\ m^2$. However, he knows about surds, and realises his square needs to have a side of $\sqrt{2}$ m, which is irrational. This means he cannot measure the length exactly, as irrational numbers have no exact fraction or decimal equivalent.

He suddenly has a brilliant idea how he can draw his square without having to measure an irrational length. Can you think how he might have done it?

HINT He started with a bigger square, with double the area.

4 Charlotte and Jack are having an argument about square roots.
Charlotte says that $\sqrt{90}$ must be irrational because all square roots are irrational.
Jack says she is wrong; $\sqrt{9} = 3$, so $\sqrt{90} = 30$.
Who is wrong? Explain how you know.

5 Write down three irrational numbers between 2 and 3.

6 Simplify these.

a $\sqrt{20}$ **c** $\sqrt{32}$ **e** $\sqrt{80}$ **g** $\sqrt{108}$

b $\sqrt{45}$ **d** $\sqrt{500}$ **f** $\sqrt{98}$ **h** $\sqrt{75}$

7 Simplify these.

a $2\sqrt{3} + 3\sqrt{3}$ **c** $3\sqrt{5} - 2\sqrt{5}$ **e** $\sqrt{18} - \sqrt{8}$ **g** $\sqrt{72} + \sqrt{18}$

b $3\sqrt{2} + \sqrt{2}$ **d** $\sqrt{20} + \sqrt{5}$ **f** $3\sqrt{7} - \sqrt{28}$ **h** $\sqrt{8} + 3\sqrt{2} - \sqrt{50}$

8 Bobby says $\sqrt{10} + \sqrt{15} = \sqrt{25} = 5$.
Jen works it out differently, but gets the same answer.
She says $\sqrt{10} + \sqrt{15} = \sqrt{2} \times \sqrt{5} + \sqrt{3} \times \sqrt{5} = (\sqrt{2} + \sqrt{3}) \times \sqrt{5} = \sqrt{5} \times \sqrt{5} = 5$.
But Ed knows that $\sqrt{10}$ and $\sqrt{15}$ are both greater than 3, so the answer must be more than 6.

a What did Bobby do wrong?

b Where did Jen go wrong?

9 Simplify:

a $\sqrt{3} \times \sqrt{12}$ **c** $\sqrt{18} \times \sqrt{2}$ **e** $\sqrt{10} \times 3\sqrt{5}$ **g** $2\sqrt{5} \times \sqrt{10}$

b $\sqrt{10} \times \sqrt{20}$ **d** $2\sqrt{3} \times \sqrt{6}$ **f** $5\sqrt{6} \times 2\sqrt{3}$ **h** $7\sqrt{7} \times 2\sqrt{7}$

10 Simplify:

a $\dfrac{\sqrt{12}}{\sqrt{3}}$ **d** $\dfrac{\sqrt{88}}{2}$ **g** $\dfrac{6\sqrt{10}}{2\sqrt{5}}$ **j** $\dfrac{2\sqrt{21}}{8\sqrt{3}}$

b $\sqrt{\dfrac{81}{16}}$ **e** $\dfrac{3\sqrt{6}}{\sqrt{3}}$ **h** $\dfrac{8\sqrt{30}}{4\sqrt{5}}$ **k** $\dfrac{2\sqrt{10}}{10\sqrt{2}}$

c $\sqrt{\dfrac{45}{20}}$ **f** $\dfrac{5\sqrt{6}}{\sqrt{2}}$ **i** $\dfrac{2\sqrt{5}}{4}$ **l** $\dfrac{6\sqrt{21}}{15\sqrt{7}}$

11 Dan says that $\dfrac{\sqrt{20}}{5} = \sqrt{4} = 2$.

Is Dan correct?
Give a reason for your answer.

12 Get Real!

Tessell Ltd make wall tiles. They make two different sizes of square tile and one rectangle, to tessellate as shown. The large square has an area of 30 cm^2, and the small square has an area of 15 cm^2.

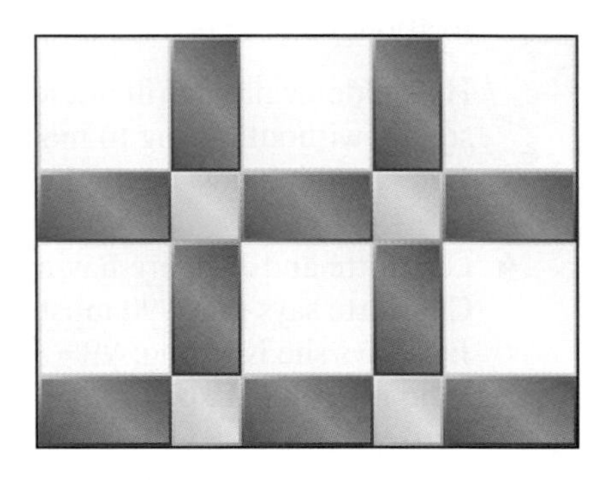

a What are the exact dimensions (length and width) of the rectangular tile?
Give your answers as surds in their simplest form.

b What is the area of the rectangular tile?
Give your answer as a surd in its simplest form.

13 a Find two irrational numbers that multiply together to make a rational number.

b Find two irrational numbers that, when divided, give the answer 2.

14 Put the numbers below into pairs with the same value.

a $2\sqrt{2} + \sqrt{2}$ **c** $\dfrac{6\sqrt{3}}{3}$ **e** $\dfrac{6\sqrt{3}}{\sqrt{3}}$ **g** $\sqrt{\dfrac{12}{2}}$

b $\sqrt{3} \times \sqrt{2}$ **d** $\sqrt{18}$ **f** $\dfrac{\sqrt{24}}{\sqrt{2}}$ **h** $\dfrac{\sqrt{72}}{\sqrt{2}}$

15 Show that:

a $\sqrt{18} \times \sqrt{2} = 6$ **d** $\sqrt{\dfrac{90}{10}} = 3$ **g** $\dfrac{\sqrt{60}}{2} = \sqrt{15}$

b $\sqrt{72} = 6\sqrt{2}$ **e** $\sqrt{2} \times \sqrt{3} \times \sqrt{4} \times \sqrt{5} \times \sqrt{6} = 12\sqrt{5}$ **h** $5\sqrt{3} - \sqrt{3} = \sqrt{48}$

c $3\sqrt{2} \times \sqrt{8} = 12$ **f** $\dfrac{3\sqrt{5} + 2\sqrt{5}}{5} = \sqrt{5}$ **i** $\sqrt{80} - \sqrt{20} = 2\sqrt{5}$

16 Write $\sqrt{15} \times \sqrt{5}$ in the form $a\sqrt{b}$, where a and b are both prime numbers.

17 a Show that $(\sqrt{12} - \sqrt{3})^2 = 3$.

b In part **a** the answer was 3, a rational number, but not many expressions of the form $(\sqrt{a} - \sqrt{b})^2$, where $\sqrt{a}$ and $\sqrt{b}$ are irrational, give a rational answer. Expand $(\sqrt{a} - \sqrt{b})^2$, and find another example of $(\sqrt{a} - \sqrt{b})^2$, where $\sqrt{a}$ and $\sqrt{b}$ are irrational, which gives a rational answer.

18 Simplify these.

a $\sqrt{2}(3+\sqrt{2})$
b $\sqrt{3}(2+\sqrt{12})$
c $(1+\sqrt{2})(2+\sqrt{2})$
d $(3-\sqrt{2})(1-\sqrt{2})$
e $(\sqrt{3}+2)(\sqrt{3}-2)$
f $(5-\sqrt{7})^2$
g $(a+\sqrt{3})(a-\sqrt{3})$
h $(c+\sqrt{d})(c-\sqrt{d})$

Explore

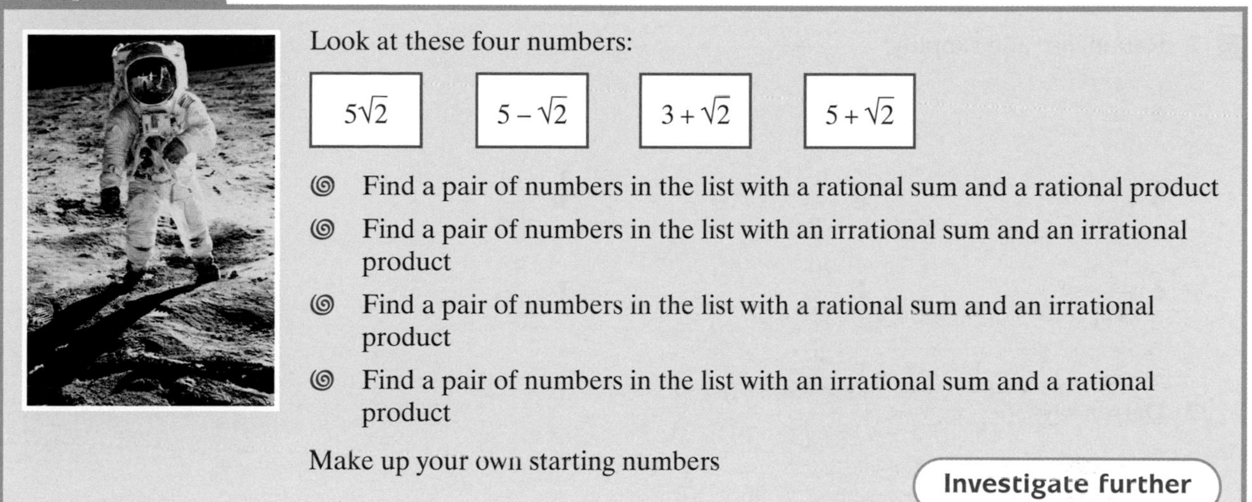

Look at these four numbers:

$5\sqrt{2}$ | $5-\sqrt{2}$ | $3+\sqrt{2}$ | $5+\sqrt{2}$

- Find a pair of numbers in the list with a rational sum and a rational product
- Find a pair of numbers in the list with an irrational sum and an irrational product
- Find a pair of numbers in the list with a rational sum and an irrational product
- Find a pair of numbers in the list with an irrational sum and a rational product

Make up your own starting numbers

Investigate further

Learn 2 Rationalising the denominator of a surd

Examples:

Rationalise and simplify these.

Rationalise means remove the square roots

a $\frac{\sqrt{3}}{\sqrt{2}}$ **b** $\frac{5}{2\sqrt{3}}$

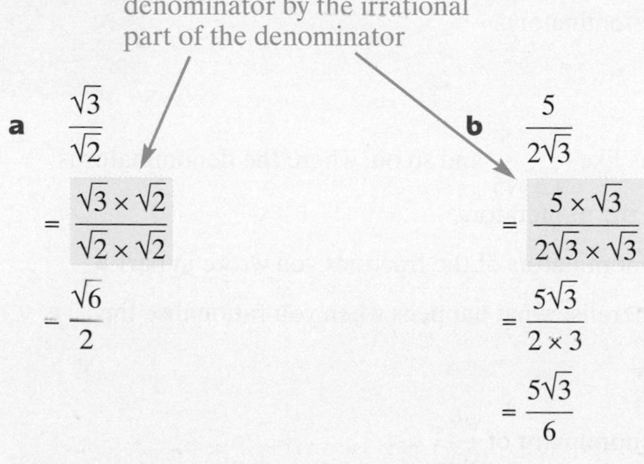

Multiply the numerator and denominator by the irrational part of the denominator

a $\frac{\sqrt{3}}{\sqrt{2}}$

$= \frac{\sqrt{3}\times\sqrt{2}}{\sqrt{2}\times\sqrt{2}}$

$= \frac{\sqrt{6}}{2}$

b $\frac{5}{2\sqrt{3}}$

$= \frac{5\times\sqrt{3}}{2\sqrt{3}\times\sqrt{3}}$

$= \frac{5\sqrt{3}}{2\times 3}$

$= \frac{5\sqrt{3}}{6}$

You can multiply surds together, for example,

$\sqrt{a}\times\sqrt{a}=a$ (as $\sqrt{a}\times\sqrt{a}=\sqrt{a\times a}=\sqrt{a^2}=a$)

$(a+\sqrt{b})(a-\sqrt{b})=a^2-b$ (as $(a+\sqrt{b})(a-\sqrt{b})=a^2-a\sqrt{b}+a\sqrt{b}-b=a^2-b$)

Apply 2

1 Simplify:

a $\frac{\sqrt{12}}{\sqrt{3}}$ **b** $\frac{\sqrt{7}}{\sqrt{28}}$ **c** $\frac{2\sqrt{3}}{\sqrt{27}}$ **d** $\frac{3\sqrt{5}}{\sqrt{20}}$

2 Rationalise and simplify:

a $\frac{5}{\sqrt{3}}$ **d** $\frac{5}{\sqrt{10}}$ **g** $\frac{3}{\sqrt{3}}$ **j** $\frac{3}{2\sqrt{3}}$

b $\frac{2}{\sqrt{7}}$ **e** $\frac{9}{\sqrt{3}}$ **h** $\frac{30}{\sqrt{15}}$ **k** $\frac{7}{4\sqrt{5}}$

c $\frac{1}{\sqrt{11}}$ **f** $\frac{20}{\sqrt{5}}$ **i** $\frac{8}{\sqrt{2}}$ **l** $\frac{9}{4\sqrt{3}}$

3 Darren says $\sqrt{\frac{6}{18}} = \sqrt{\frac{1}{3}} = \sqrt{\frac{1 \times \sqrt{3}}{3 \times \sqrt{3}}} = \frac{\sqrt{3}}{3\sqrt{3}} = \frac{1}{3}$

Pablo says $\sqrt{\frac{6}{18}} = \frac{\sqrt{6}}{\sqrt{18}} = \frac{\sqrt{6}}{\sqrt{9} \times \sqrt{2}} = \frac{\sqrt{6}}{3\sqrt{2}} = \frac{\sqrt{2}}{\sqrt{2}} = 1$

a Find Darren's mistake.

b Find Pablo's mistake.

c Work out the correct answer.

4 Get Real!

A room has a length which is exactly $\sqrt{2}$ times its width. The length is 10 m.

a What is the width? Give your answer as a surd (with a rational denominator).

b What is the area?

5 a Write five fractions like $\frac{2}{\sqrt{2}}, \frac{5}{\sqrt{5}}$ and so on, where the denominator is the square root of the numerator.

b Rationalise the denominators of the fractions you wrote in part **a**.

c Use algebra to generalise what happens when you rationalise the denominator of $\frac{a}{\sqrt{a}}$

d Rationalise the denominator of $\frac{ab}{\sqrt{a}}$

6 Copy and find your way through this maze, only occupying spaces where the answer is correct.

Start ➡	$\frac{2}{\sqrt{3}} = \frac{2\sqrt{3}}{3}$	$\frac{2}{\sqrt{5}} = \frac{\sqrt{5}}{2}$	$\frac{\sqrt{4}}{3} = \frac{2}{3}$	**End**
$\frac{1}{\sqrt{2}} = \sqrt{2}$	$\frac{4}{\sqrt{6}} = \frac{2\sqrt{6}}{3}$	$\frac{a}{\sqrt{2a}} = \sqrt{a}$	$\frac{6}{\sqrt{3}} = 2\sqrt{3}$	$\frac{2}{1+\sqrt{4}} = \frac{2}{5}$
$\frac{4}{\sqrt{2}} = 2\sqrt{2}$	$\frac{1}{\sqrt{5}} = \frac{\sqrt{5}}{5}$	$\frac{7}{\sqrt{2}} = 7\sqrt{2}$	$\frac{a}{\sqrt{b}} = \frac{a\sqrt{b}}{b}$	$\frac{7}{\sqrt{7}} = \sqrt{7}$
$\frac{3}{\sqrt{3}} = \sqrt{3}$	$\frac{1}{5\sqrt{2}} = \frac{5\sqrt{2}}{2}$	$\frac{2a}{\sqrt{2a}} = 2\sqrt{a}$	$\frac{3}{2\sqrt{3}} = \frac{\sqrt{3}}{3}$	$\frac{12}{\sqrt{3}} = 4\sqrt{3}$
$\frac{3a}{\sqrt{a}} = 3\sqrt{a}$	$\frac{5}{\sqrt{2}} = \frac{5\sqrt{2}}{2}$	$\frac{11}{\sqrt{11}} = \frac{\sqrt{11}}{11}$	$\frac{4}{\sqrt{36}} = \frac{2}{3}$	$\frac{5}{\sqrt{10}} = \frac{\sqrt{10}}{2}$
$\frac{3}{\sqrt{2}} = \sqrt{2}$	$\frac{9}{\sqrt{6}} = \frac{3\sqrt{6}}{2}$	$\frac{a}{\sqrt{50}} = \frac{a\sqrt{2}}{10}$	$\frac{9}{\sqrt{3}} = 3\sqrt{3}$	$\frac{1}{\sqrt{50}} = \frac{\sqrt{50}}{25}$

Explore

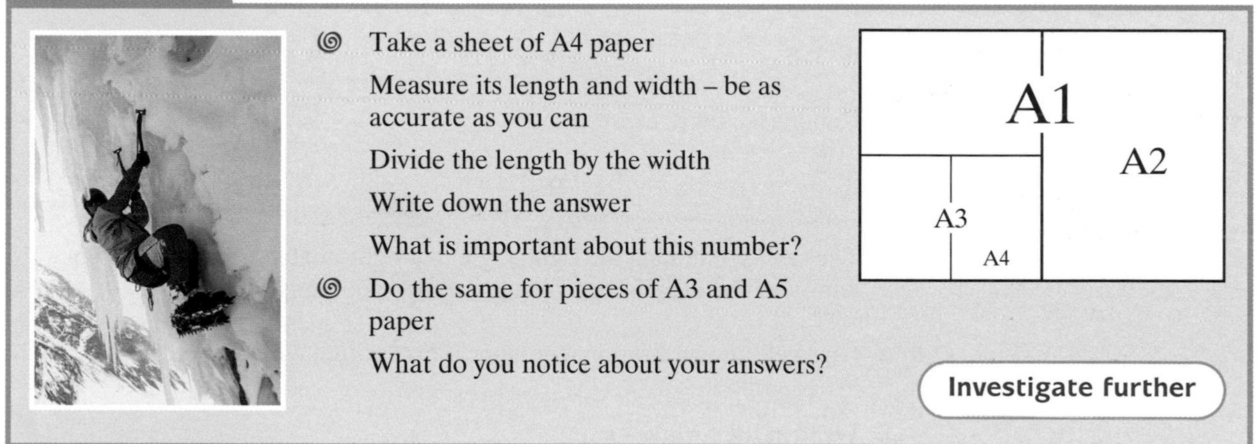

- Take a sheet of A4 paper

  Measure its length and width – be as accurate as you can

  Divide the length by the width

  Write down the answer

  What is important about this number?
- Do the same for pieces of A3 and A5 paper

  What do you notice about your answers?

Investigate further

Explore

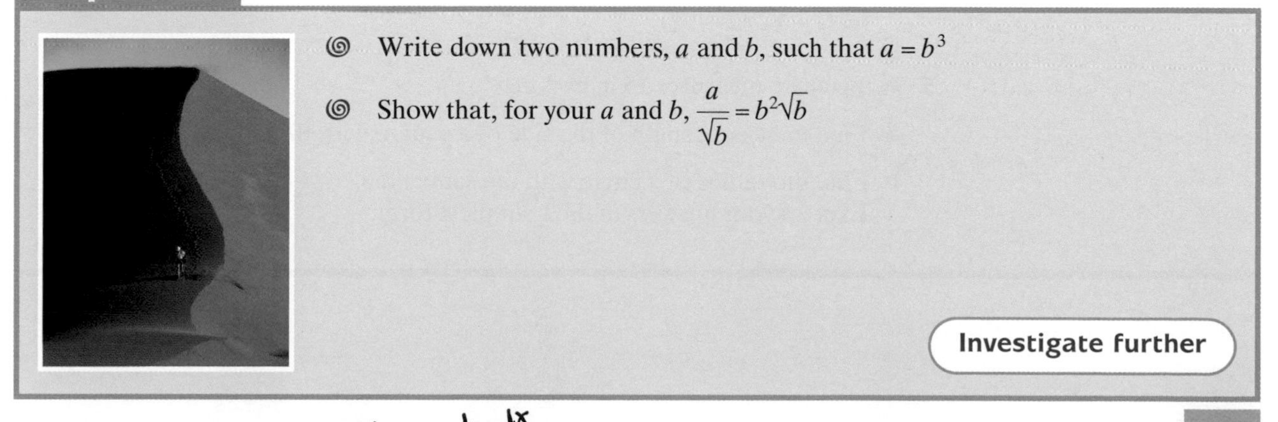

- Write down two numbers, a and b, such that $a = b^3$
- Show that, for your a and b, $\frac{a}{\sqrt{b}} = b^2\sqrt{b}$

Investigate further

Surds

ASSESS

The following exercise tests your understanding of this chapter, with the questions appearing in order of increasing difficulty.

1 **a** Which of these numbers are rational and which are irrational? How can you tell?

$\sqrt{5}$ $\quad\sqrt{9}$ $\quad\dfrac{\sqrt{16}}{7}$ $\quad\sqrt{\dfrac{16}{7}}$ $\quad\dfrac{\sqrt{63}}{\sqrt{7}}$

b Square these numbers.

i $\sqrt{8}$ **iv** $\sqrt{x}$ **vii** $\sqrt{3} \times \sqrt{8}$

ii $\sqrt{13}$ **v** $2\sqrt{3}$ **viii** $\sqrt{y} \times \sqrt{z}$

iii $\sqrt{25}$ **vi** $5\sqrt{a}$

2 **a** Write, in their simplest form:

i $\sqrt{8}$ **iii** $\sqrt{98}$ **v** $5\sqrt{50}$

ii $\sqrt{45}$ **iv** $4\sqrt{18}$

b Express these as square roots of a single number, for example, $3\sqrt{7} = \sqrt{9} \times \sqrt{7} = \sqrt{63}$.

i $2\sqrt{3}$ **iii** $4\sqrt{7}$ **v** $10\sqrt{2}$

ii $3\sqrt{5}$ **iv** $2\sqrt{11}$

3 **a** Rationalise these expressions.

i $\dfrac{2}{\sqrt{2}}$ **ii** $\dfrac{3}{\sqrt{5}}$ **iii** $\dfrac{4}{\sqrt{7}}$

b The sides containing the right angle in a right-angled triangle are 7 cm and 5 cm. What is the *exact* length of the hypotenuse?

4 **a** One well-known irrational number is not written with a $\sqrt{}$ sign. What number is it?

b Write in the form $a + b\sqrt{c}$:

i $(\sqrt{2} + 3)^2$ **iv** $(\sqrt{3} + \sqrt{2})(\sqrt{3} - \sqrt{2})$

ii $(\sqrt{5} - 4)^2$ **v** $(2\sqrt{5} + 3\sqrt{7})(3\sqrt{5} - 2\sqrt{7})$

iii $(1 + \sqrt{2})(2 - \sqrt{2})$

5 A rectangle measures 15 m by 3 m.

a Find the exact length of the side of a square with the same area.

b Find the radius of a circle with the same area. Express your answers in their simplest form.

9 Representing data

OBJECTIVES

E **Examiners would normally expect students who get an E grade to be able to:**

Construct a pie chart

D **Examiners would normally expect students who get a D grade to be able to:**

Construct a stem-and-leaf diagram (ordered)

Construct a frequency diagram

Interpret a time series graph

B **Examiners would normally expect students who get a B grade also to be able to:**

Construct and interpret a cumulative frequency diagram

Use a cumulative frequency diagram to estimate the median and interquartile range

Construct and interpret a box plot

Compare two sets of data using box plots

Construct a time series and plot the moving averages

Use the trend line to estimate other values

A **Examiners would normally expect students who get an A grade also to be able to:**

Construct and interpret a histogram with unequal class intervals

What you should already know ...

- Measures of average including mean, median and mode
- Accurate use of ruler and protractor
- Construct and interpret a pictogram, bar chart and pie chart
- Interpret a stem-and-leaf diagram
- Interpret a time series graph

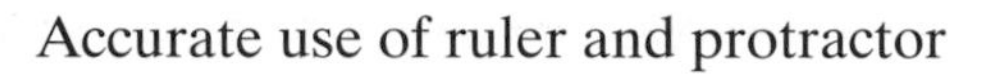

VOCABULARY

Pie chart – in a pie chart, frequency is shown by the angles (or areas) of the sectors of a circle

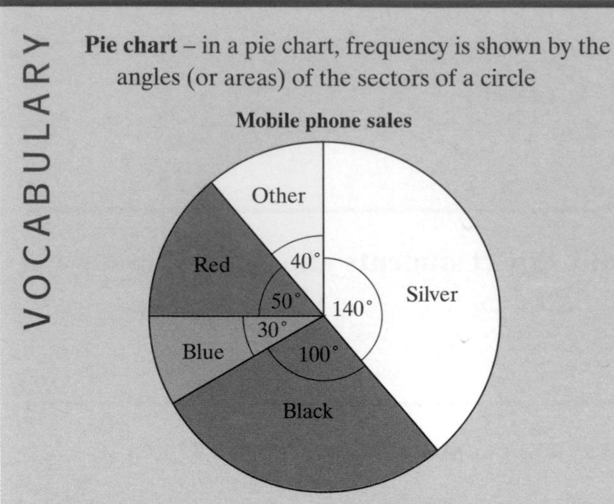

Frequency diagram – a graphical method of showing how many results or observations fall into each category in a survey or experiment

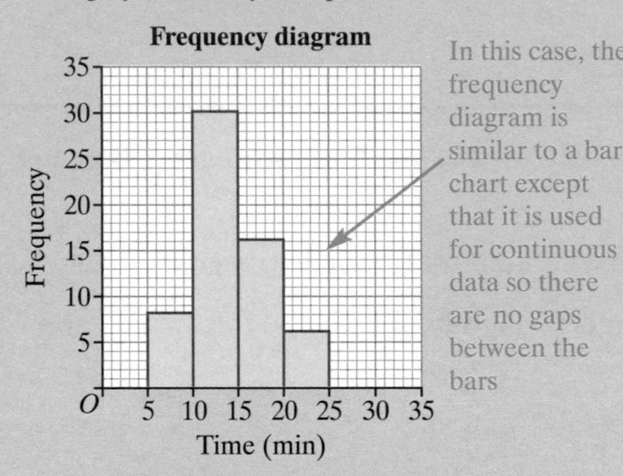

Stem-and-leaf diagram – a way of arranging data using a key to explain the 'stem' and 'leaf' so that 3|4 represents 34

Number of minutes to complete a task

Stem (tens)	Leaf (units)
1	6 8 1 9 7
2	7 8 2 7 7 2 9
3	4 1 6

Key: 3|4 represents 34 minutes

Ordered stem-and-leaf diagram – a stem-and-leaf diagram where the data is placed in order

Number of minutes to complete a task

Stem (tens)	Leaf (units)
1	1 6 7 8 9
2	2 2 7 7 7 7 8 9
3	1 4 6

Key: 3|4 represents 34 minutes

Back-to-back stem-and-leaf diagram – a stem-and-leaf diagram used to represent two sets of data

Number of minutes to complete a task

Leaf (units) Girls	Stem (tens)	Leaf (units) Boys
7 7 6 5 4 2 2	1	1 6 7 8 9
7 6 4 3 2 1	2	2 2 7 7 7 7 8 9
7 0	3	1 4 6

Key: 3|2 represents 23 minutes

Key: 3|4 represents 34 minutes

Line graph – a line graph is a series of points joined with straight lines

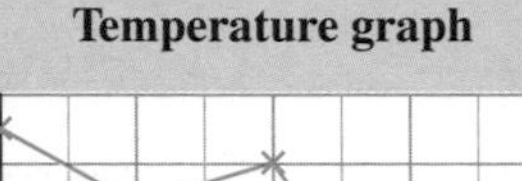

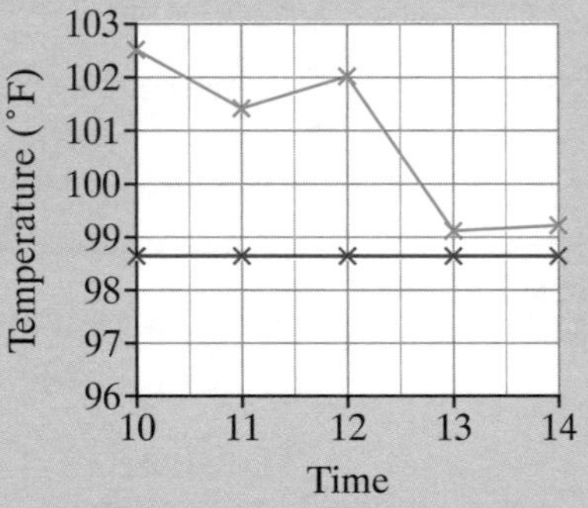

Index number – an index number is used to compare measurements over a period of time. A common index is the price index which compares prices over a period of time. The base year is given a value of 100 (representing 100%) and the price index shows the increase (or decrease) since the base year. A price index of 120 represents a 20% increase and a price index of 90 represents a 10% decrease

Time series – a graph of data recorded at regular intervals

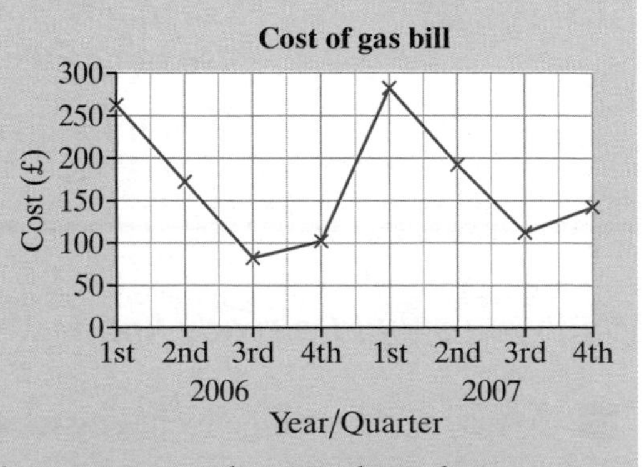

Moving average – used to smooth out the fluctuations in a time series, for example, a four-point moving average is found by averaging successive groups of four readings

The four-point moving averages can be plotted on the graph as shown

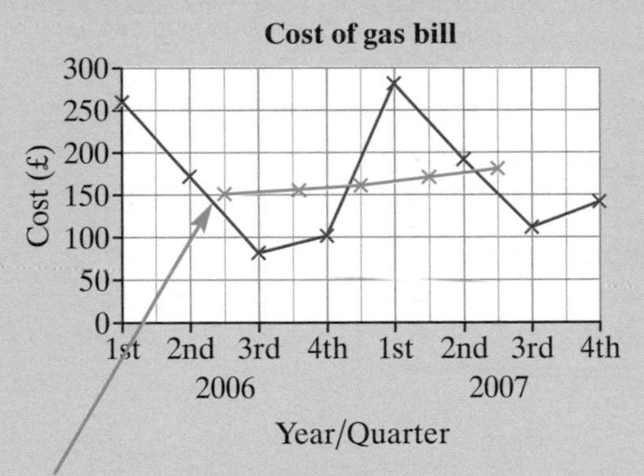

The first four-point moving average is plotted in the 'middle' of the first four points, and so on

Cumulative frequency diagram – a cumulative frequency diagram can be used to find an estimate for the mean and quartiles of a set of data; find the cumulative frequency by adding the frequencies in turn to give a 'running total'

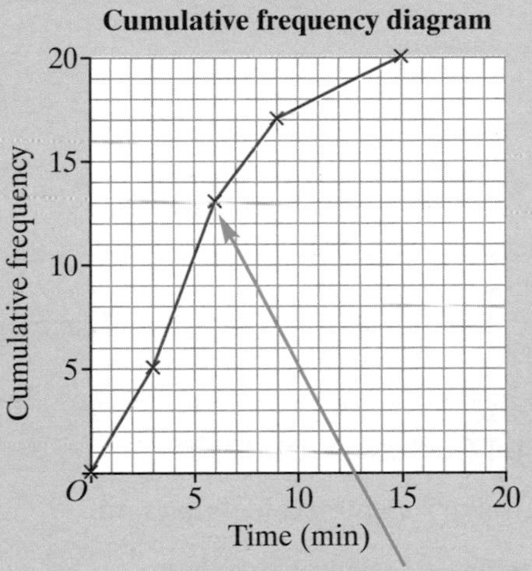

The cumulative frequencies are plotted at the end of the interval to which they relate

Median – the middle value when all the values have been arranged in order of size; for an even set of numbers, the median is the mean of the two middle values

Lower quartile – the value 25% of the way through the data

Upper quartile – the value 75% of the way through the data

Interquartile range – the difference between the upper quartile and the lower quartile

Interquartile range = upper quartile − lower quartile

Box plot or **box and whisker plot** – used to show how the data is distributed

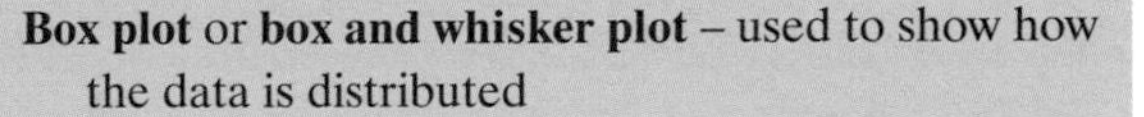

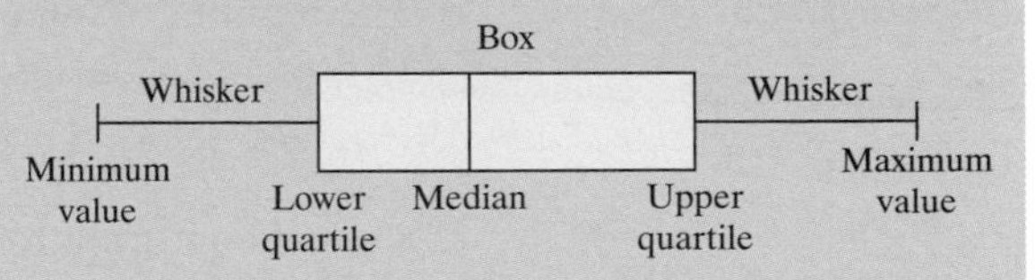

Histogram – a histogram is similar to a bar chart except that the *area* of the bar represents the frequency

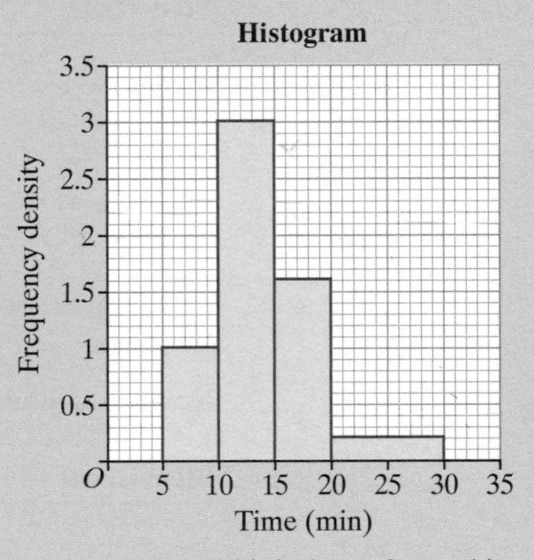

Frequency polygon – this is drawn from a histogram (or bar chart) by joining the midpoints of the tops of the bars with straight lines to form a polygon

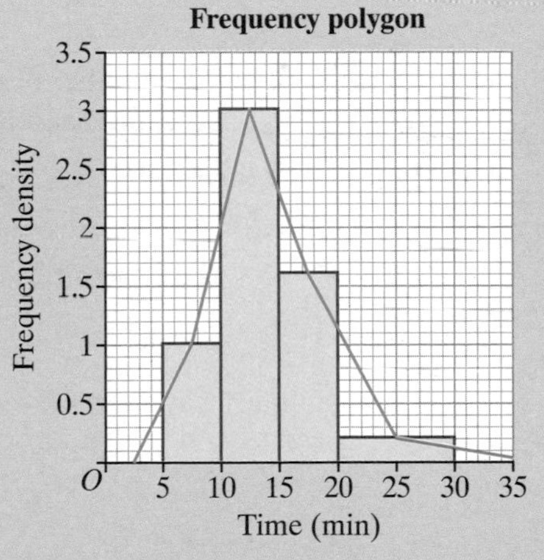

Frequency density – in a histogram, the area of the bars represents the frequency and the height represents the frequency density

$$\text{Frequency density (bar height)} = \frac{\text{frequency}}{\text{class width}}$$

Learn 1 Stem-and-leaf diagrams and pie charts

Examples:

a The number of minutes taken to complete an exercise was recorded for 15 boys in a class.

16, 27, 28, 22, 34, 18, 11, 19, 27, 31, 27, 36, 22, 17, 29

Show the information as a stem-and-leaf diagram.

Number of minutes to complete a task

Stem (tens)	Leaf (units)
1	6 8 1 9 7
2	7 8 2 7 7 2 9
3	4 1 6

In this case the number 6 stands for 16 (1 ten and 6 units)

In this case the number 6 stands for 36 (3 tens and 6 units)

Key: 3|4 represents 34 minutes

It is useful to provide an ordered stem-and-leaf diagram to find the median and range.

Number of minutes to complete a task

Stem (tens)	Leaf (units)
1	1 6 7 8 9
2	2 2 7 7 7 8 9
3	1 4 6

Here the leaves (units) are all arranged numerically

Key: 3|4 represents 34 minutes

b The number of minutes taken to complete an exercise was recorded for 15 boys and 15 girls in a class.

Boys: 16, 27, 28, 22, 34, 18, 11, 19, 27, 31, 27, 36, 22, 17, 29
Girls: 12, 23, 22, 17, 30, 16, 15, 14, 17, 37, 26, 24, 21, 12, 27

Show the information as a back-to-back stem-and-leaf diagram.

Here the leaves (units) are all arranged numerically from the right-hand side

Number of minutes to complete a task

Leaf (units) Girls	Stem (tens)	Leaf (units) Boys
7 7 6 5 4 2 2	1	1 6 7 8 9
7 6 4 3 2 1	2	2 2 7 7 7 8 9
7 0	3	1 4 6

Key: 3|2 represents 23 minutes

Key: 3|2 represents 32 minutes

The boys' data has already been recorded in an ordered stem-and-leaf diagram

This diagram is called a back-to-back (ordered) stem-and-leaf diagram.

c The following distribution shows the sales of mobile phones by colour.

Colour	Silver	Black	Blue	Red	Other
Frequency	14	10	3	5	4

Show this information as a pie chart.

In a pie chart, the frequency is represented by the angles (or areas) of the sectors of a circle.

The pie chart needs to be drawn to represent 36 phones. There are 360° in a full circle, so each phone will be shown by 360° ÷ 36 = 10°

Colour	**Frequency**	
Silver	14	$14 \times 10° = 140°$
Black	10	$10 \times 10° = 100°$
Blue	3	$3 \times 10° = 30°$
Red	5	$5 \times 10° = 50°$
Other	4	$4 \times 10° = 40°$
Total	36	360°

Draw your pie chart as large as possible and remember to label the sectors or provide a key.

Check that the sum of the angles does add up to 360°

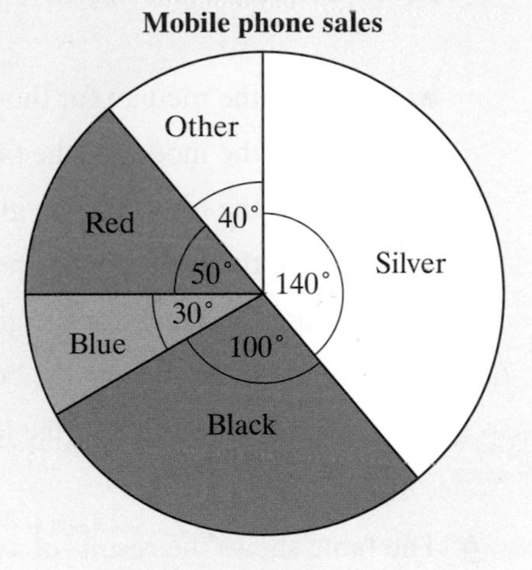

Apply 1

1 The prices paid for some takeaway food are shown below.

£3.64 £4.15 £5.22 £5.88 £4.21 £4.55 £3.75 £4.78 £5.05 £4.52 £4.60

Copy and complete the stem-and-leaf diagram to show this information.

Prices paid for takeaway food

Stem (£)	**Leaf (pence)**
3	
4	
5	

Key: 4 | 15 represents £4.15

2 The marks obtained in a test were recorded as follows.

8 20 9 21 18 22 19 13 22 24 14 9 25 10 19 20 17 14 12

a Show this information in an ordered stem-and-leaf diagram.

b What was the highest mark in the test?

c Write down the median of the marks in the test.

d Write down the range of the marks in the test.

3 The number of minutes taken to complete an exercise was recorded for 15 boys and 15 girls in this back-to-back stem-and-leaf diagram.

Number of minutes to complete a task

Leaf (units) Girls	Stem (tens)	Leaf (units) Boys
7 7 6 5 4 2 2	1	1 6 7 8 9
7 6 4 3 2 1	2	2 2 7 7 7 8 9
7 0	3	1 4 6

Key : 3|2 represents 23 minutes   Key : 3|4 represents 34 minutes

a Calculate the median for the girls.

b Calculate the mode for the boys.

c Calculate the mean for the girls.

d Calculate the mean for the boys.

e Calculate the range for the girls.

f Calculate the range for the boys.

g What can you say about the length of time to complete the exercise by the girls and the boys?

4 The table shows the results of a survey to find students' favourite colours. Draw a pie chart to show the information.

Colour	Tally
Red	𝍸 𝍸 𝍸
Blue	𝍸 𝍸 𝍸 𝍸 I
Green	𝍸 IIII
Yellow	𝍸 III
Black	II

5 Students at a college were asked to choose their favourite film. Their choices are shown in the pie chart.

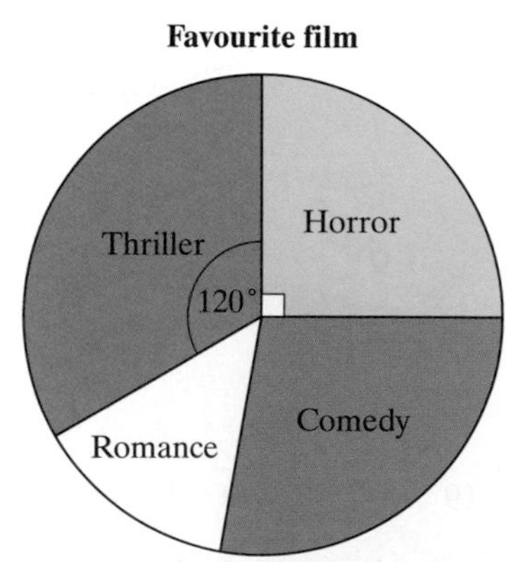

A total of 135 students chose horror films.

a How many students were included in the survey?

b How many students chose thrillers?

Twice as many students chose 'Comedy' as chose 'Romance'.

c How many students chose 'Comedy'?

Explore

Collect some information from the students in your class, and show it in a stem-and-leaf diagram. You might wish to collect:

- Height, arm length, head circumference ... in metric or imperial units
- Estimate of the weight of ten mathematics books in metric or imperial units
- Time to complete a task in minutes and seconds
- Time spent on homework in hours and minutes
- Time to run or walk a certain distance in minutes and seconds
- Weekly pocket money or weekly wages in pounds and pence

Alternatively, collect some other information that can be shown in a stem-and-leaf diagram

Investigate further

Learn 2 Frequency diagrams, line graphs and time series

Examples:

a 50 people were asked how long they had to wait for a train. The table below shows the results.

Time, t (minutes)	Frequency
$5 \leqslant t < 10$	16
$10 \leqslant t < 15$	22
$15 \leqslant t < 20$	11
$20 \leqslant t < 25$	1

Draw a frequency diagram to represent the data.

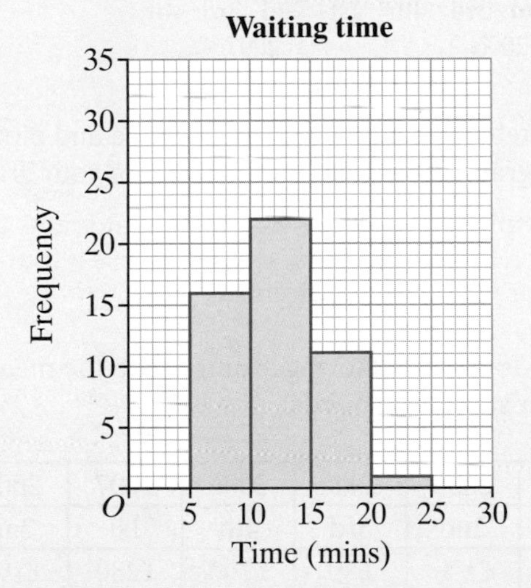

A frequency diagram is a graphical method of showing how many results fall into each category. In this example, the frequency diagram is similar to a bar chart except that it is used for continuous data so there are no gaps between the bars

b The table shows the temperature of a patient at different times during the day.

Time	10.00	11.00	12.00	13.00	14.00
Temperature (°F)	102.5	101.3	102	99.1	99.2

Draw a line graph to show this information.

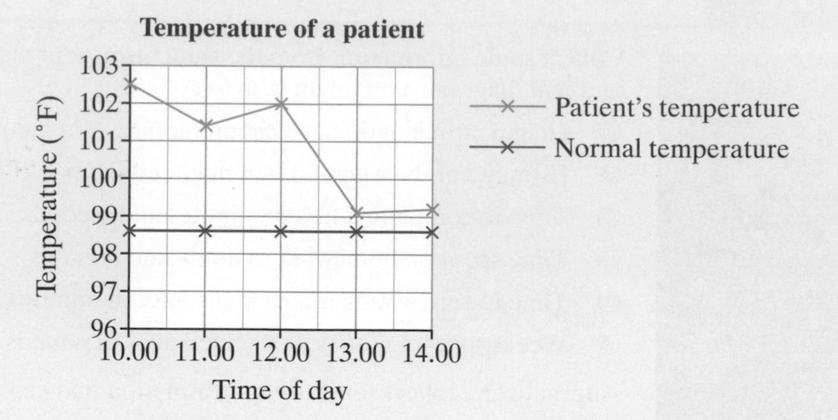

In this example, the temperature is expected to fall to the normal body temperature of 98.8°F. However, other line graphs can fluctuate over time.

c The table shows the cost of gas bills at the end of every three months.

Year	2006	2006	2006	2006	2007	2007	2007	2007
Quarter	1st	2nd	3rd	4th	1st	2nd	3rd	4th
Cost	£260	£170	£80	£100	£280	£190	£110	£140

i Draw a time series to show this information.

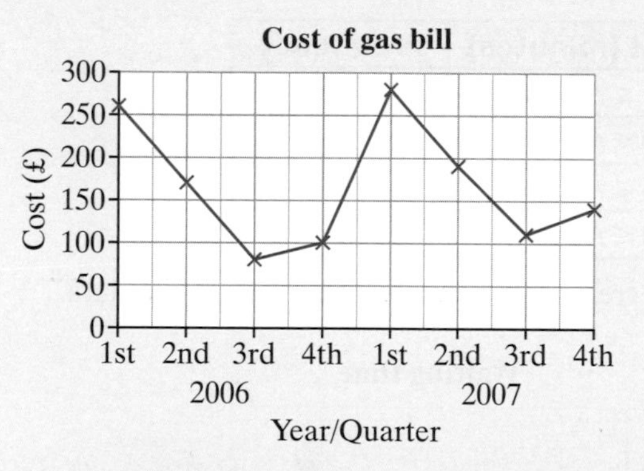

ii Use the table to find the moving average and plot this on your graph. Comment on the trend of your graph.

In this example, the cost follows a cycle that seems to be repeating every four quarters, so you can use a four-point moving average to find any trend

To find the four-point moving average take the mean of each four successive points.

To find the second moving average, miss out the first reading and include the fifth reading

Year	2006	2006	2006	2006	2007	2007	2007	2007
Quarter	1st	2nd	3rd	4th	1st	2nd	3rd	4th
Cost	£260	£170	£80	£100	£280	£190	£110	£140

The first four-point moving average $= \dfrac{£260 + £170 + £80 + £100}{4} = £152.50$

The second four-point moving average $= \dfrac{£170 + £80 + £100 + £280}{4} = £157.50$

The third four-point moving average $= \frac{£80 + £100 + £280 + £190}{4} = £162.50$

The fourth four-point moving average $= \frac{£100 + £280 + £190 + £110}{4} = £170.00$

The fifth four-point moving average $= \frac{£280 + £190 + £110 + £140}{4} = £180.00$

Plotting the four-point moving averages on the graph:

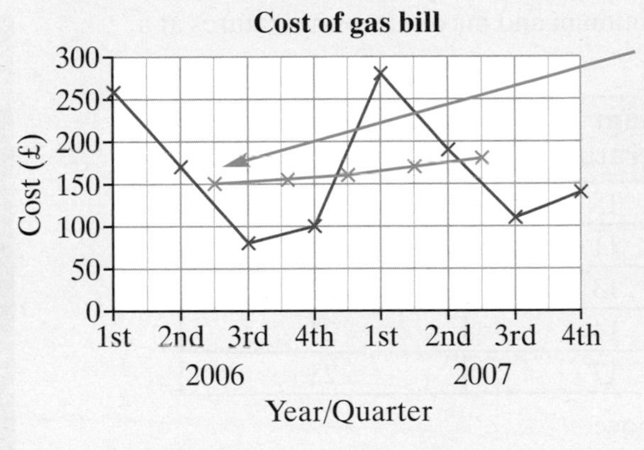

The first four-point moving average is plotted in the 'middle' of the first four points, and so on

From this graph you can see that the trend is upwards.

Apply 2

1 The table shows the time spent in a local shop by 60 customers.

Time, t (minutes)	Frequency
$5 \leqslant t < 10$	8
$10 \leqslant t < 15$	30
$15 \leqslant t < 20$	16
$20 \leqslant t < 25$	6

Draw a frequency diagram to represent the data.

2 The frequency diagram shows the ages of 81 people in a factory.
Copy and complete this table to show this information.

Age, y (years)	Frequency
$20 \leqslant y < 30$	
$30 \leqslant y < 40$	
$40 \leqslant y < 50$	
$50 \leqslant y < 60$	

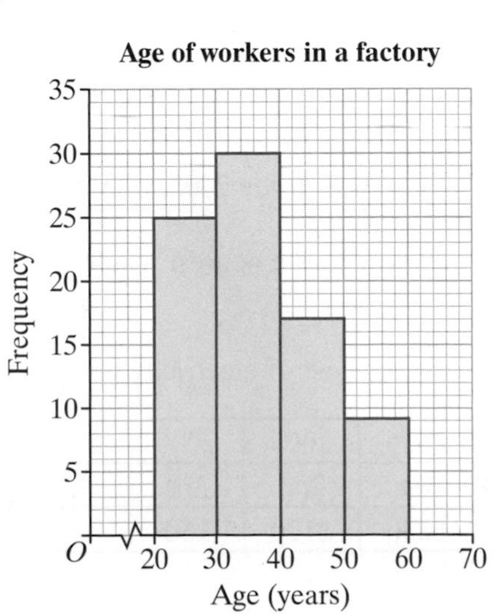

3 The table shows the pressure in millibars (mb) over five days at a seaside resort.

Day	Pressure (mb)
Monday	1018
Tuesday	1022
Wednesday	1028
Thursday	1023
Friday	1019

Draw a line graph to show the pressure each day.

4 The table shows the minimum and maximum temperatures at a seaside resort.

Day	Minimum temperature (°C)	Maximum temperature (°C)
Monday	15	19
Tuesday	11	21
Wednesday	13	22
Thursday	14	23
Friday	17	23

Draw a line graph to show:

a the minimum temperatures

b the maximum temperatures.

Use your graph to find:

c the day on which the lowest temperature was recorded

d the day on which the highest temperature was recorded

e the biggest difference between the daily minimum and maximum temperatures.

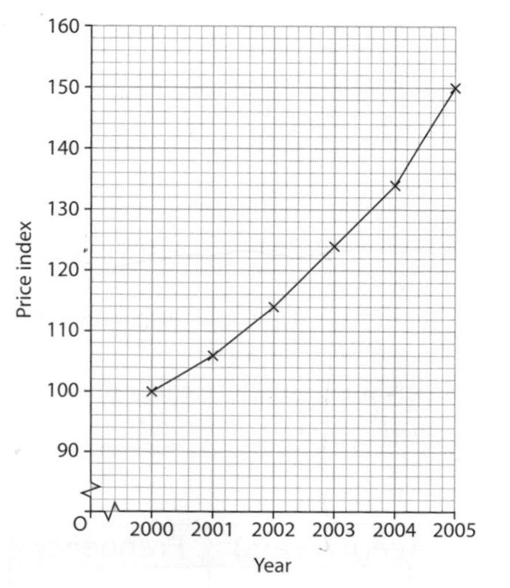

5 The graph shows the price index for petrol from the year 2000 to the year 2005.

a What is the price index for petrol in 2003?

A litre of petrol cost 60p in 2000.

b What is the cost of a litre of petrol in 2005?

c Taking the base year to be 2002, what would be the new price index for 2004?
Give your answer to the nearest whole number.

6 The table shows the cost of electricity bills at the end of every three months.

Year	2006	2006	2006	2006	2007	2007	2007	2007
Quarter	1st	2nd	3rd	4th	1st	2nd	3rd	4th
Cost	£230	£120	£50	£80	£215	£120	£25	£55

a Copy this table and calculate the four-point moving averages.
The first two have been done for you.

Year	2006	2006	2006	2006	2007	2007	2007	2007
Quarter	1st	2nd	3rd	4th	1st	2nd	3rd	4th
Cost	£230	£120	£50	£80	£215	£120	£25	£55
Four-point moving average		£120.00	£116.25					

b Show this information on a graph.

c What can you say about the trend?

7 The table shows the number of students at college present during morning and afternoon registration.

Day	Mon	Mon	Tue	Tue	Wed	Wed	Thu	Thu	Fri	Fri
Session	a.m.	p.m.	a.m.	p.m.	a.m.	p.m.	a.m.	p.m.	a.m.	p.m.
Number	220	210	243	215	254	218	251	201	185	152

a Copy this table and calculate the two-point moving averages.
The first four have been done for you.

Day	Mon	Mon	Tue	Tue	Wed	Wed	Thu	Thu	Fri	Fri
Session	a.m.	p.m.	a.m.	p.m.	a.m.	p.m.	a.m.	p.m.	a.m.	p.m.
Number	220	210	243	215	254	218	251	201	185	152
Two-point moving average	215	226.5	229	234.5						

b Show this information on a graph.

c What can you say about the trend?

8 Simon keeps a note of his termly exam results.

Year	2005	2006	2006	2006	2007	2007	2007
Session	Autumn	Spring	Summer	Autumn	Spring	Summer	Autumn
Exam result (%)	86	93	70	83	93	67	77

a Use the information to calculate the three-point moving averages.

b Show this information on a graph.

c Simon says that his performance is improving
Is he correct?
Give a reason for your answer.

9 The graph shows the quarterly sales at a shop.

a Use the graph to calculate the four-point moving averages.

b Copy the graph and plot your four-point moving averages.

c What can you say about the sales trend?

d Use the moving averages to predict the likely sales for July 2006.

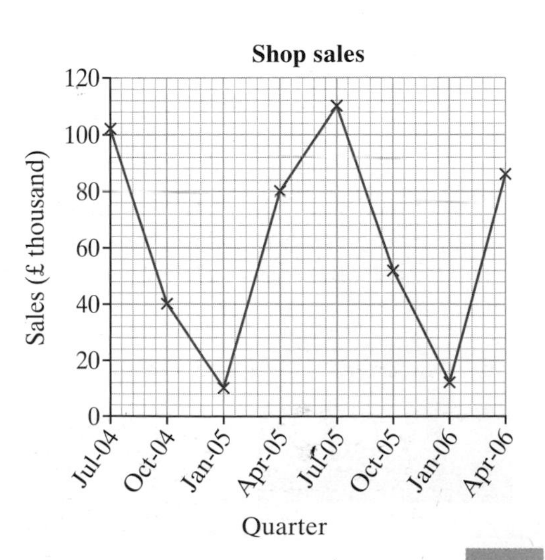

10 The table shows a company's quarterly profits (£ million) recorded at the end of each quarter in 2005 and 2006.

Year	Quarter	Profit (£ million)	Four-point moving average
2005	March	1.2	
	June	1.5	1.35
	September	1.9	
	December	0.8	
2006	March	1.1	
	June	1.3	
	September	2.0	
	December	1.1	

a Calculate the four-point moving averages. The first one has been done for you.

b Show this information on a graph.

c Does the graph suggest that profits are going up or down? Give a reason for your answer.

d Use the trend line to predict the likely profit for March 2007.

11 Copy and complete the table. Identify the most appropriate moving average for each item. One of them has been done for you.

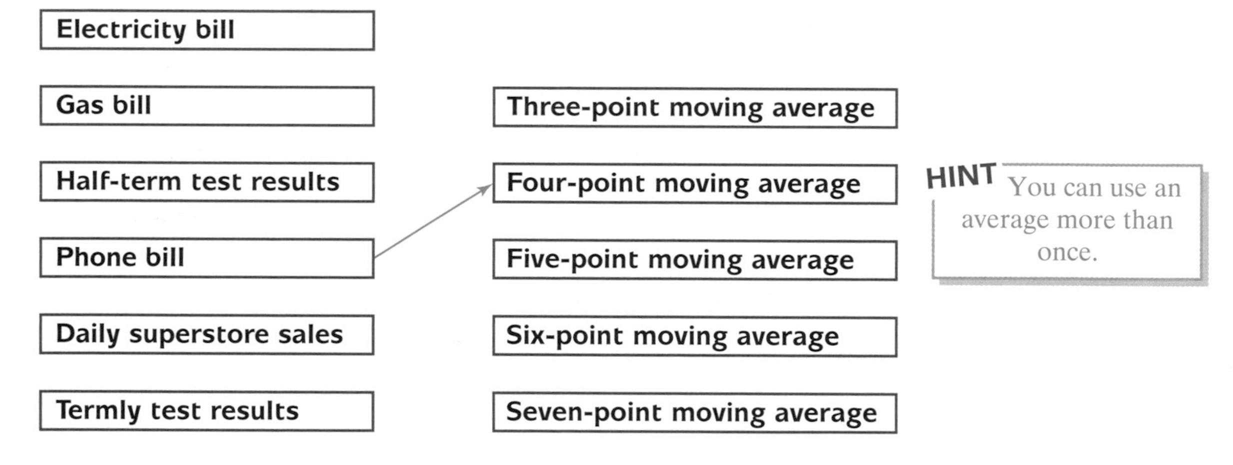

HINT You can use an average more than once.

Explore

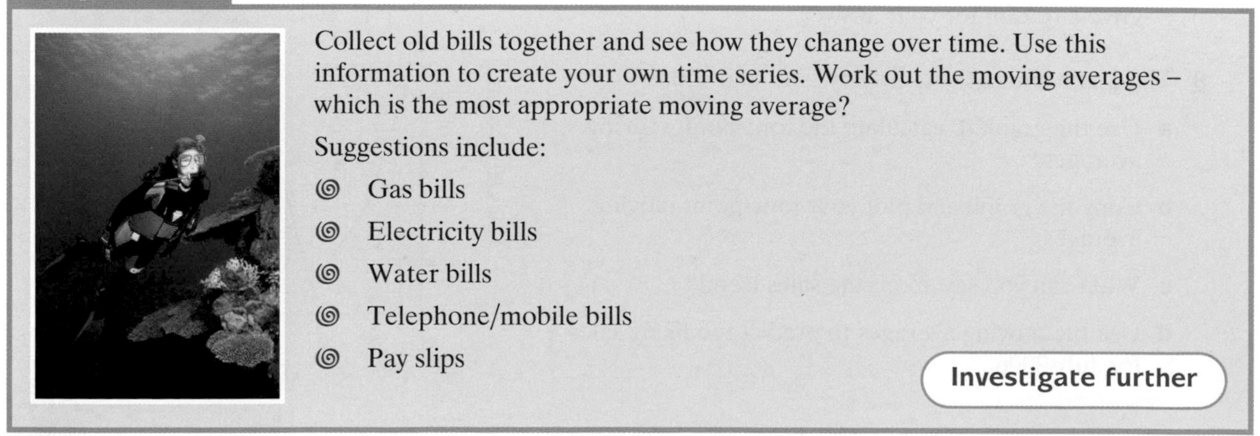

Collect old bills together and see how they change over time. Use this information to create your own time series. Work out the moving averages – which is the most appropriate moving average?

Suggestions include:

- Gas bills
- Electricity bills
- Water bills
- Telephone/mobile bills
- Pay slips

Investigate further

Explore

Collect information over a period of two weeks that you can use to create your own time series. Work out the moving averages – which is the most appropriate moving average?

Suggestions for data include:

- Time taken to complete homework
- Time taken to travel to work/school/college
- Time spent sleeping/eating/exercising
- Homework/examination or test results over time

Investigate further

Learn 3 Cumulative frequency diagrams

Example: The table shows the times taken to complete 20 telephone calls.

Time (min)	Frequency
$0 \leqslant t < 3$	5
$3 \leqslant t < 6$	8
$6 \leqslant t < 9$	4
$9 \leqslant t < 15$	3

The group $3 \leqslant t < 6$ means greater than or equal to 3 and less than 6

The value of 6 minutes will not be included in this interval

Show this information on a cumulative frequency diagram.
Use your diagram to estimate the median and interquartile range.

To find cumulative frequencies, you add the frequencies in turn to give you a 'running total'.

Add a third column for the cumulative frequency.

Time (min)	Frequency	Cumulative frequency
$0 \leqslant t < 3$	5	5
$3 \leqslant t < 6$	8	13
$6 \leqslant t < 9$	4	17
$9 \leqslant t < 15$	3	20

5 calls took less than 3 minutes

$5 + 8 = 13$ calls took less than 6 minutes

Check that the final cumulative frequency is the same as the total number of calls

Now you can draw the cumulative frequency diagram.

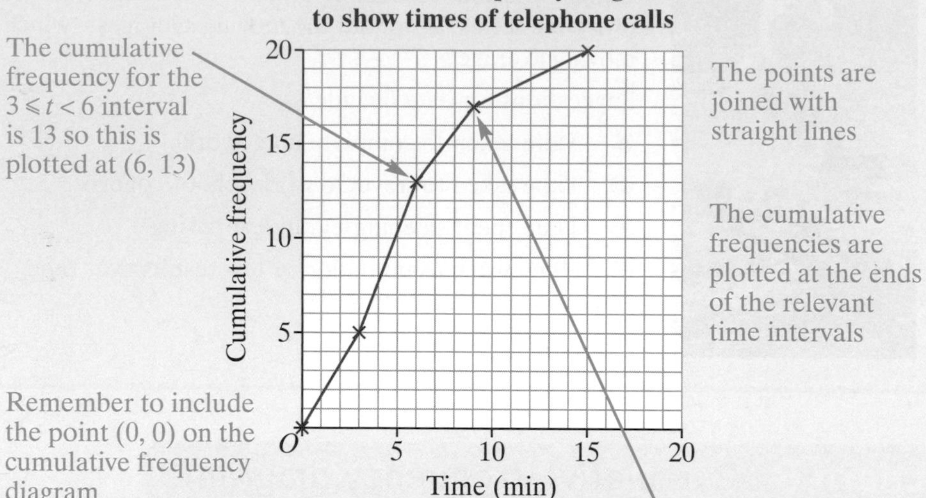

The cumulative frequency can be divided by four to find the quartiles and the median as follows.

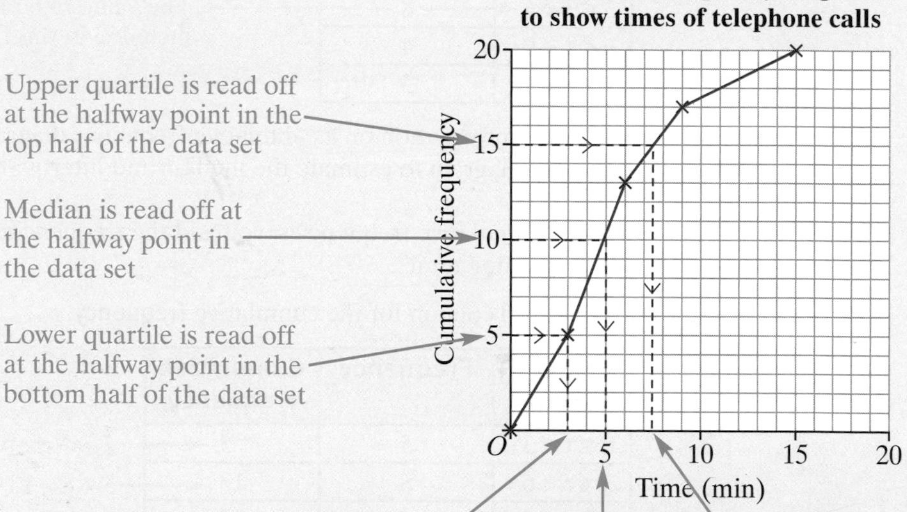

Median = 5

Interquartile range = UQ − LQ = 7.5 − 3 = 4.5

The interquartile range is not affected by very large or very small values, which have a disproportionate effect on the range

Apply 3

1 The table shows the distances travelled to work by 40 commuters.

Distance (miles)	Frequency
$0 < d \leqslant 2$	11
$2 < d \leqslant 4$	16
$4 < d \leqslant 8$	10
$8 < d \leqslant 16$	3

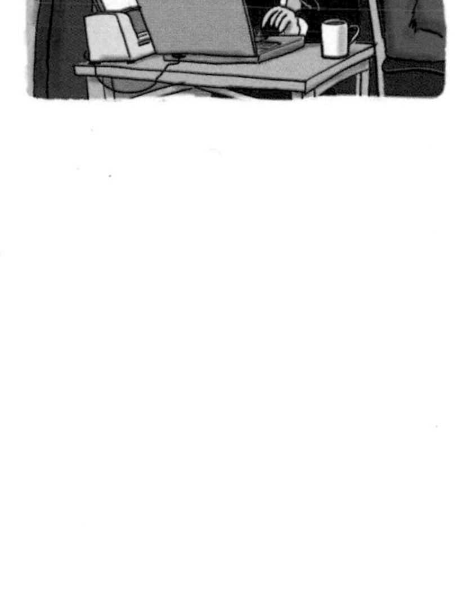

a Draw a cumulative frequency diagram to show this data.

b Use your diagram to estimate:

i the median distance

ii the interquartile range.

2 The table shows the heights of 50 students in a class.

Height (cm)	Frequency
$150 < h \leqslant 155$	4
$155 < h \leqslant 160$	7
$160 < h \leqslant 165$	18
$165 < h \leqslant 170$	11
$170 < h \leqslant 175$	6
$175 < h \leqslant 180$	4

a Show this information on a cumulative frequency diagram.

b Use your diagram to estimate:

i the median height

ii the limits between which the middle 50% of the heights lie.

3 The table shows the wages of workers in a factory.

Wages (£)	Frequency
$0 < x \leqslant 100$	120
$100 < x \leqslant 150$	165
$150 < x \leqslant 200$	182
$200 < x \leqslant 250$	197
$250 < x \leqslant 300$	40
$300 < x \leqslant 500$	6

a Draw a cumulative frequency diagram to show this data.

b Use your diagram to estimate:

i the median wage

ii the interquartile range

iii the number of workers whose wage is below £175

iv the number of workers whose wage is above £420.

HINT First find the number who earn £420 or less and then subtract from the total number of workers.

4 Peter records the waiting times (to the nearest minute) at a post office and puts the times in a table.

Time (min)	Frequency
1–3	23
4–6	17
7–9	8
10–15	2

a Show this information on a cumulative frequency diagram.

b Use your diagram to estimate:

i the median waiting time

ii the interquartile range

iii the percentage of people who waited over twelve minutes.

5 The cumulative frequency diagram shows the times taken by 60 students to complete an arithmetic test.

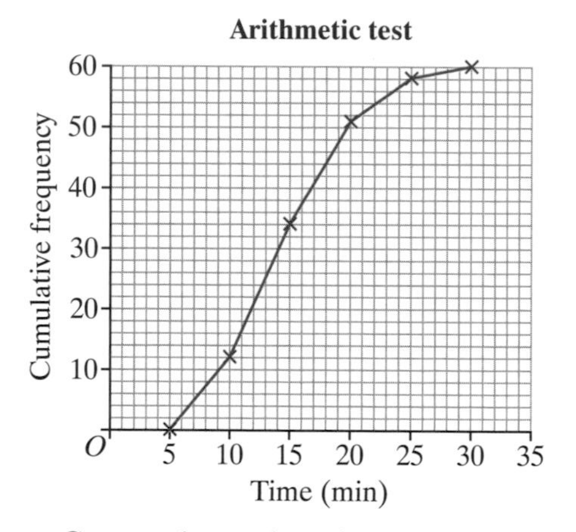

a Copy and complete the table to show the cumulative frequencies.

Time (min)	Cumulative frequency
$5 < t \leqslant 10$	
$10 < t \leqslant 15$	
$15 < t \leqslant 20$	
$20 < t \leqslant 25$	
$25 < t \leqslant 30$	

b Use your table to complete the frequency distribution.

Time (min)	Frequency
$5 < t \leqslant 10$	
$10 < t \leqslant 15$	
$15 < t \leqslant 20$	
$20 < t \leqslant 25$	
$25 < t \leqslant 30$	

6 The table shows the arm spans of 25 students.

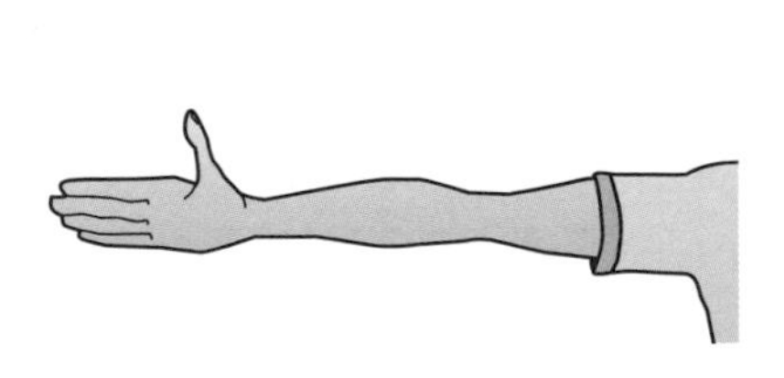

Arm span (cm)	Frequency
140–150	0
150–160	3
160–170	11
170–180	9
180–190	2

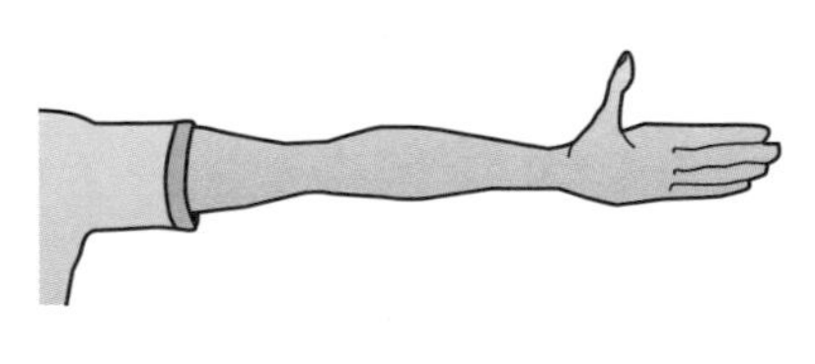

a Show this information on a cumulative frequency diagram.

b Use your diagram to estimate:

i the median

ii the interquartile range.

Another class of 25 students has this distribution.

Median	178
Interquartile range	26

c What can you say about the two classes?
Give reasons for your answer.

7 The table shows the number of words per paragraph in a newspaper.

Words	Number of paragraphs
0	0
1–10	15
11–20	31
21–30	45
31–40	25
41–50	4

a Show this information on a cumulative frequency diagram.

b Use your diagram to estimate:

i the median number of words

ii the interquartile range.

This table shows the number of words per paragraph in a magazine.

Words	Number of paragraphs
0	0
1–10	20
11–20	46
21–30	36
31–40	13
41–50	5

c Draw a cumulative frequency diagram on the same graph as part **a**.
What can you say about the two distributions?
Give reasons for your answer.

8 Match the frequency diagram with the cumulative frequency diagram.

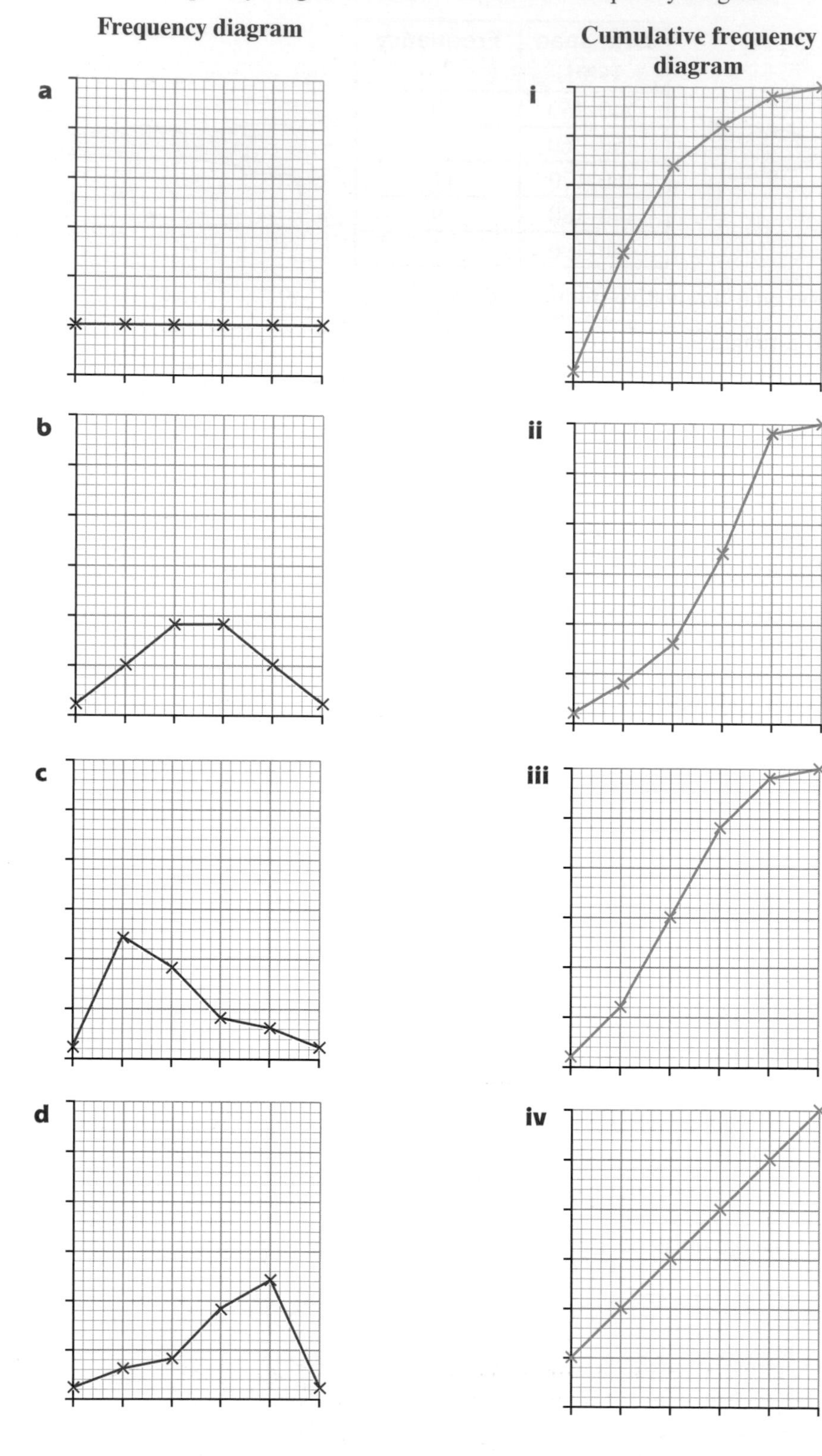

Explore

Collect data from your work/school/family/internet about these variables:

- Height
- Arm length
- Head circumference
- Running time over a fixed distance
- Pulse rate
- Heart rate
- Time taken to complete homework
- Time taken to travel to work/school/college
- Time spent sleeping/eating/exercising

Use your data to draw cumulative frequency diagrams
Where do you lie in each distribution?

Investigate further

Learn 4 Box plots

Examples:

a Peter keeps a note of his sales over the past eleven days.

Sun	Mon	Tue	Wed	Thu	Fri	Sat	Sun	Mon	Tue	Wed
28	2	8	11	16	15	30	25	4	10	12

Draw a box plot for the data.

To draw a box plot you need:

- the minimum and maximum values
- the lower and upper quartiles
- the median.

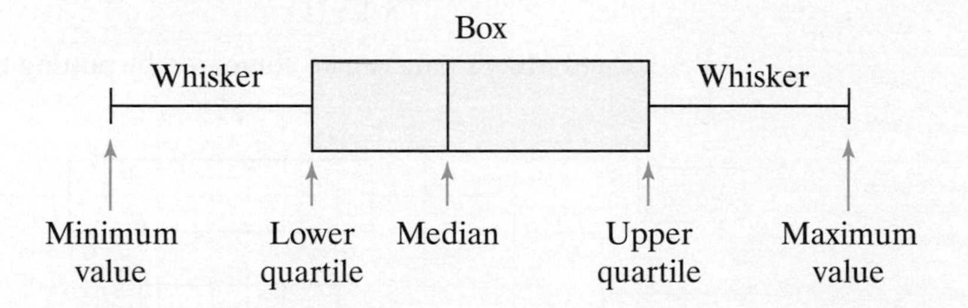

Rearranging the data in order:

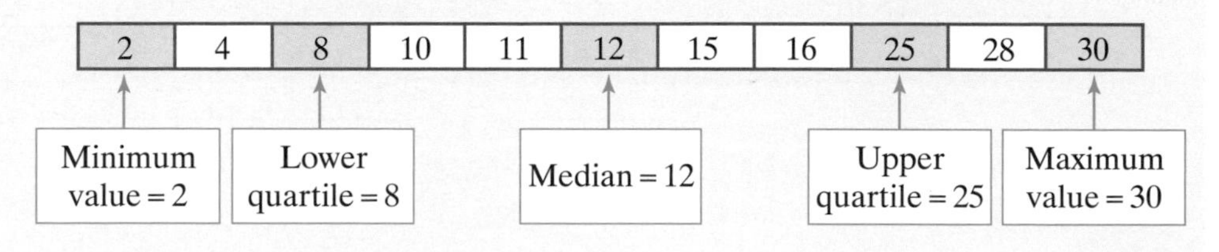

The box plot looks like this:

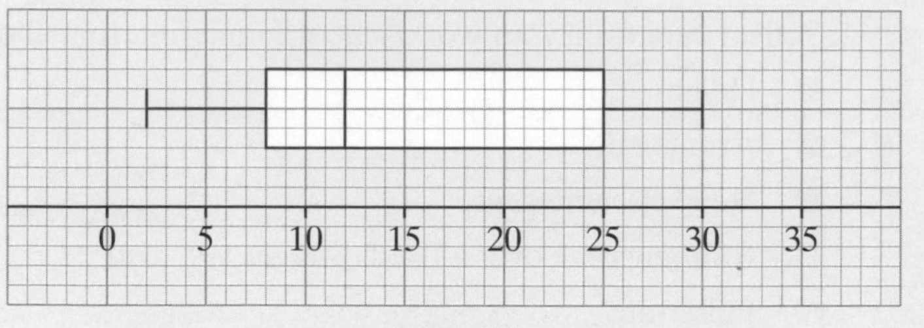

Sales (£)

b Mary also keeps a note of her sales over the past eleven days.

Sun	Mon	Tue	Wed	Thu	Fri	Sat	Sun	Mon	Tue	Wed
26	3	10	18	14	21	19	31	6	11	17

Draw a box plot for the data.

Rearranging the data in order:

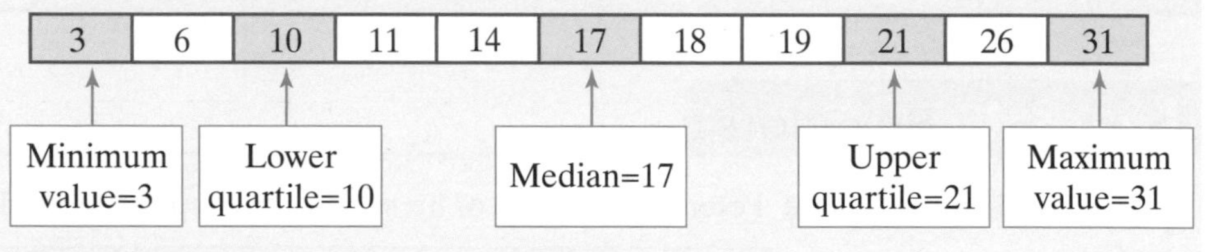

3	6	10	11	14	17	18	19	21	26	31

Mary's box plot:

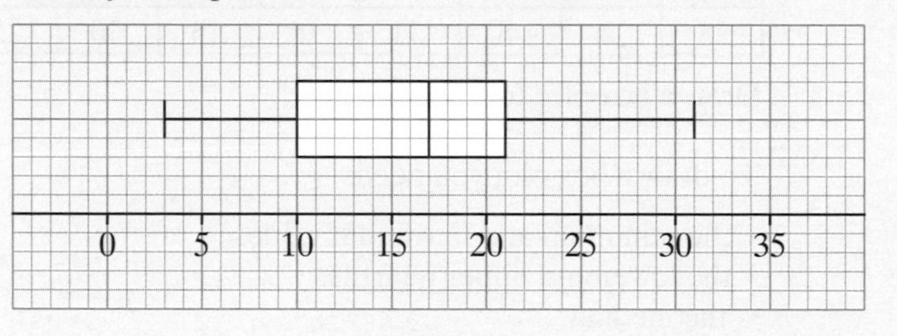

Sales (£)

Peter's and Mary's data can be compared by putting the two box plots on the same axis.

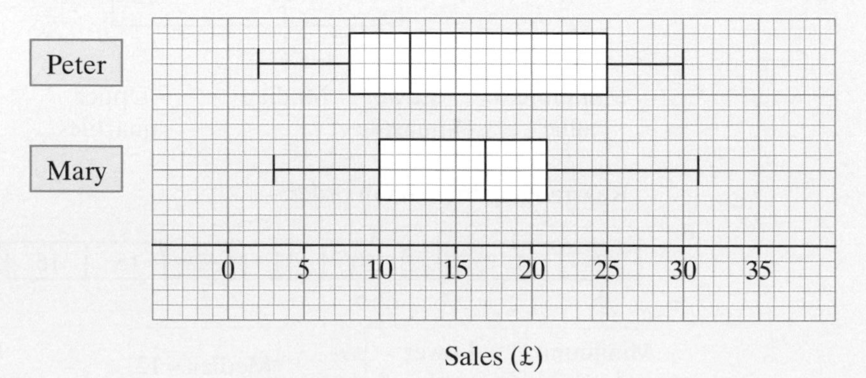

Sales (£)

Information from graphs should always be interpreted in terms of the given data

Measure	Description	Interpretation
Median	Mary's median is greater	Mary's average (median) sales are greater than Peter's average (median) sales
IQR	Peter's interquartile range is greater	Taking the middle 50% of the sales, Peter's sales are more spread out than Mary's sales
Range	The ranges are the same	Over the whole range, the spread of sales is the same (£28) for both Mary and Peter

Box plots can also be used for highlighting and comparing information from cumulative frequency diagrams.

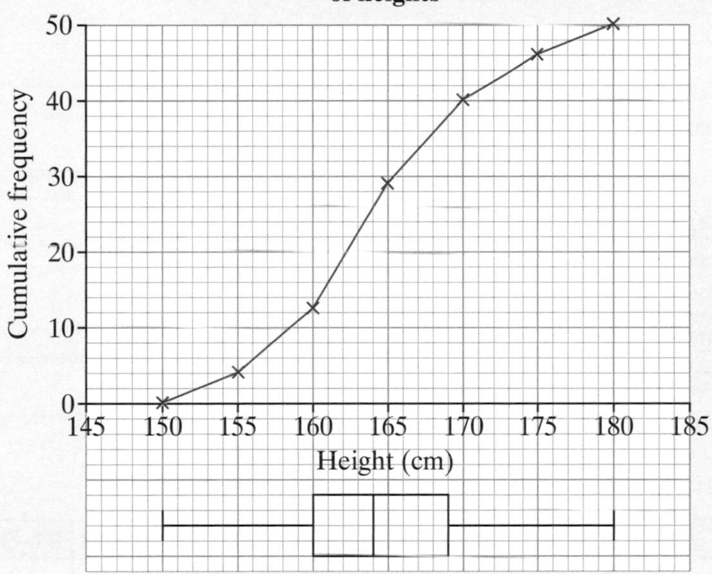

From the cumulative frequency diagram:

- the minimum value = 150
- the maximum value = 180
- the lower quartile = 160
- the upper quartile = 169
- the median = 164

Apply 4

1 Draw box plots of this data.

a

23	7	7	11	28	17	9	21	8	18	25	29	5	13	25

b

13	5	13	17	20	15	3	7	11	8	9

c

12	5	14	15	20	24	4	6	11	6	9	15

2 These box plots show the ages of all of the people in two villages.

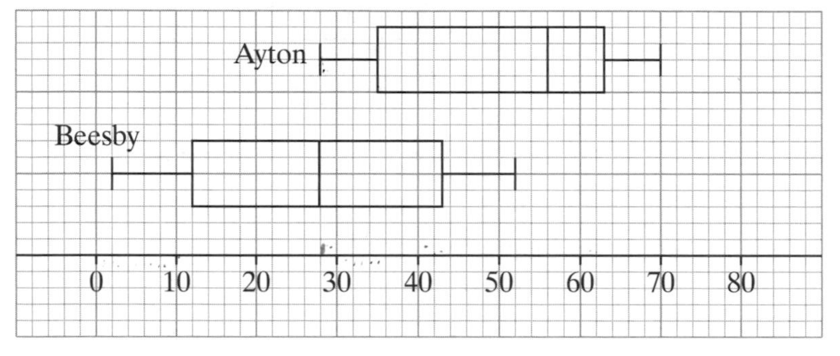

Age (years)

a Copy and complete the table for the two villages.

	Ayton	Beesby
Minimum age		
Maximum age		
Lower quartile		
Upper quartile		
Median		

b Compare and contrast the ages of people in the two villages.

3 The examination results for 60 students are shown in the table.

1–10	11–20	21–30	31–40	41–50	51–60	61–70	71–80	81–90	91–100
1	1	2	3	5	9	10	14	12	3

a Show this information on a cumulative frequency diagram.

b Use the diagram to estimate:

i the median

ii the lower quartile

iii the upper quartile.

c Draw a box plot for the data.

4 Match each cumulative frequency diagram with a box plot.

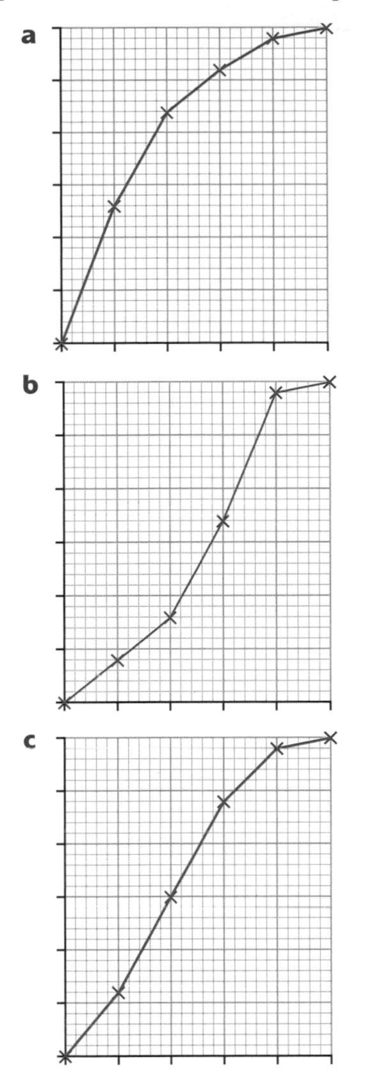

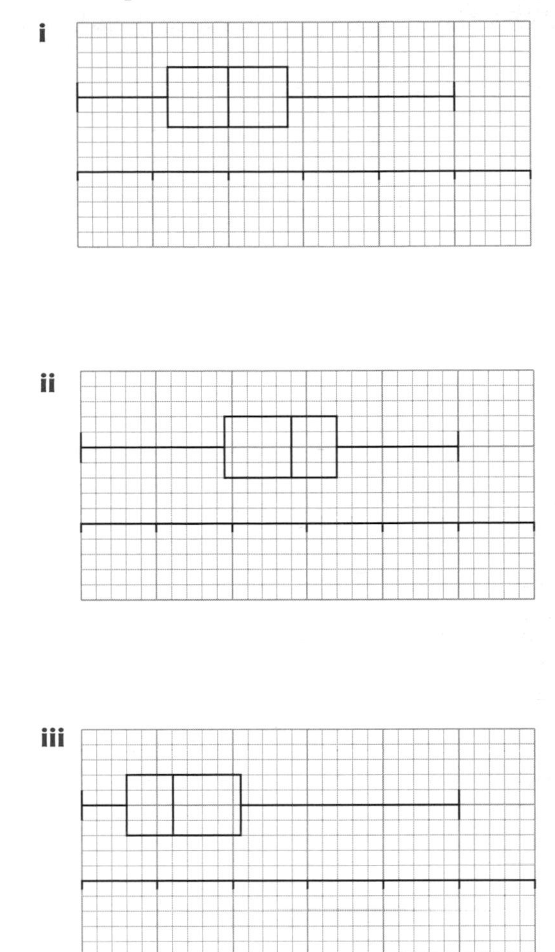

5 The cumulative frequency diagram shows the waiting times (to the nearest minute) at a main post office.

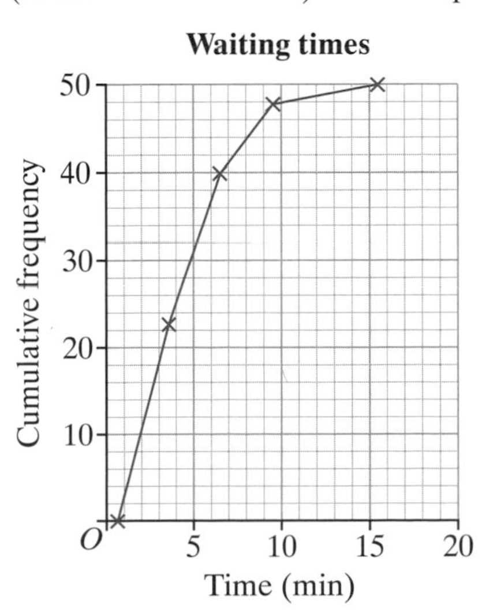

a Use the diagram to draw a box plot for the data.

b This box plot shows the waiting times at a village post office. Write down two differences between the waiting times at the two post offices.

0 5 10 15

Waiting time (min)

Explore

Collect data from your work/school/family/internet about these variables:

- Height
- Arm length
- Head circumference
- Running time over a fixed distance
- Pulse rate
- Heart rate
- Time taken to complete homework
- Time taken to travel to work/school/college
- Time spent sleeping/eating/exercising

Use your data to draw box plots. Where do you lie in each distribution?

Investigate further

Learn 5 Histograms

Examples:

The table shows the waiting times in an accident and emergency department.

Time (min)	Frequency
$5 \leqslant t < 10$	5
$10 \leqslant t < 15$	15
$15 \leqslant t < 20$	8
$20 \leqslant t < 30$	2

In a histogram, the area of the bar represents the frequency

Frequency = class width × height of bar

or

$$\text{Height} = \frac{\text{frequency}}{\text{class width}}$$

Show the information on:

a a histogram
b a frequency polygon.

a Add additional columns for class width and height (frequency density).

Time (min)	Frequency	Class width	Height $= \dfrac{\text{frequency}}{\text{class width}}$
$0 \leqslant t < 5$	0	5	Height $= 0 \div 5 = 0$
$5 \leqslant t < 10$	5	5	Height $= 5 \div 5 = 1$
$10 \leqslant t < 15$	15	5	Height $= 15 \div 5 = 3$
$15 \leqslant t < 20$	8	5	Height $= 8 \div 5 = 1.6$
$20 \leqslant t < 30$	2	10	Height $= 2 \div 10 = 0.2$

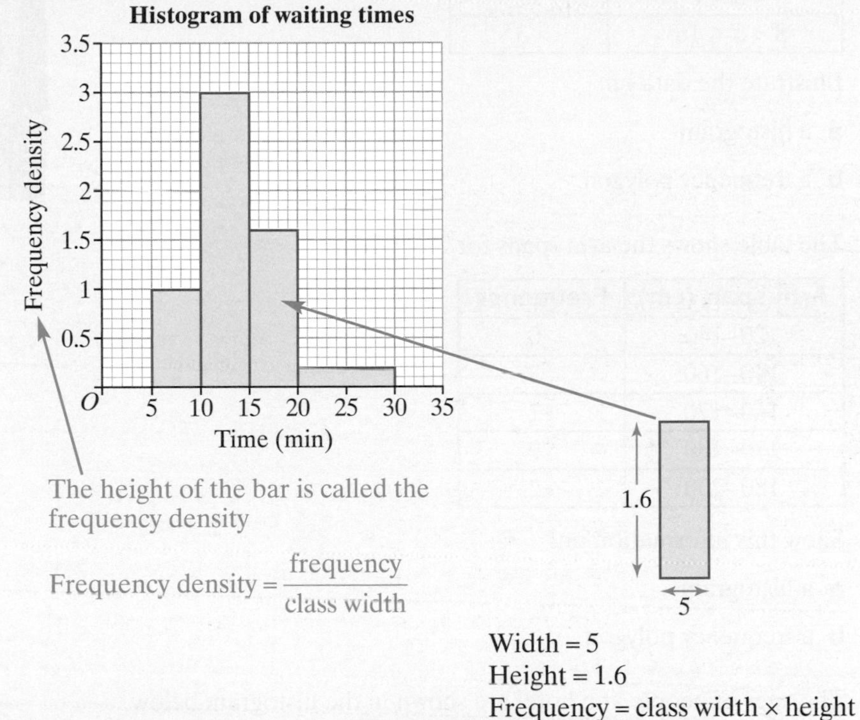

The height of the bar is called the frequency density

$$\text{Frequency density} = \frac{\text{frequency}}{\text{class width}}$$

Width = 5
Height = 1.6
Frequency = class width × height
= 5 × 1.6
= 8

The bar represents a frequency of 8.

b A frequency polygon can be drawn from a histogram (or bar chart) by joining the midpoints of the tops of the bars with straight lines to form a polygon.

The area of the polygon is equal to the total frequency.

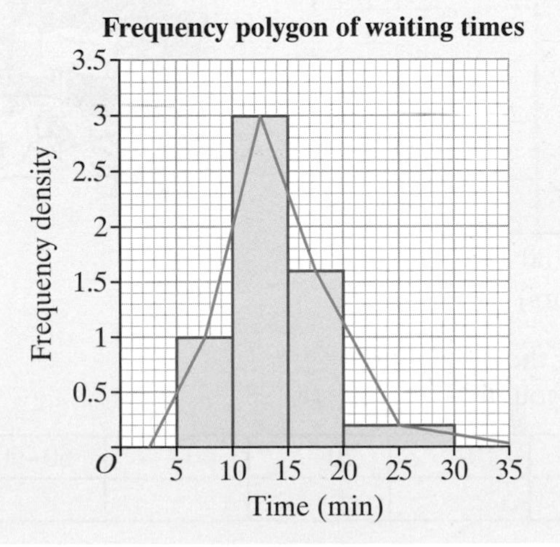

It is usual to extend the lines to the horizontal axis as shown

Apply 5

1 The table shows the distances travelled to work by 36 commuters.

Distance (miles)	Frequency
$0 < d \leqslant 1$	0
$1 < d \leqslant 2$	6
$2 < d \leqslant 4$	16
$4 < d \leqslant 8$	11
$8 < d \leqslant 16$	3

Illustrate the data on:

a a histogram

b a frequency polygon.

2 The table shows the arm spans for 25 students.

Arm span (cm)	Frequency
120–140	0
140–160	3
160–170	11
170–180	9
180–200	2

Show this information on:

a a histogram

b a frequency polygon.

3 The ages of people at a hotel are shown in the histogram below.

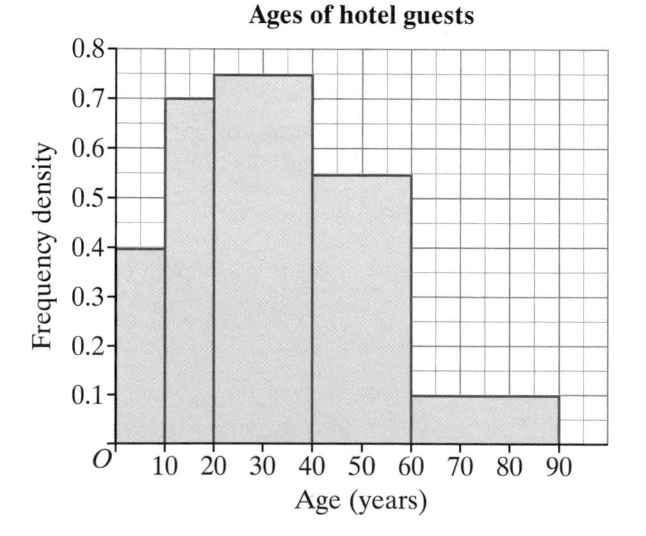

Copy and complete the table for the distribution.
The first one has been done for you.

Age (years)	0–	10–	20–	40–	60–90
Number of guests	4				

4 The weights of 60 students in a year group were recorded.

Weight (kg)	Frequency
$40 \leqslant w < 42$	1
$42 \leqslant w < 45$	9
$45 \leqslant w < 50$	22
$50 \leqslant w < 55$	18
$55 \leqslant w < 65$	8
$65 \leqslant w < 75$	2

Show this information on:

a a histogram

b a frequency polygon.

5 Light bulbs are tested to see how long they last.
The table shows the results of 60 tests.

Time (hours)	Frequency
$600 \leqslant x < 700$	3
$700 \leqslant x < 800$	15
$800 \leqslant x < 850$	20
$850 \leqslant x < 900$	12
$900 \leqslant x < 1000$	8
$1000 \leqslant x < 1200$	2

a Draw a histogram to show this information.

b Use your histogram to estimate the median life of a light bulb.

HINT Draw a vertical line dividing the area of the histogram into two equal parts.

6 The table shows the wages of workers in a factory.

Wages (£)	Frequency
$0 \leqslant x < 100$	120
$100 \leqslant x < 150$	165
$150 \leqslant x < 200$	182
$200 \leqslant x < 250$	197
$250 \leqslant x < 300$	40
$300 \leqslant x < 500$	6

Show this information on:

a a histogram

b a frequency polygon.

Explore

Collect data from your work/school/family about these variables:

- Height
- Arm length
- Head circumference
- Running time over a fixed distance
- Pulse rate
- Heart rate
- Time taken to complete homework
- Time taken to travel to work/school/college
- Time spent sleeping/eating/exercising

Use your data to draw histograms. Where do you lie in each distribution?

Investigate further

Representing data

ASSESS

The following exercise tests your understanding of this chapter, with the questions appearing in order of increasing difficulty.

1 a The list shows the average gestation times, to the nearest day, of some animals. Show this data using a suitable stem-and-leaf diagram.

Animal	Gestation time (days)
Common opossum	13
Marine turtle	55
Grass lizard	42
Emperor penguin	63
House mouse	19
Royal albatross	79
Australian skink	30
Falcon	29
Hawk	44
Swan	30

Animal	Gestation time (days)
Python	61
Thrush	14
Wren	16
Spiny lizard	63
Alligator	61
Dog	63
Finch	12
Ostrich	42
Pheasant	22

b This data shows the track times in minutes and seconds on some of Luciano's CDs. Show the data using a suitable stem-and-leaf diagram.

3.21 2.29 2.25 2.49 2.57 3.30 3.19 2.25 3.34 2.45 2.44 3.34

3.19 3.30 2.10 3.00 2.44 2.25 2.54 3.43 2.22 2.54 2.55 2.24

3.35 2.10 3.55 3.07 2.54 2.08 3.22 3.33 2.43 3.50 2.22 2.57

2 a This information gives the waiting times at J A Williams' dental practice.

Waiting time (minutes)	Frequency
$0 \leqslant t < 3$	9
$3 \leqslant t < 6$	17
$6 \leqslant t < 9$	12
$9 \leqslant t < 12$	6
$12 \leqslant t < 15$	2
$15 \leqslant t < 18$	1

i Draw a cumulative frequency diagram for this data.

Use your diagram to find:

ii how many patients waited less than four minutes

iii how many patients waited more than ten minutes.

iv Use your diagram to estimate the values of the median and the interquartile range.

b The pie charts below show how 200 students travelled to college one day during winter and one day during summer.

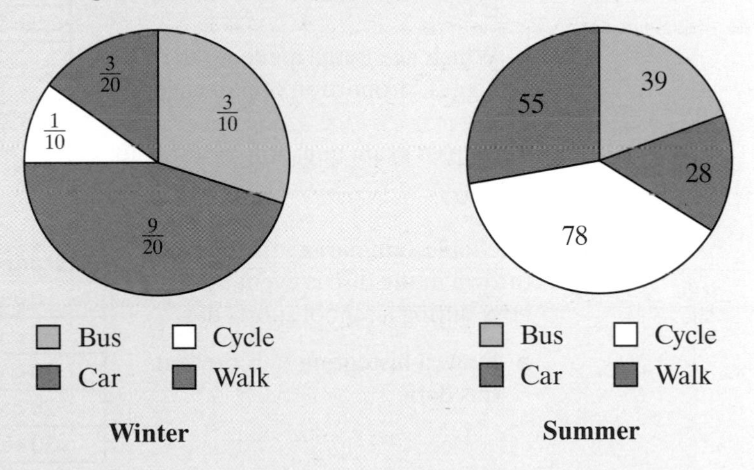

i Which method of transport was most popular on the winter day?

ii Calculate the number of students who travelled by bus on the winter day.

iii Which method of transport was most popular on the summer day?

iv Calculate the fraction of students who travelled by bus on the summer day.

3 a Draw a box plot to represent this data:

1 3 5 6 8 10 11 12 14 15 15

b Draw a box plot to represent the information given in question **2**.

4 BROLLIES Я US have issued their sales figures for the past two years. (Figures are to the nearest £1000.) Draw a time series of these figures and include:

a a six-point moving average graph

b a twelve-point moving average graph.

c What conclusions do you draw from the graphs for the periods:

i January–April in Year 2

ii June–August in Year 2?

	Year 1	Year 2
Jan	210	200
Feb	204	190
Mar	165	158
Apr	144	140
May	126	130
Jun	96	110
Jul	72	104
Aug	51	76
Sep	153	150
Oct	155	182
Nov	178	183
Dec	195	199

5 The distribution of ages of passengers on a train is given in the table.

a Draw a histogram to illustrate the data.

b How many people were on the train?

c Which age range made up the biggest proportion of passengers?

Age range (years)	Frequency
$0 \leqslant a < 5$	11
$5 \leqslant a < 10$	32
$10 \leqslant a < 15$	28
$15 \leqslant a < 20$	35
$20 \leqslant a < 30$	45
$30 \leqslant a < 50$	58
$50 \leqslant a < 65$	115
$65 \leqslant a < 90$	76

Try a real past exam question to test your knowledge:

6 The table summarises the distance thrown in the discus event by 20 boys during a school sports day.

a Draw a histogram to represent this data.

Distance, x (m)	Number of boys
$0 < x \leqslant 5$	1
$5 < x \leqslant 10$	0
$10 < x \leqslant 20$	9
$20 < x \leqslant 30$	5
$30 < x \leqslant 35$	4
$35 < x \leqslant 40$	1

The distances thrown in the discus event by 20 girls are represented by the histogram opposite.

b Write down two comparisons between the distances thrown by the boys and the girls.

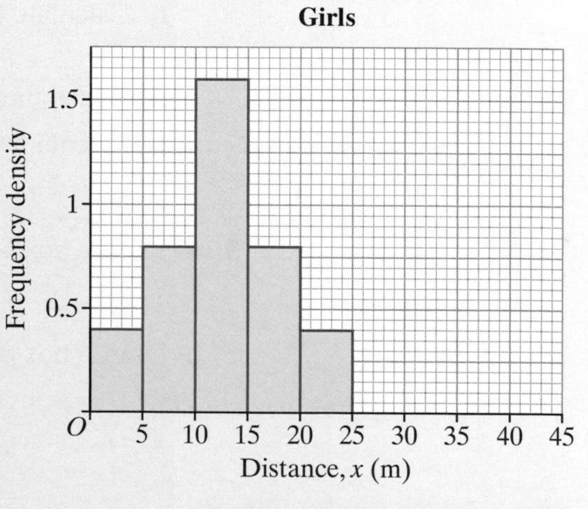

Spec B, Higher Paper, Nov 04

10 Scatter graphs

OBJECTIVES

D **Examiners would normally expect students who get a D grade to be able to:**

Draw a scatter graph by plotting points on a graph

Interpret the scatter graph

Draw a line of best fit on the scatter graph

C **Examiners would normally expect students who get a C grade also to be able to:**

Interpret the line of best fit

Identify the type and strength of the correlation

What you should already know ...

- Use coordinates to plot points on a graph
- Draw graphs including labelling axes and giving a title

VOCABULARY

Coordinates – a system used to identify a point; an x-coordinate and a y-coordinate give the horizontal and vertical positions

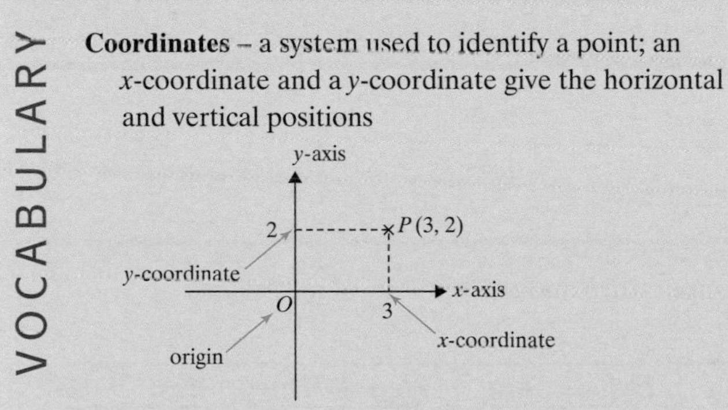

Correlation – a measure of the relationship between two sets of data; correlation is measured in terms of type and strength

Strength of correlation

The strength of correlation is an indication of how close the points lie to a straight line (perfect correlation)

Strong correlation **Weak correlation**

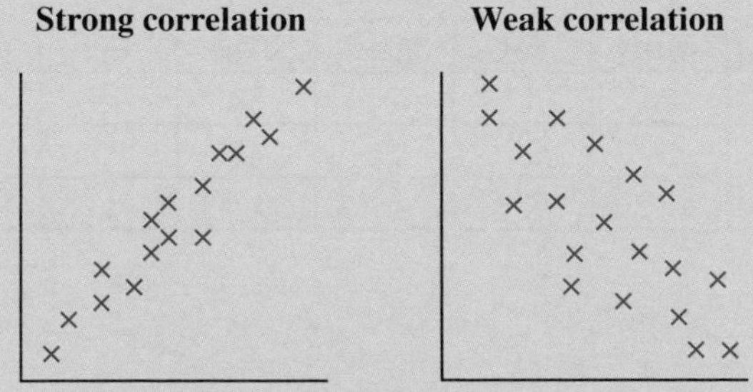

Correlation is usually described in terms of strong correlation, weak correlation or no correlation

Type of correlation

Positive correlation

In positive correlation an increase in one set of variables results in an increase in the other set of variables

Negative correlation

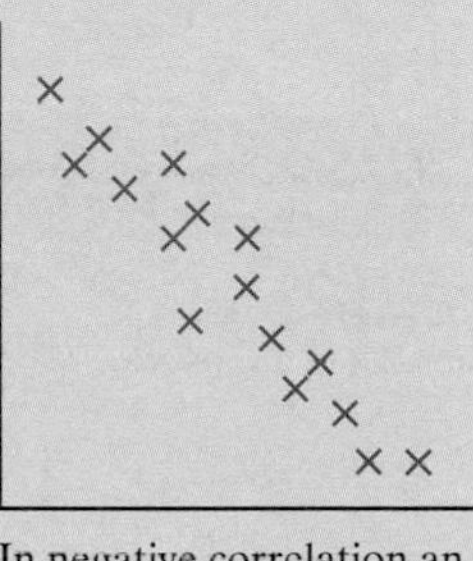

In negative correlation an increase in one set of variables results in a decrease in the other set of variables

Zero or no correlation

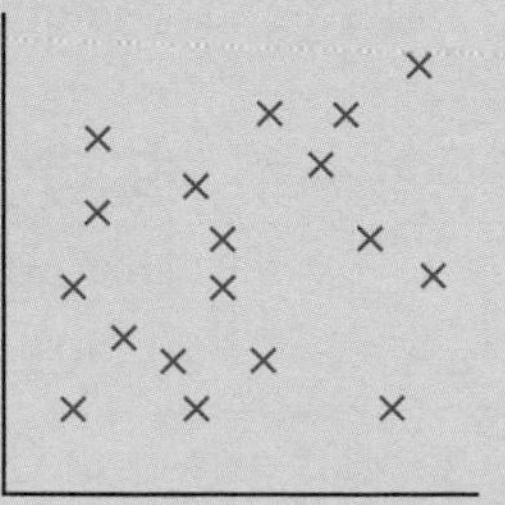

Zero or no correlation is where there is no obvious relationship between the two sets of data

Line of best fit – a line drawn to represent the relationship between two sets of data. Ideally it should only be drawn where the correlation is strong, for example,

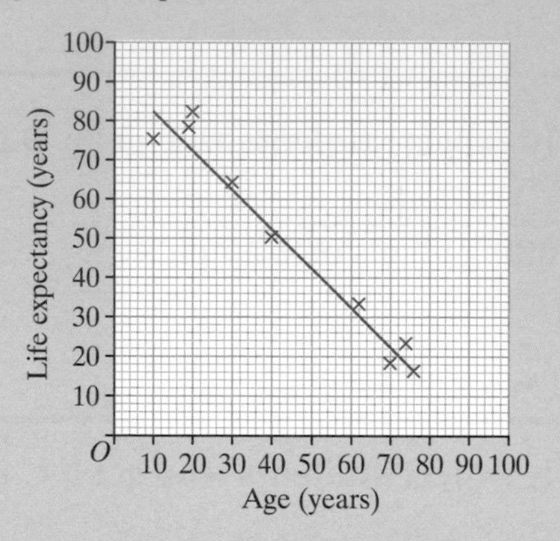

Outlier – a value that does not fit the general trend, for example,

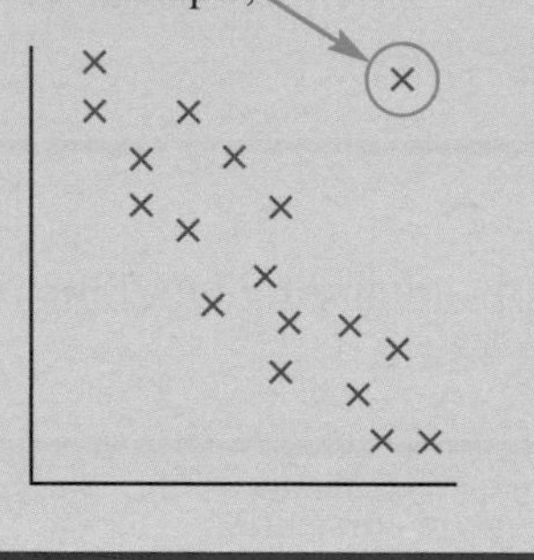

Scatter graph – a graph used to show the relationship between two sets of variables, for example, temperature and ice cream sales

Learn 1 Scatter graphs

Example:

A shopkeeper notes the temperature and the number of ice creams sold each day.

	Sun	Mon	Tue	Wed	Thu	Fri	Sat
Temperature (°C)	20	26	17	24	30	15	18
Ice cream sales	35	39	27	36	45	25	32

Plot this information on a scatter graph and describe the correlation between temperature and ice cream sales.

The information can be plotted as a series of coordinate pairs.

	Sun	Mon	Tue	Wed	Thu	Fri	Sat
Temperature (°C)	20	26	17	24	30	15	18
Ice cream sales	35	39	27	36	45	25	32
	(20, 35)	(26, 39)	(17, 27)	(24, 36)	(30, 45)	(15, 25)	(18, 32)

Before drawing a scatter graph, you need to choose carefully the scales on the axes.

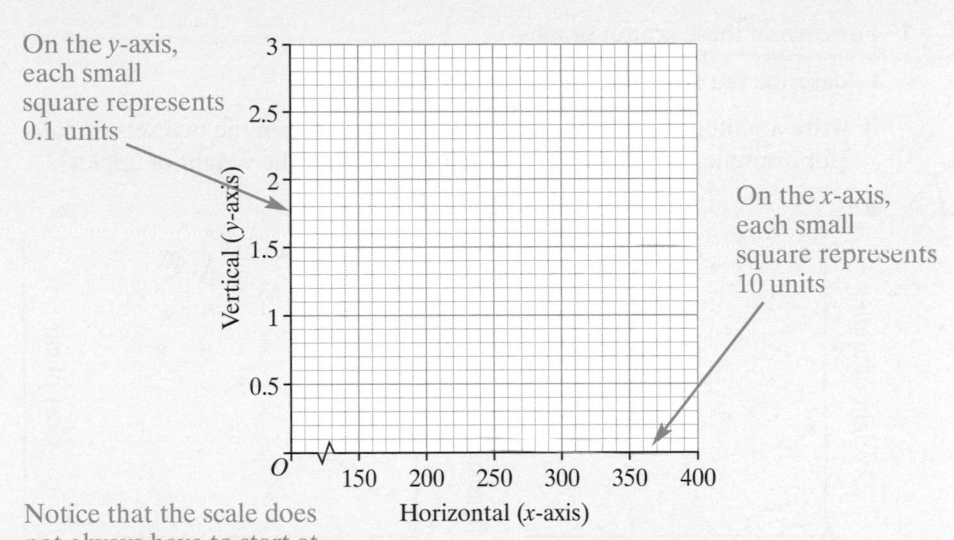

Notice that the scale does not always have to start at zero; a jagged line is often used to show the scale does not start at O

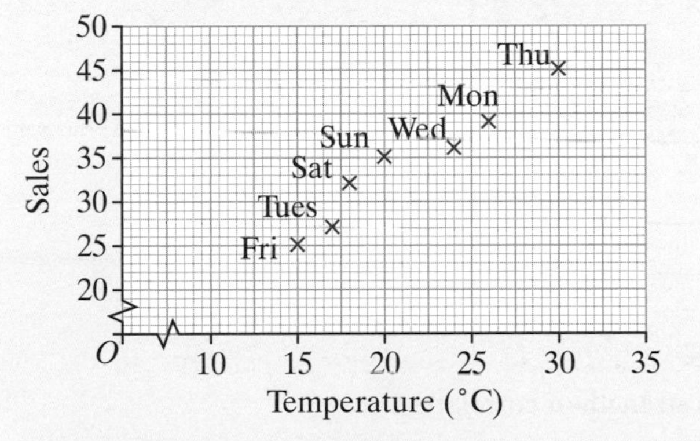

Plot the first variable on the horizontal axis

You can see from the graph that as the temperature rises, the sales of ice cream increase. There is a link between the temperature and the sales of ice cream.

Correlation measures the relationship between two sets of data

It is measured in terms of the **strength** and **type** of correlation

Therefore, there is **strong positive** correlation between the ice cream sales and the temperature.

Apply 1

1 For each of these scatter graphs:

i describe the type and strength of correlation

ii write a sentence explaining the relationship between the two sets of data (for example, the higher the rainfall, the heavier the weight of apples).

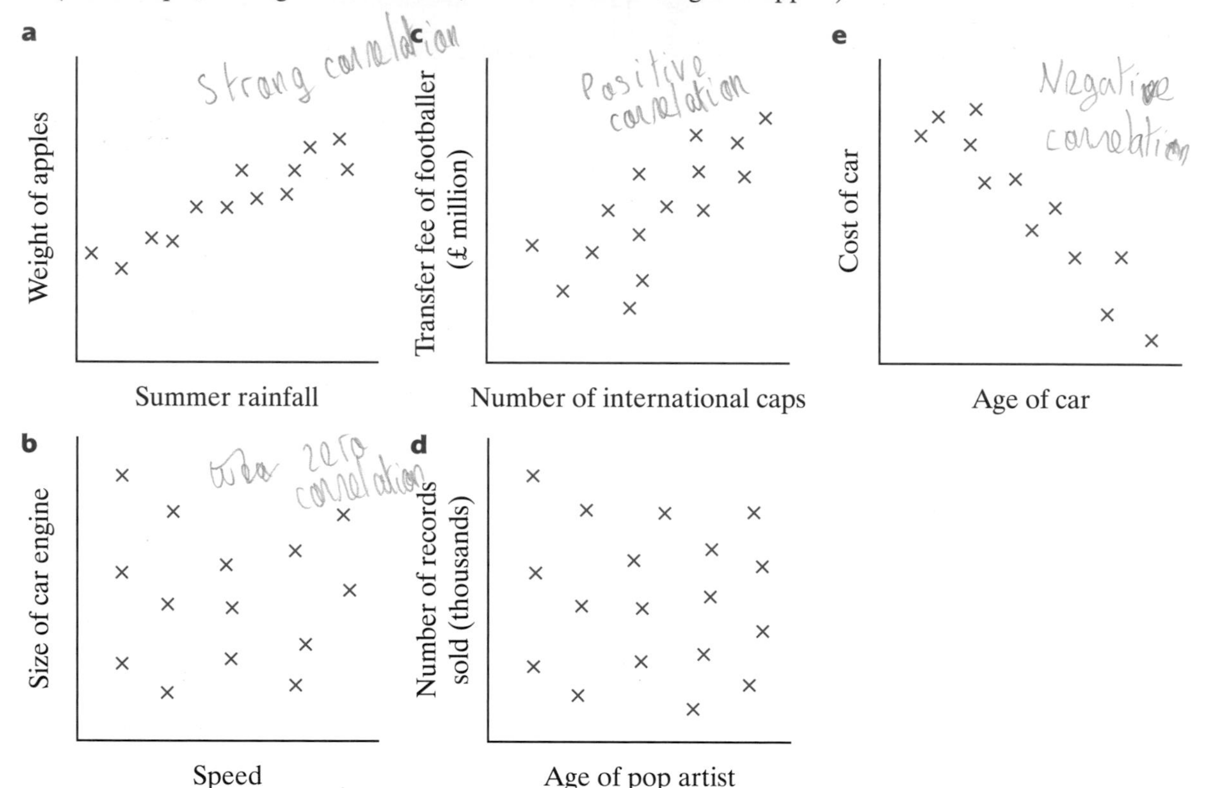

2 For each of the following:

i describe the type and strength of correlation

ii write a sentence explaining the relationship between the two sets of data.

a The hours of sunshine and the income from hiring deckchairs.

b The number of cars on one particular road and the average speed of cars.

c The distance travelled and the money spent on petrol.

d The height of a person and the number of children in his or her family.

3 The table shows the ages and arm spans of seven students in a school.

Age (years)	16	13	13	10	18	10	15
Arm span (inches)	62	57	59	57	64	55	61

a Draw a scatter graph of the results.

b Describe the type and strength of correlation.

c Write a sentence explaining the relationship between the two sets of data.

4 The table shows the ages and second-hand values of seven cars.

Age of car (years)	2	1	4	7	10	9	8
Value of car (£)	4200	4700	2800	1900	400	1100	2100

a Draw a scatter graph of the results.

b Describe the type and strength of correlation.

c Write a sentence explaining the relationship between the two sets of data.

5 The table shows the daily rainfall and the number of sunbeds sold at a resort on the south coast.

Amount of rainfall (mm)	0	1	2	5	6	9	11
Number of sunbeds sold	380	320	340	210	220	110	60

a Draw a scatter graph of the results.

b Describe the type and strength of correlation.

c Write a sentence explaining the relationship between the two sets of data.

6 For each graph, write down two variables that might fit the relationship.

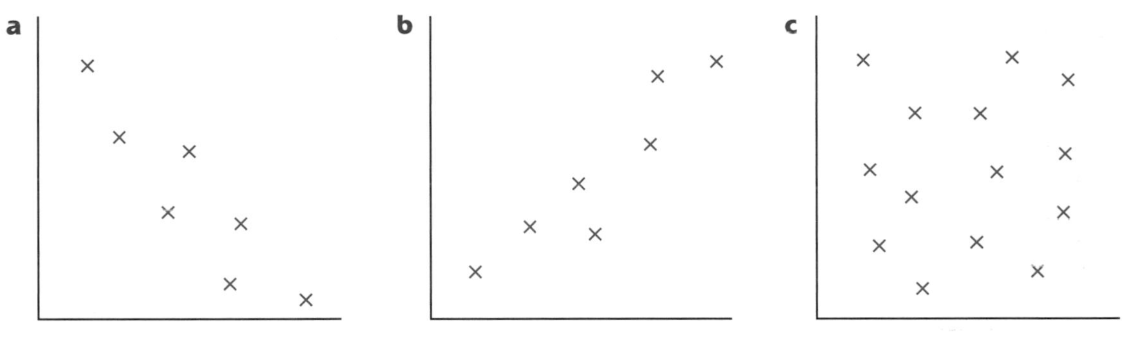

7 The scatter graph shows the ages and shoe sizes of a group of people.

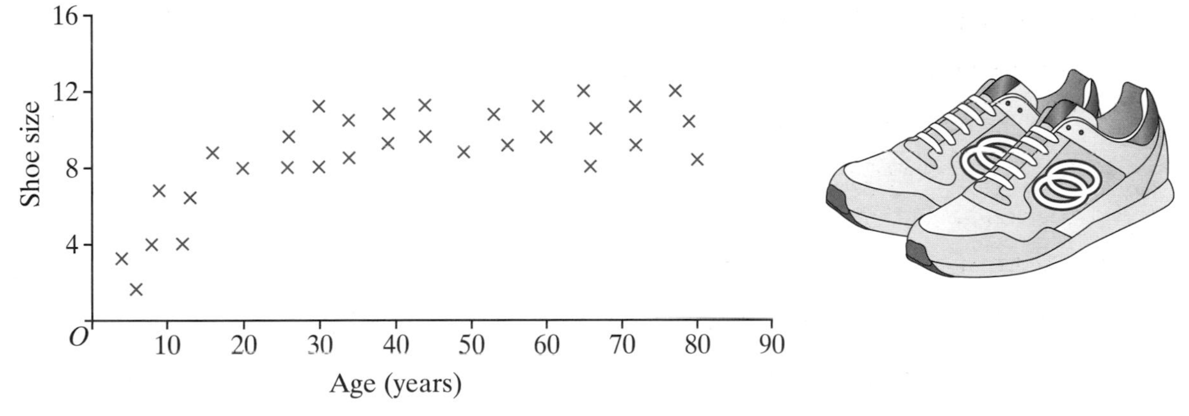

a Describe the correlation.

b Give a reason for your answer.

8 Get Real!

Steve is investigating the fat content and the calorie values of food at his local fast-food restaurant.
He collects the following information.

	Fat (g)	Calories
Hamburger	9	260
Cheeseburger	12	310
Chicken nuggets	24	420
Fish sandwich	18	400
Medium fries	16	350
Medium cola	0	210
Milkshake	26	1100
Breakfast	46	730

a Draw a scatter graph of the results.

b Describe the type and strength of correlation.

c Does the relationship hold for all the different foods?
Give a reason for your answer.

Explore

- Undertake some research of your own into the fat content and calorie values for food in a local restaurant
- You can find this information from the restaurant itself or on the internet

Investigate further

Learn 2 Lines of best fit

Example:

Use the graph to draw a line of best fit and estimate the likely sales of ice cream for a temperature of 28°C.

Temperature against ice cream sales

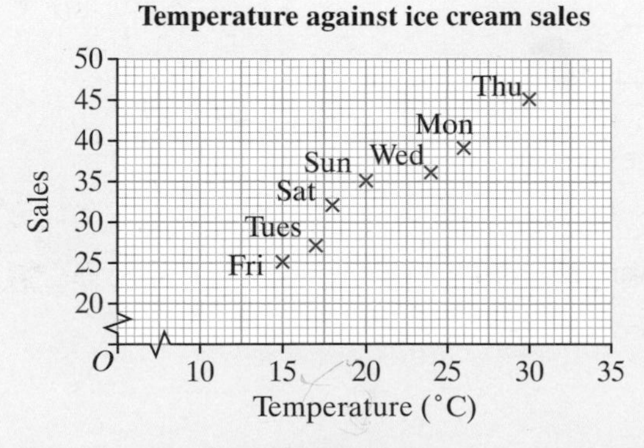

Drawing the line of best fit, you can use the graph to estimate the likely sales of ice creams for a temperature of 28°C.

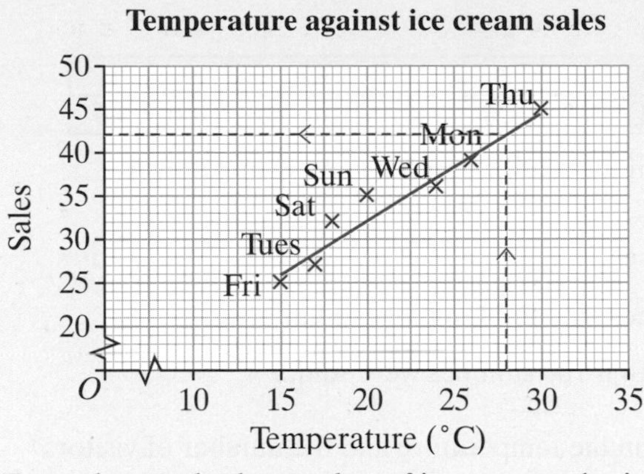

For some graphs, a straight line is not possible and a curve of best fit may be appropriate

From the graph, the number of ice cream sales is 42.

However, care should be taken when making use of such methods. For example, you will notice that the sales of ice creams were slightly higher on Saturday and Sunday.

You might also use your line of best fit to estimate the temperature given the number of ice cream sales – but it is not likely that you would need to do this.

Apply 2

1 The table shows the ages and arm spans of seven students in a school.

Age (years)	16	13	13	10	18	10	15
Arm span (inches)	62	57	59	57	64	55	61

a Draw a line of best fit on the scatter graph you drew in Apply **1**, question **3**.

b Use your line of best fit to estimate:

i the arm span of an 11-year-old student

ii the age of a student with an arm span of 61 inches.

2 The table shows the ages and second-hand values of seven cars.

Age of car (years)	2	1	4	7	10	9	8
Value of car (£)	4200	4700	2800	1900	400	1100	2100

a Draw a line of best fit on the scatter graph you drew in Apply **1**, question **4**.

b Use your line of best fit to estimate:

i the value of a car if it is 7.5 years old

ii the age of a car if its value is £3700.

3 The table shows the daily rainfall and the number of sunbeds sold at a resort on the south coast.

Amount of rainfall (mm)	0	1	2	5	6	9	11
Number of sunbeds sold	380	320	340	210	220	110	60

a Draw a line of best fit on the scatter graph you drew in Apply **1**, question **5**.

b Use your line of best fit to estimate:

i the number of sunbeds sold for 4 mm of rainfall

ii the amount of rainfall if 100 sunbeds were sold.

4 Sally collects information on the temperature and the number of visitors to a museum.

Temperature (°C)	15	25	16	18	19	22	24	23	17	20	26	20
Number of visitors	720	180	160	620	510	400	310	670	720	530	180	420

a Draw a scatter graph and a line of best fit.

b Use your line of best fit to estimate:

i the number of visitors if the temperature is 21°C

ii the temperature if 350 people visit the museum.

c Sally is sure that two of the data pairs are incorrect. Identify these two pairs on your graph.

5 The graph shows the line of best fit for the relationship between house prices in 2000 and house prices in 2006.

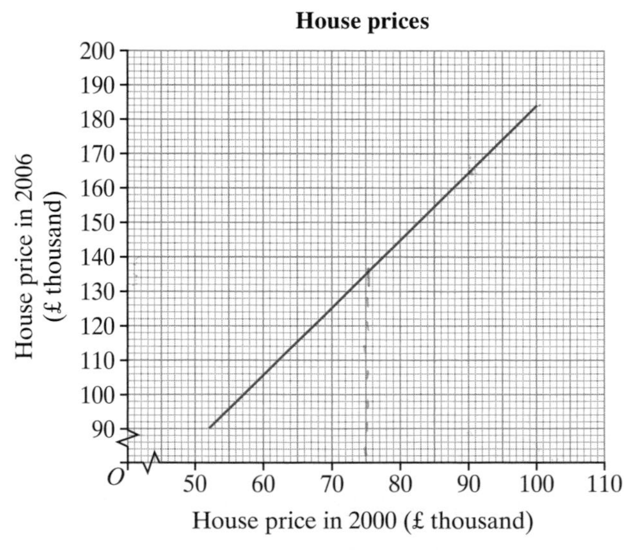

a Copy and complete the table, giving estimates for the missing values.

House price in 2000 (£ thousand)	70	75	90	100		
House price in 2006 (£ thousand)	125				105	155

b Andrea says her house price was £80 000 in 2000 and £170 000 in 2006. Is she correct? Give a reason for your answer.

c Find an estimate for the 2006 price of a house priced £55 000 in 2000.

d Find an estimate for the 2000 price of a house priced £180 000 in 2006.

6 Jenny collects information on the maximum speed and engine size of various motorbikes.
Her results are shown in the table.

Engine size (cc)	50	250	350	270	400	440	600	800	950	900	1200	1000
Speed (kph)	70	120	140	150	180	190	220	250	270	260	270	240

a Draw a scatter graph.

b What do you notice about the correlation between speed and engine size?

c Draw a line of best fit and use this to estimate:

i the engine size if your speed is 170 kph

ii the engine size if your speed is 250 kph.

d Which of these results is likely to be the most reliable?
Give a reason for your answer.

7 Readings of two variables, A and B, are shown in the table.

A	1	2	3	4	5	6	7	0.8	2.1	3.2	3.9	5.1	6.2	7.1
B	1.8	8.8	20	33	48	73	95	2	9	18	31	49	72	98

a Draw a scatter graph.

b What can you say about the correlation between the two sets of data?

c Draw a curve of best fit and use this to estimate:

i the value of B if $A = 2.5$

ii the value of A if $B = 64$.

d Express the relationship between A and B.

Explore

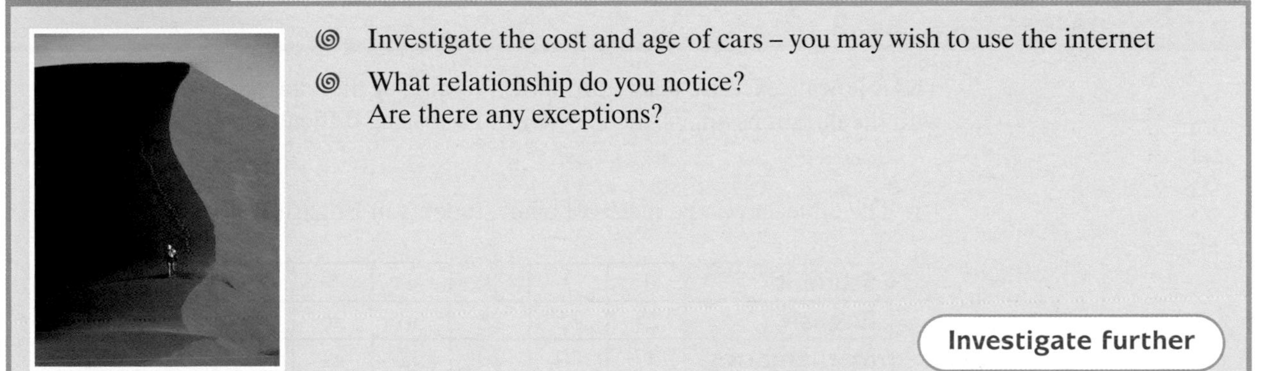

- Investigate the cost and age of cars – you may wish to use the internet
- What relationship do you notice?
  Are there any exceptions?

Investigate further

Explore

- Investigate house prices and the number of bedrooms – you may wish to use a local newspaper
- What relationship do you notice? Are there any exceptions?

Investigate further

Explore

- Investigate body lengths – for example, your arm span and your height are closely related
- Collect data from your friends and family to see what relationships you can find

Investigate further

Scatter graphs

ASSESS

The following exercise tests your understanding of this chapter, with the questions appearing in order of increasing difficulty.

1 The table shows the marks of eight students in English and mathematics.

Student	1	2	3	4	5	6	7	8
English	25	35	28	30	36	44	15	21
Mathematics	27	40	29	32	41	48	17	20

Draw a scatter graph and comment on the relationship between the marks in the two subjects.

2 Mr Metcalf, the maths teacher, told his class they had a test in a week's time. He also asked them to record how many hours of TV they watched during the week before the test.
When he had marked the test he showed the class a scatter graph of the data in the table.

Student	1	2	3	4	5	6	7	8	9	10
TV watched (hours)	4	7	9	10	13	14	15	20	21	25
Test mark	92	90	74	30	74	66	95	38	35	30

a Draw a scatter graph and comment on the relationship between the marks in the test and the amount of TV watched.

b Two students do not seem to 'fit the trend'.
Identify the students and give a possible reason for their results.

3 The tables show the relationship between the areas (in thousands of km^2) of some countries and their populations (in millions), all to 2 s.f.

	Area (km^2)	**Population (millions)**
Monaco	0.0020	0.030
Malta	0.32	0.40
Jersey	0.12	0.090
Netherlands	42	16
UK	250	60
Germany	360	83
Italy	300	58
Switzerland	41	7.3
Andorra	0.47	0.068
Denmark	43	5.4

	Area (km^2)	**Population (millions)**
France	550	60
Austria	84	8.2
Turkey	780	67
Greece	130	11
Spain	500	40
Ireland	70	3.9
Latvia	65	2.4
Sweden	450	8.9
Norway	320	4.5
Iceland	100	0.28

Draw a scatter graph of this data and comment on the graph.

4 **a** Draw a suitable scatter graph to illustrate this data, which shows the relationship between the distances jumped in long jump trials and the leg lengths of the jumpers.

Leg length (cm)	71	73	74	75	76	79	82
Distance jumped (m)	3.2	3.1	3.3	4.1	3.9	4	4.8

b Draw a line of best fit on the graph.

c Use your line of best fit to estimate:

i the leg length of an athlete who jumped a distance of 3.5 m

ii the distance jumped by an athlete with a leg length of 83 cm.

d Explain why one of these estimates is more reliable than the other.

Try a real past exam question to test your knowledge:

5 The scatter graph shows the height and trunk diameter of each of eight trees.

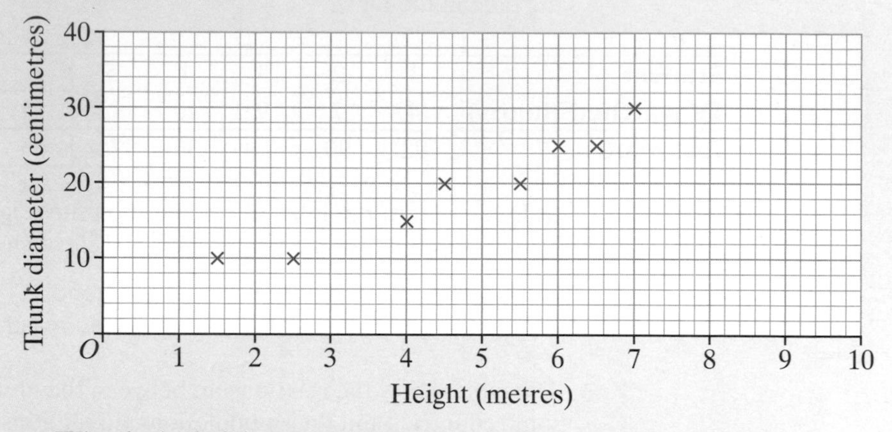

a What is the height of the tallest tree?

b Draw a line of best fit through the points on the scatter graph.

c Describe the relationship shown in the scatter graph.

d i Estimate the height of a tree with trunk diameter 35 centimetres.

ii Comment on the reliability of your estimate.

Spec B, Int Paper 1, Mar 02

11 Properties of polygons

OBJECTIVES

C

Examiners would normally expect students who get a C grade to be able to:

Classify a quadrilateral by geometric properties

Calculate exterior and interior angles of a regular polygon

What you should already know ...

- Recall and use properties of angles at a point, angles on a straight line, perpendicular lines, and opposite angles at a vertex
- Distinguish between acute, obtuse, reflex and right angles
- Use parallel lines, alternate angles and corresponding angles
- Prove that the angle sum of a triangle is 180°
- Prove that the exterior angle of a triangle is equal to the sum of the interior opposite angles
- Use angle properties of equilateral, isosceles and right-angled triangles
- Understand simple congruence
- Understand and use angle properties of quadrilaterals

VOCABULARY

Equilateral triangle – a triangle with 3 equal sides and 3 equal angles – each angle is 60°

Isosceles triangle – a triangle with 2 equal sides and 2 equal angles; the equal angles are called **base angles**

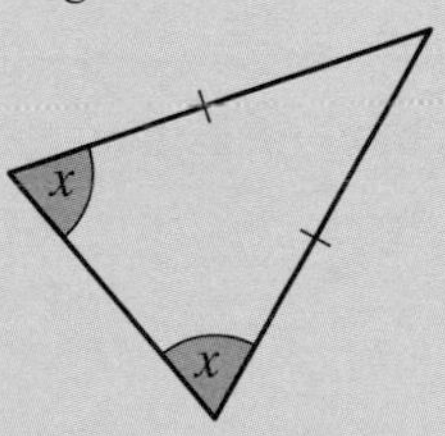

Square – a quadrilateral with four equal sides and four right angles

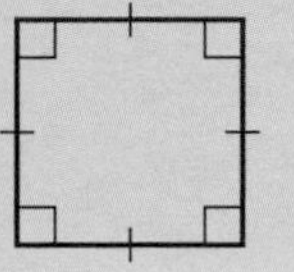

Rectangle – a quadrilateral with four right angles, and opposite sides equal in length

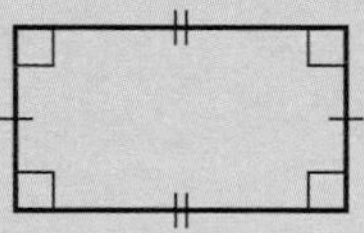

Kite – a quadrilateral with two pairs of equal adjacent sides

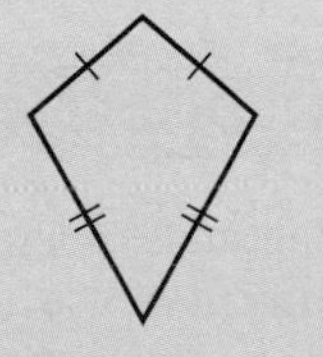

Trapezium (pl. **trapezia**) – a quadrilateral with one pair of parallel sides

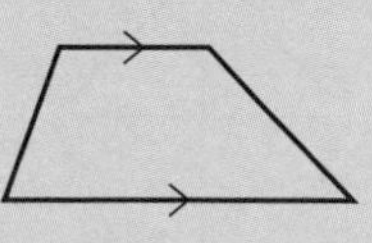

Isosceles trapezium – a quadrilateral with one pair of parallel sides. Non-parallel sides are equal

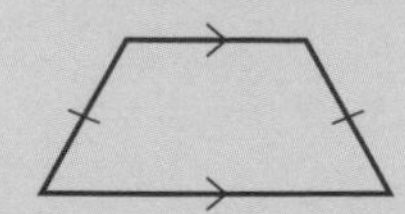

Parallelogram – a quadrilateral with opposite sides equal and parallel

Rhombus – a quadrilateral with four equal sides and opposite sides parallel

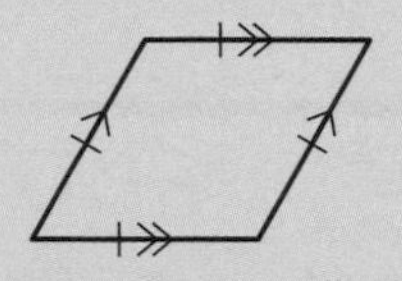

Polygon – a closed two-dimensional shape made from straight lines

Pentagon – a polygon with five sides

Hexagon – a polygon with six sides

Heptagon – a polygon with seven sides

Octagon – a polygon with eight sides

Nonagon – a polygon with nine sides

Decagon – a polygon with ten sides

Regular polygon – a polygon with all sides and all angles equal

Regular pentagon

Irregular polygon – a polygon whose sides and angles are not all equal (they do not all have to be different)

Interior angle – an angle inside a polygon

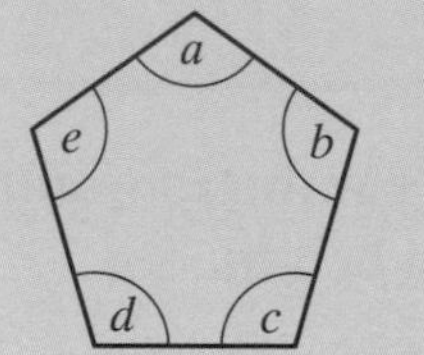

a, b, c, d and *e* are interior angles

Exterior angle – the angle between one side of a polygon and the extension of the adjacent side

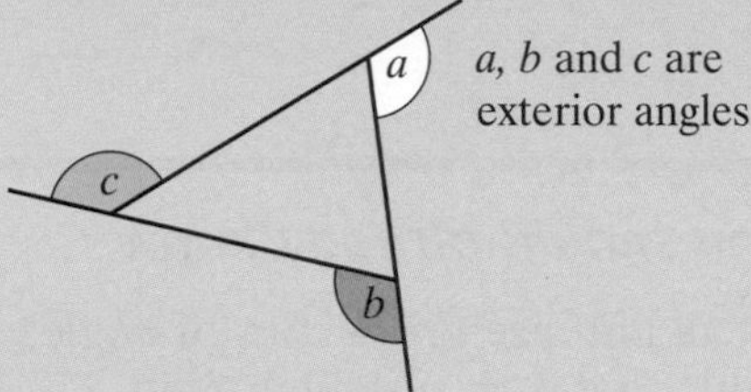

Convex polygon – a polygon with no interior reflex angles

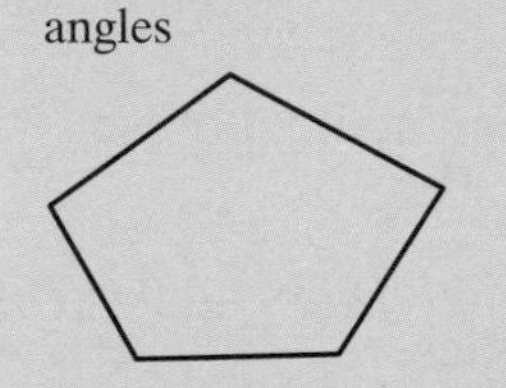

Concave polygon – a polygon with at least one interior reflex angle

Tessellation – a pattern where one or more shapes are fitted together repeatedly leaving no gaps

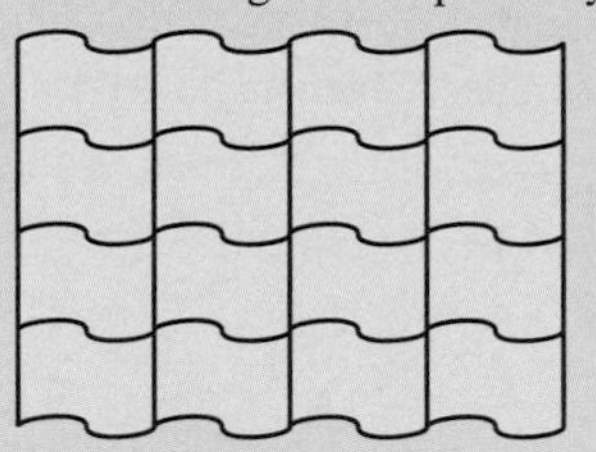

Learn 1 Diagonal properties of quadrilaterals

Example: The diagonals of a quadrilateral are different lengths, but one is the perpendicular bisector of the other.
What is the mathematical name of the quadrilateral?

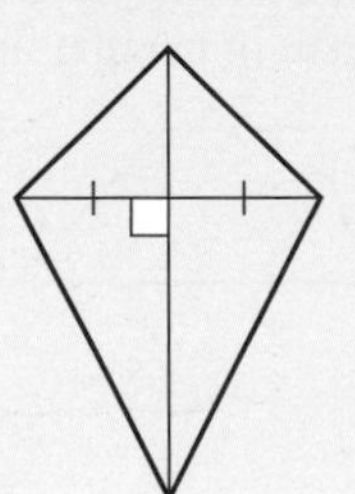

The quadrilateral is a kite.

Apply 1

1 Make an accurate drawing of each quadrilateral listed in the table, then draw the diagonals.
Use your diagrams to complete a copy of the table.

Shape	Are the diagonals equal? (Yes/No)	Do the diagonals bisect each other? (Both/One only/No)	Do the diagonals cross at right angles? (Yes/No)	Do the diagonals bisect the angles of the quadrilateral? (Yes/Two only/No)
Square				
Kite				
Parallelogram				
Trapezium				
Rectangle				
Rhombus				

2 Rajesh says that he has drawn a quadrilateral. Its diagonals are equal.
What quadrilaterals might he have drawn?
(Use the table from question **1** to help you.)

3 Michelle says that the diagonals of a rectangle bisect the corner angles.
So angles a and c are both 45°, and angle b must be 90°.
Is she right? Explain your answer.

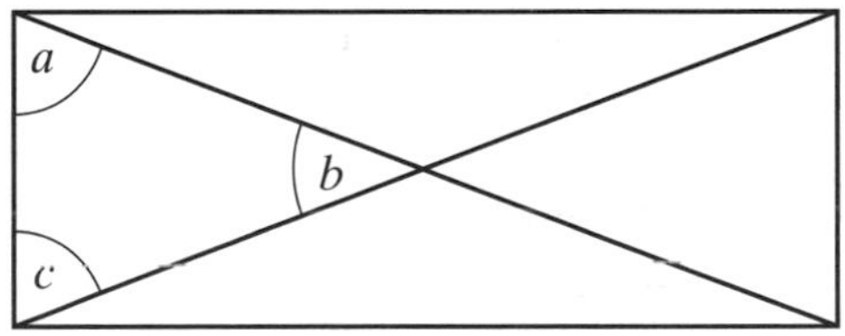

4 The diagram below shows a rhombus ABCD. AC and BD are the diagonals.
Angle ADB = 32°.

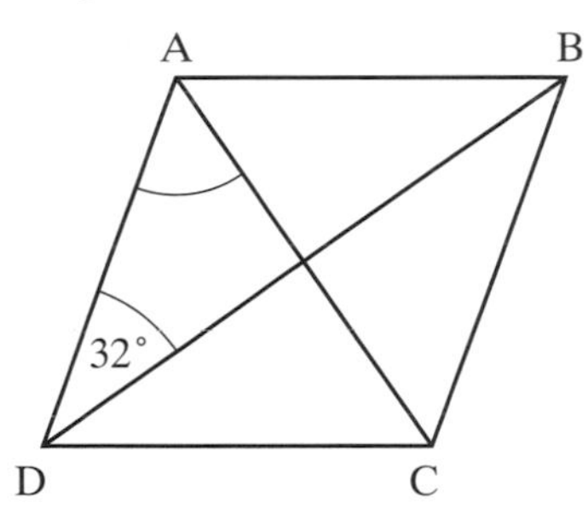

Not drawn accurately

Calculate angle DAC.

5 Calculate the angles marked with letters in the diagrams below.
You will need to use the properties of diagonals.
Give a reason for each of the angles.

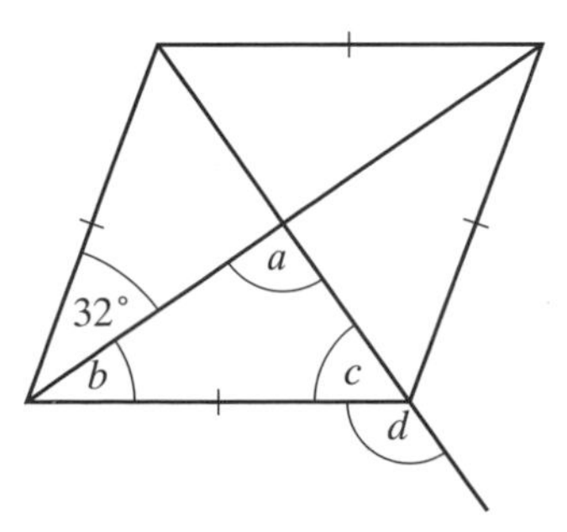

Not drawn accurately

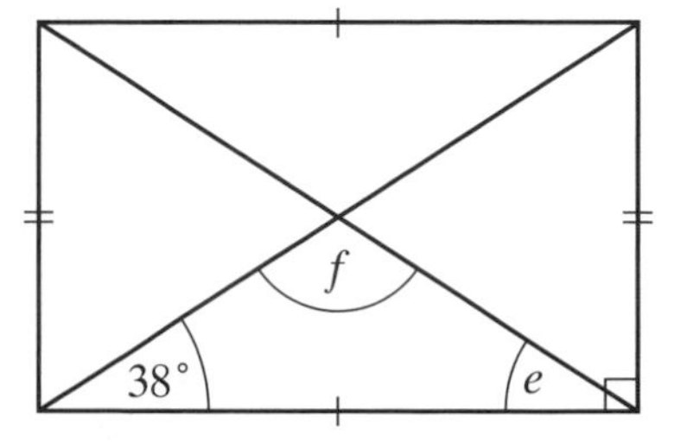

6 Get Real!

A builder has two types of wall tile. One is an isosceles trapezium, the other is a rhombus. He creates this tessellation.

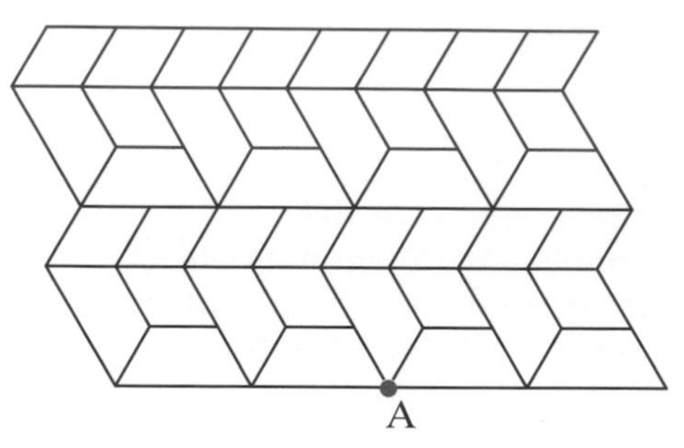

a By looking at the angles around point A, calculate the angles of the trapezium.

b Now calculate the angles of the rhombus.

7 Copy and complete this table. The top line has been done for you.

Shape	Number of different length sides (at most)	Number of right angles (at least)	Pairs of opposite sides parallel	Diagonals must be equal	Diagonals bisect each other	Diagonals cross at right angles
Square	1	4	Both	Yes	Yes	Yes
Rectangle						
Trapezium						
Rhombus						
Parallelogram						
Kite						
Isosceles trapezium						

8 In the diagram, EF is parallel to GH, and AB is parallel to CD.
IJ is perpendicular to AB, and IK is equal to JK.
Calculate the angles a to f, giving reasons for your answers.

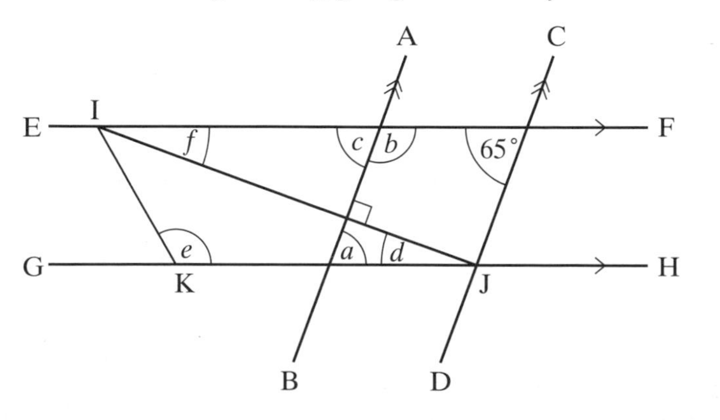

Not drawn accurately

9 Matt, Tess and Sam all draw kites with one angle of 76° and one angle of 60°.
All three kites are different.
Matt's kite has two obtuse angles. Sam's kite has a larger angle than the other two kites.
What are the angles of each kite? Draw diagrams to help.

Explore

- Roger draws a quadrilateral
  He uses it to make this tessellation pattern

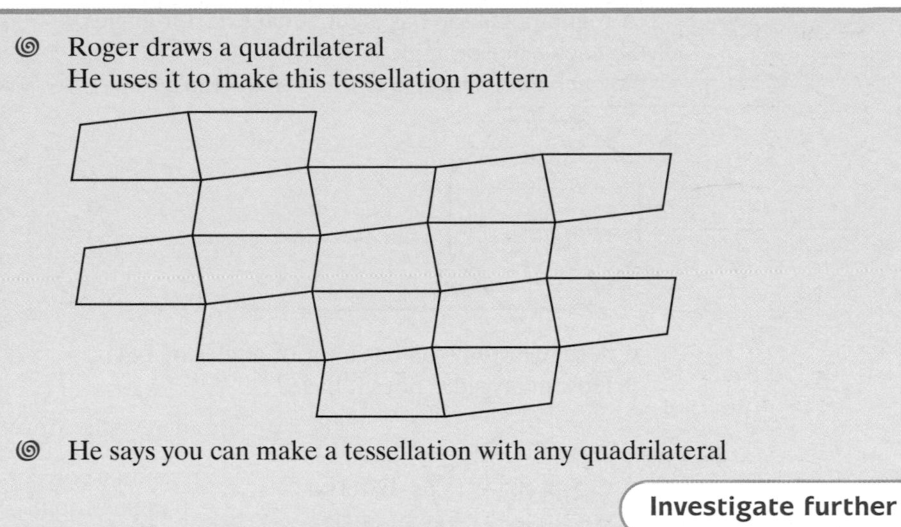

- He says you can make a tessellation with any quadrilateral

Investigate further

Learn 2 Angle properties of polygons

Examples:

a Calculate the sum of the interior angles of a:

i pentagon **ii** hexagon.

i The pentagon can be divided into three triangles by drawing diagonals from a start point.

Start point

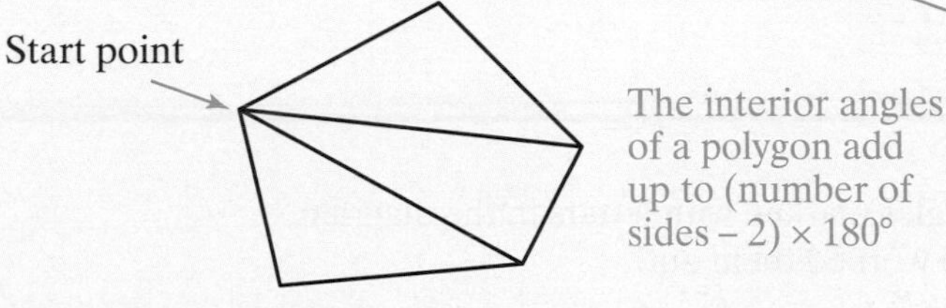

The sum of the angles is

$(5 - 2) \times 180° = 540°$

ii The hexagon can be divided into four triangles in the same way.

Start point

The interior angles of a polygon add up to (number of sides – 2) × 180°

The sum of the angles is

$(6 - 2) \times 180° = 720°$

b Calculate the interior angle of a regular octagon.

To calculate the interior angle of a regular octagon you can use one of the following methods.

Either:
An octagon has eight sides.
The sum of the angles can be found by dividing the octagon into six triangles as shown.
So the sum of the angles is $(8 - 2) \times 180° = 1080°$.
A regular octagon has all angles equal, so each angle is $1080° \div 8 = 135°$.

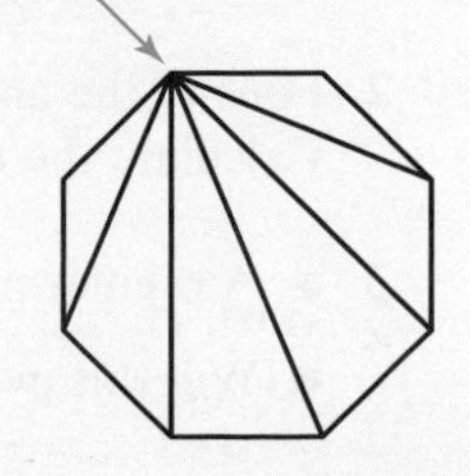

Or:

A regular octagon has eight equal exterior angles.
So each exterior angle is 360° ÷ 8 = 45°.
So each interior angle is 180° – 45° = 135°.

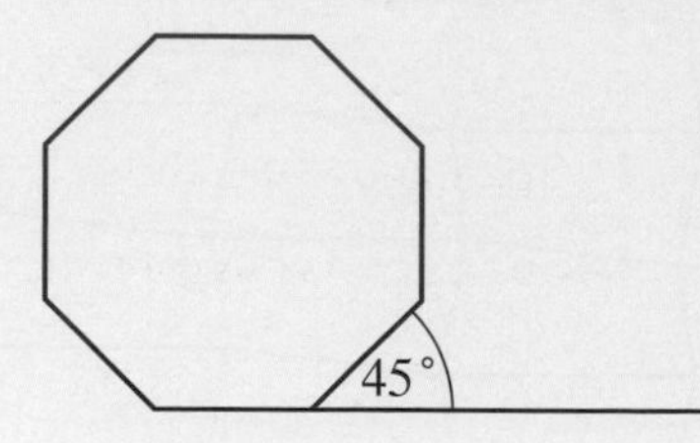

c A regular polygon has interior angles of 144°.
How many sides does it have?

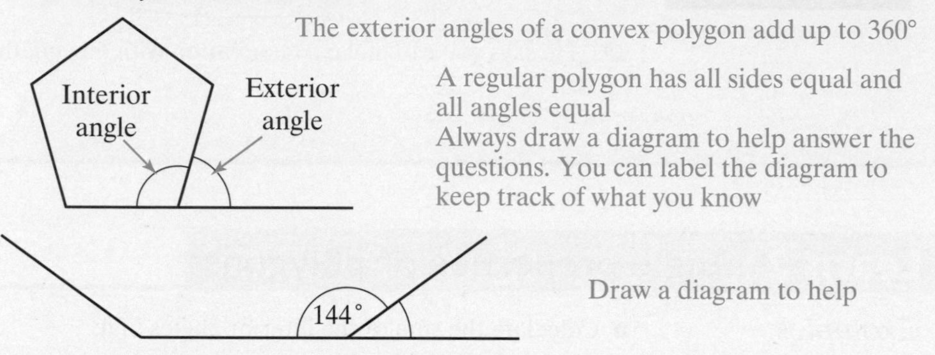

Each exterior angle must be 180° – 144° = 36°.
Exterior angles add up to 360°, so there must be 360 ÷ 36 = 10 exterior angles, and 10 sides.

Apply 2

1 Calculate the angles marked with letters in the diagram.
Explain how you worked them out.

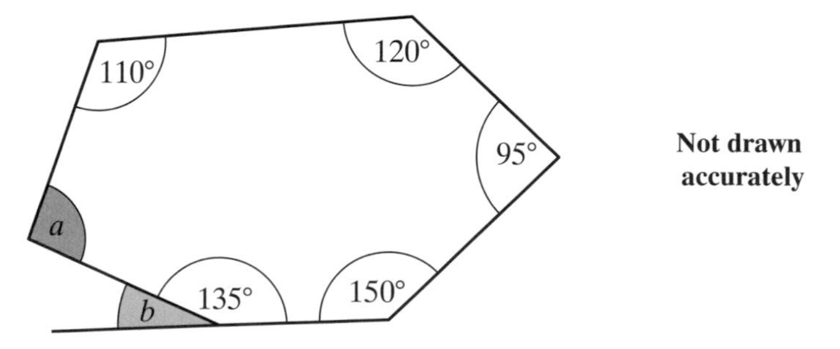

Not drawn accurately

2 Four of the angles of a pentagon are 110°, 130°, 102° and 97°.
Calculate the fifth angle.

3 a A regular polygon has an exterior angle of 40°. How many sides has it?

b Will this polygon tessellate?

4 Some of these quadrilaterals are possible, and others are not.
If possible draw:

a a kite with a right angle

b a kite with two right angles

c a trapezium with two right angles

d a trapezium with only one right angle

e a triangle with a right angle

f a triangle with two right angles

g a pentagon with one right angle

h a pentagon with two right angles

i a pentagon with three right angles

j a pentagon with four right angles.

5 Elena says that the interior angles of a decagon add up to $10 \times 180° = 1800°$.
Is she right? Give a reason for your answer.

6 Get Real!
A company makes containers as shown.
The top is in the shape of a regular octagon.

a What is the size of each interior angle?

b When the company packs them into a box, will they tessellate? If not, what shape will be left between them?

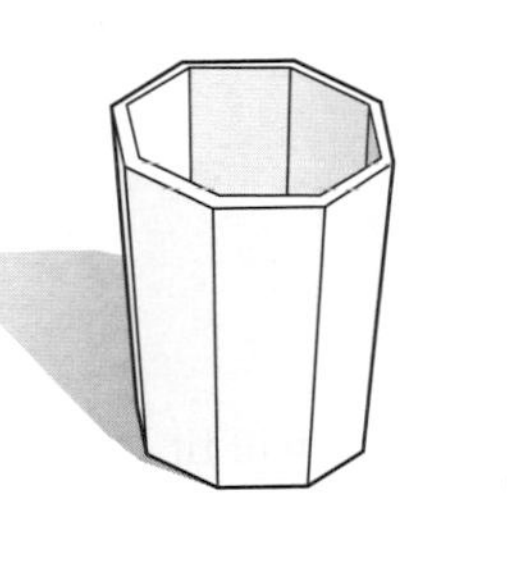

7 The diagrams show how you can draw an equilateral triangle and a regular hexagon inside a circle by dividing the angle at the centre into equal parts.

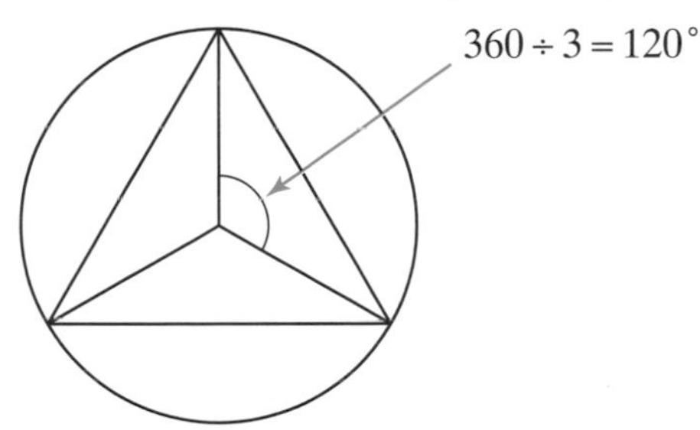

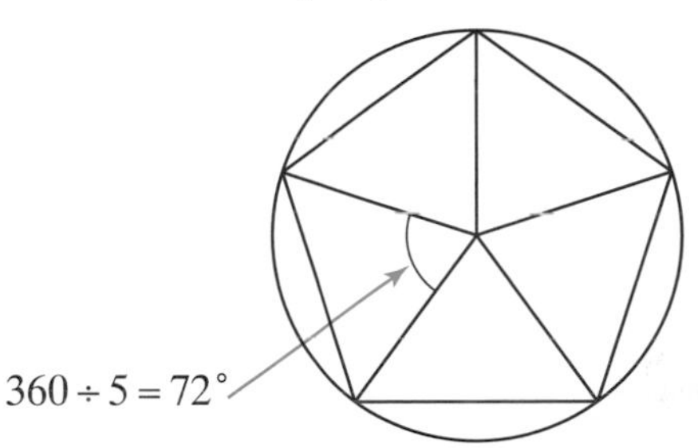

Use the same method to draw a regular hexagon and a regular nonagon inside a circle.

8 The diagram shows a regular pentagon ABCDE and a regular hexagon DEFGHI.

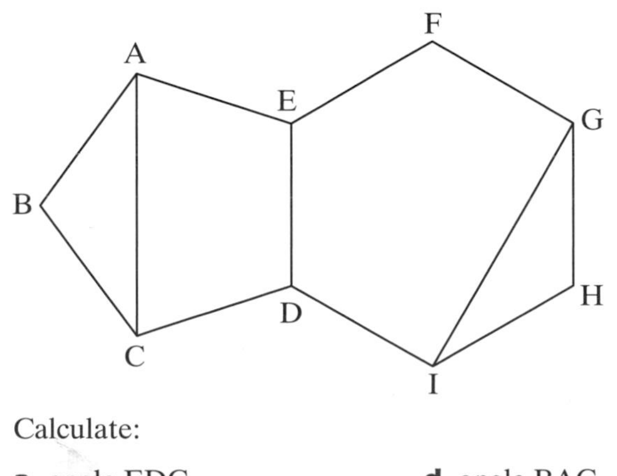

Calculate:

a angle EDC

b angle EDI

c obtuse angle CDI

d angle BAC

e angle CAE

f angle HIG

g angle DIG

9 Show that a convex polygon cannot have more than three acute angles.

> **HINT** Think about the sum of the exterior angles.

10 Show that if a convex polygon has more than six sides, then at least one of the sides has an obtuse angle at both ends.

11 Penny fits regular pentagons together in a circular arrangement. Part of the arrangement is shown in the diagram.

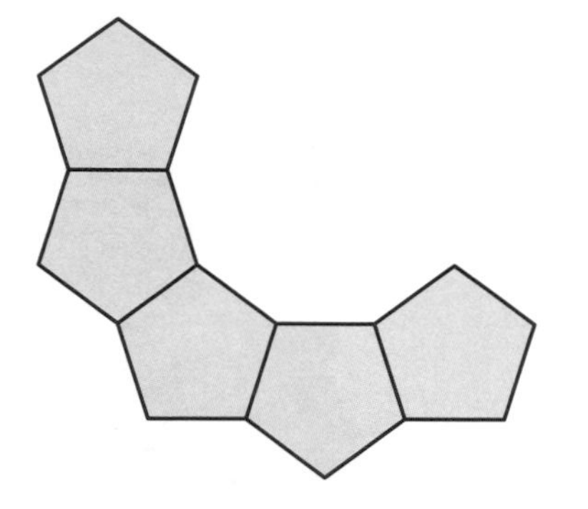

Show that exactly ten pentagons will fit together in this way before meeting up.

Explore

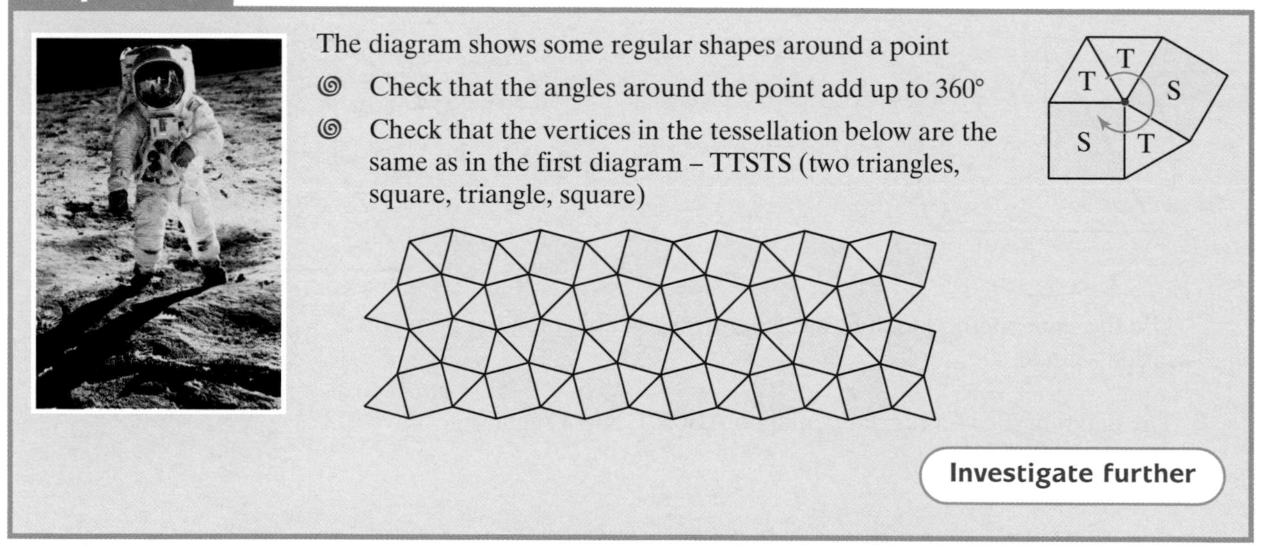

The diagram shows some regular shapes around a point

- Check that the angles around the point add up to 360°
- Check that the vertices in the tessellation below are the same as in the first diagram – TTSTS (two triangles, square, triangle, square)

Investigate further

Properties of polygons

ASSESS

The following exercise tests your understanding of this chapter, with the questions appearing in order of increasing difficulty.

1 **a** Draw a quadrilateral that can be cut into two pieces by drawing a straight line across it.

b Draw a quadrilateral that can be cut into three pieces by drawing a straight line across it.

2 Find the values of the marked angles in these diagrams.

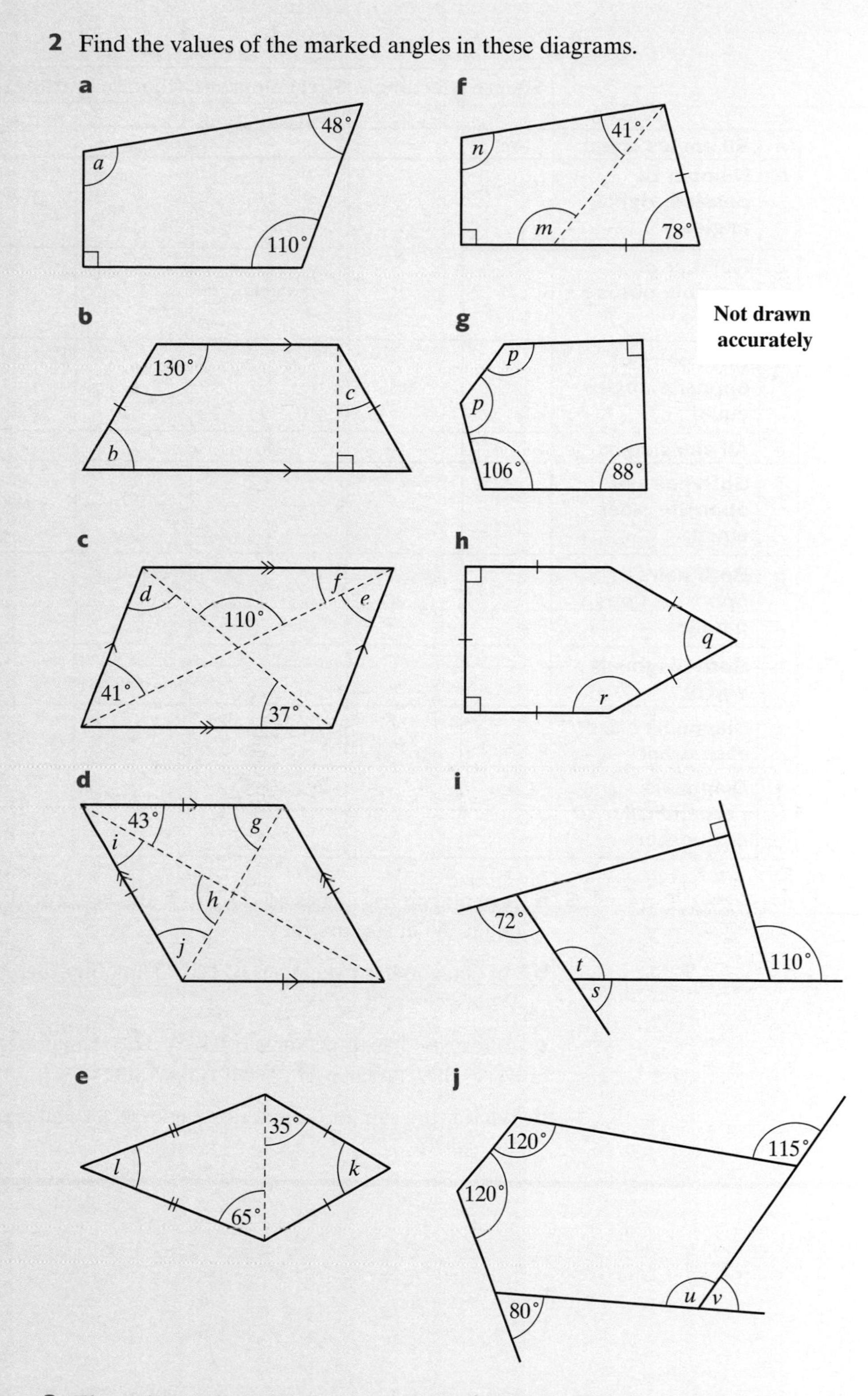

3 The only regular polygons that tessellate on their own are those whose interior angles divide exactly into 360°. Which ones are they?

4 Copy and complete the table.

		Square	Rectangle	Parallelogram	Rhombus	Trapezium	Isosceles trapezium	Kite
a	All angles equal	Yes						
b	Number of possible right angles	Exactly 4				0 or 2		1 or 2
c	Number of possible obtuse angles	0		Exactly 2				
d	Both pairs of opposite angles equal		Yes					
e	All sides equal			No				
f	Both pairs of opposite sides equal					No		
g	Both pairs of opposite sides parallel						No	
h	Both diagonals equal							No
i	Diagonals bisect each other		Yes					
j	Diagonals perpendicular to each other			No				

5 **a** A pentagon has angles of 115°, 165° and 80°. The other two angles are equal. What size are they?

b An octagon has five angles of 120°. The other three angles are equal. What size are they?

c A hexagon has three angles of 125°. The remaining three angles are w, $w + 120°$ and $w - 30°$. What is the value of w?

d What is the sum of the interior angles of a regular polygon with 102 sides?

12 Indices and standard form

OBJECTIVES

D **Examiners would normally expect students who get a D grade to be able to:**

Use the terms square, positive square root, negative square root, cube and cube root

Recall integer squares from 2×2 to 15×15 and the corresponding square roots

Recall the cubes of 2, 3, 4, 5 and 10

C **Examiners would normally expect students who get a C grade also to be able to:**

Use index notation and index laws for positive and negative powers such as $w^3 \times w^5$ and $\frac{w^3}{w^7}$

B **Examiners would normally expect students who get a B grade also to be able to:**

Use index notation and index laws for positive and negative powers such as $3w^3y \times 2w^5y^2$ and $\frac{8w^5z}{2w^3z^2}$

Convert between numbers in ordinary and standard index form

Use standard index form with and without a calculator

A **Examiners would normally expect students who get an A grade also to be able to:**

Use index notation and index laws for fractional powers such as $16^{\frac{1}{4}}$

A* **Examiners would normally expect students who get an A* grade also to be able to:**

Use index notation and index laws for fractional powers such as $16^{\frac{3}{4}}$

What you should already know ...

- Calculate squares and square roots (with and without the use of a calculator)
- Calculate cubes and cube roots (with and without the use of a calculator)
- Use function keys on a calculator for powers, roots and reciprocals

VOCABULARY

Square number – a square number is the outcome when a whole number is multiplied by itself

Cube number – a cube number is the outcome when a whole number is multiplied by itself then multiplied by itself again

Square root – the square root of a number such as 16 is a number whose outcome is 16 when multiplied by itself

Cube root – the cube root of a number such as 125 is a number whose outcome is 125 when multiplied by itself then multiplied by itself again

Index or **power** or **exponent** – the index tells you how many times the base number is to be multiplied by itself

5^3 (3 = Index, 5 = Base)

So $5^3 = 5 \times 5 \times 5$

Indices – the plural of index

Standard form – standard form is a shorthand way of writing very large and very small numbers; standard form numbers are always written as:

$$A \times 10^n$$

10^n: A power of 10

A must be at least 1 but less than 10

Learn 1 Rules of indices

Examples:

a Work out 5^3

b Work out:

i $6^3 \times 6^2$ **ii** $a^3 \times a^2$ **iii** $6a^3 \times 3a^2$

c Work out:

i $\frac{2^5}{2^2}$ **ii** $\frac{a^5}{a^2}$ **iii** $\frac{15a^5}{3a^2}$

d Work out:

i $(3^5)^2$ **ii** $(a^5)^2$ **iii** $(6a^5)^2$

a 5^3 (3 = Index (or power or exponent), 5 = Base)

So $5^3 = 5 \times 5 \times 5 = 125$

The index (or power or exponent) tells you how many times the base number is to be multiplied by itself

You can use the x^y button for indices on your calculator

	i Number	ii Algebra	iii Higher algebra
b	$6^3 \times 6^2$ $= 6^{(3+2)}$ $= 6^5$	$a^3 \times a^2$ $= a^{(3+2)}$ $= a^5$	$6a^3 \times 3a^2$ $= 6 \times a^3 \times 3 \times a^2$ $= 6 \times 3 \times a^3 \times a^2$ $= 18 \times a^{(3+2)}$ $= 18a^5$
c	$\frac{2^5}{2^2}$ $= 2^5 \div 2^2$ $= 2^{(5-2)}$ $= 2^3$	$\frac{a^5}{a^2}$ $= a^5 \div a^2$ $= a^{(5-2)}$ $= a^3$	$\frac{15a^5}{3a^2}$ $= \frac{15}{3} \times a^5 \div a^2$ $= 5 \times a^{(5-2)}$ $= 5a^3$
d	$(3^5)^2$ $= 3^{(5\times2)}$ $= 3^{10}$	$(a^5)^2$ $= a^{(5\times2)}$ $= a^{10}$	$(6a^5)^2$ $= 6^2 \times (a^5)^2$ $= 36 \times a^{(5\times2)}$ $= 36 \times a^{10}$ $= 36a^{10}$

Rules of indices

In general

$a^m \times a^n = a^{m+n}$

$a^m \div a^n = a^{m-n}$

$(a^m)^n = a^{m \times n}$

$a^{-m} = \frac{1}{a^m}$

$a^0 = 1$

$a^{\frac{1}{n}} = \sqrt[n]{a}$

(that is, the nth root of a so $a^{\frac{1}{2}} = \sqrt{a}$ and $a^{\frac{1}{3}} = \sqrt[3]{a}$ etc)

Apply 1

1 Find the value of each of the following.

a 1.5^2
b 10^3
c $\sqrt{225}$
d $\sqrt[3]{1}$
e $\sqrt[3]{-64}$

2 Calculate:

a $3^2 + 4^2$
b $2^3 \times 3^2$
c $10^3 - \sqrt{100}$
d $\sqrt{225} - \sqrt[3]{125}$
e $\sqrt{5^2 + 12^2}$
f $\sqrt{3^2 \times 5^2}$

3 Neil says -3^2 is 9.
Andrea says -3^2 is -9.
Who is correct?
Give a reason for your answer.

4 Write the following numbers in index notation.

a $5 \times 5 \times 5 \times 5$
b $2 \times 2 \times 2 \times 2 \times 2 \times 2 \times 2$
c 13×13
d 8
e $\frac{1}{5}$
f $\frac{1}{25}$

5 Find the value of each of the following.

a 7^2
b 2^5
c 3^4
d 5^1
e 4^6
f 12^0
g 1^2
h 1^{100}
i 3^{-1}
j 2^{-3}
k 4^{-6}
l 100^{-1}

6 Find the value of each of the following.

a 2^6
b 2^{10}
c 3^5
d 8^6
e 9^4
f $2^{11} - 5^3$
g $2^6 + 6^2$
h $5^5 \times 10^{-4}$
i $10^8 - 10^6$

7 Simplify the following numbers, giving your answers in index form.

a $5^6 \times 5^2$
b $12^8 \times 12^3$
c $\dfrac{4^7}{4^3}$
d $7^{10} \div 7^5$
e $3^7 \div 3^{10}$
f $(9^2)^5$
g $\dfrac{4^2 \times 4^3}{4^6}$

8 Calculate:

a $49^{\frac{1}{2}}$
b $121^{\frac{1}{2}}$
c $64^{\frac{1}{3}}$
d $8^{\frac{2}{3}}$
e $-8^{\frac{2}{3}}$
f $(-8)^{\frac{2}{3}}$
g $32^{\frac{2}{5}}$
h $4^{-\frac{1}{2}}$
i $(-125)^{\frac{2}{3}}$
j $1^{\frac{1}{3}}$
k $1^{-\frac{1}{3}}$

9 Say whether these statements are true or false.
Give a reason for your answer.

a $6^2 = 12$
b $1^3 = 1$
c $1^{-\frac{1}{2}} = -1$
d $16^{-\frac{1}{2}} = -4$
e $\dfrac{2^{10}}{4^5} = 1$
f $3^4 + 3^5 = 3^9$
g $10^{50} \times 10^{50} = 10^{100}$
h $(-216)^{\frac{1}{3}} = -6$
i $1\,000\,000^0 = 0$

10 Use your calculator to work out each of the following.

a 60^2 **b** $16^{-0.5}$ **c** $729^{1.5}$ **d** $256^{-0.25}$

11 Put these in order, starting with the smallest.

$64^{\frac{1}{3}}$, $64^{\frac{1}{4}}$, $(1/64)^{\frac{1}{2}}$, $64^{-\frac{1}{3}}$

12 Simplify the following, giving your answers in index form.

a $x^6 \times x^2$
b $e^8 \times e^3$
c $\dfrac{a^7}{a^3}$
d $p^{10} \div p^5$
e $q^7 \div q^{10}$
f $(b^2)^5$

13 Simplify the following.

a $2x^2 \times 3x^5$
b $\dfrac{3a^6}{6a^2}$
c $5c^2 \times 2c^7$
d $(4b^2)^3$
e $\dfrac{c^6 \times c^9}{c^5}$
f $\dfrac{5c^2 \times 2c^7}{c^6}$

HINT
Write $2x^2 \times 3x^5$ as $2 \times x^2 \times 3 \times x^5$
$= 2 \times 3 \times x^2 \times x^5$

14 **a** Find the product of $7xy^2$ and $3x^4y^3$.

b Write down five other expressions that give the same product as your answer in part **a**.

HINT Write $7xy^2$ as $7x \times y^2$.

15 The number 64 can be written as 8^2 in index form.
Write down five other ways that 64 can be written in index form.

Explore

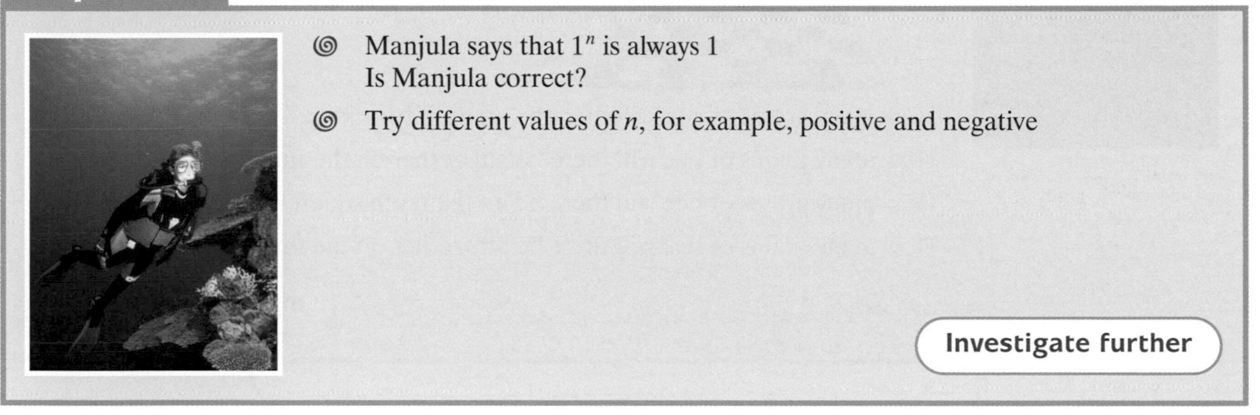

- Manjula says that 1^n is always 1
  Is Manjula correct?
- Try different values of n, for example, positive and negative

Investigate further

Explore

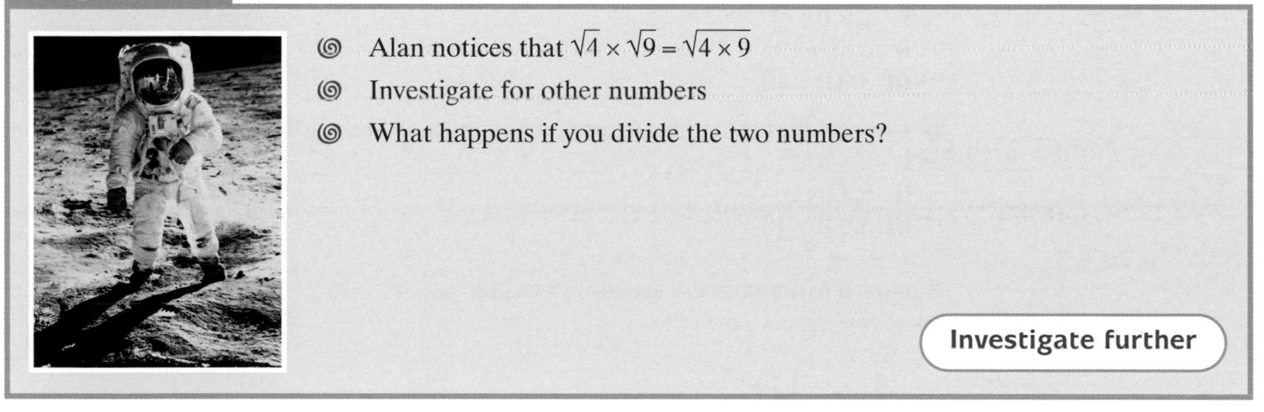

- Alan notices that $\sqrt{4} \times \sqrt{9} = \sqrt{4 \times 9}$
- Investigate for other numbers
- What happens if you divide the two numbers?

Investigate further

Explore

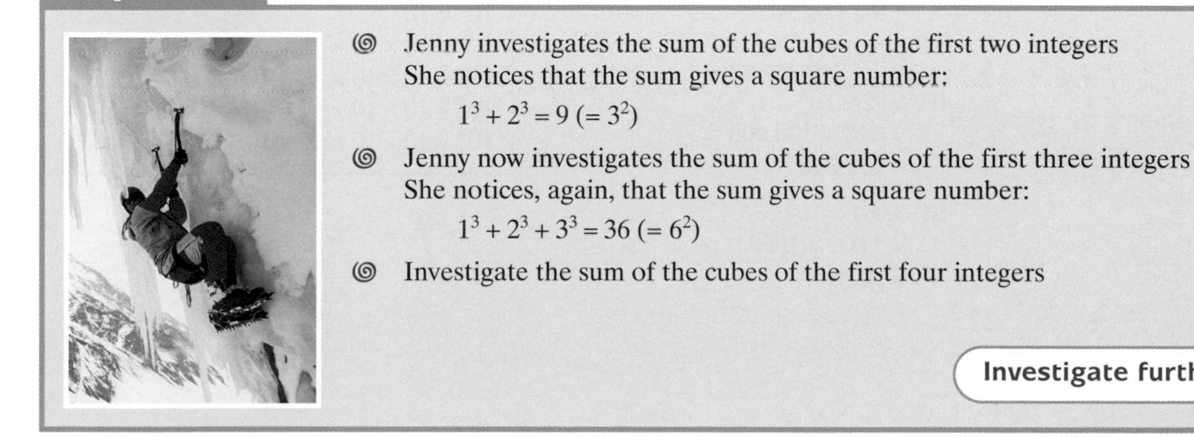

- Jenny investigates the sum of the cubes of the first two integers
  She notices that the sum gives a square number:
  $1^3 + 2^3 = 9\ (= 3^2)$
- Jenny now investigates the sum of the cubes of the first three integers
  She notices, again, that the sum gives a square number:
  $1^3 + 2^3 + 3^3 = 36\ (= 6^2)$
- Investigate the sum of the cubes of the first four integers

Investigate further

Explore

- One grain of rice is placed on the first square of a chessboard
- Two grains of rice are placed on the second square of a chessboard
- Four grains of rice are placed on the third square of a chessboard
- Eight grains of rice are placed on the fourth square of a chessboard etc

How many grains of rice will there be on the fifth square?

How many grains of rice will there be altogether on the first five squares?

How many grains of rice will there be on the tenth square?

How many grains of rice will there be altogether on the first ten squares?

Investigate further

Learn 2 Standard form

Examples:

a Convert these standard form numbers into ordinary form.

i 2×10^2
ii 6.82×10^5
iii 3.001×10^3

b Convert these standard form numbers into ordinary form.

i 2×10^{-2}
ii 6.82×10^{-5}
iii 3.001×10^{-3}

Standard form numbers are always written as:

$$A \times 10^n$$

A power of 10 (10^n)

A number between 1 and 10 (A)

a i 2×10^2
$= 2 \times 100$
$= 200$

ii 6.82×10^5
$= 6.82 \times 100\,000$
$= 682\,000$

iii 3.001×10^3
$= 3.001 \times 1000$
$= 3001$

$10^1 = 10$
$10^2 = 10 \times 10 = 100$
$10^3 = 10 \times 10 \times 10 = 1000$
$10^4 = 10 \times 10 \times 10 \times 10 = 10\,000$
$10^5 = 10 \times 10 \times 10 \times 10 \times 10 = 100\,000$

b i 2×10^{-2}
$= 2 \times 0.01$
$= 0.02$

ii 6.82×10^{-5}
$= 6.82 \times 0.00001$
$= 0.0000682$

iii 3.001×10^{-3}
$= 3.001 \times 0.001$
$= 0.003001$

$10^{-1} = \frac{1}{10^1} = \frac{1}{10} = 0.1$
$10^{-2} = \frac{1}{10^2} = \frac{1}{100} = 0.01$
$10^{-3} = \frac{1}{10^3} = \frac{1}{1000} = 0.001$
$10^{-4} = \frac{1}{10^4} = \frac{1}{10\,000} = 0.0001$
$10^{-5} = \frac{1}{10^5} = \frac{1}{100\,000} = 0.00001$

Remember that multiplying by 10^{-1} is the same as dividing by 10^1, and multiplying by 10^{-2} is the same as dividing by 10^2, etc

c Write the following in standard form.

Write your number in the form $A \times 10^n$ where A is a number between 1 and 10

i 65 000
$A = 6.5$, so $65\,000 = 6.5 \times 10\,000 = 6.5 \times 10^4$

ii 0.000000572
$A = 5.72$, so $0.000000572 = 5.72 \times 0.0000001 = 5.72 \times 10^{-7}$

Writing	Reading
Input the number $\mathbf{3.2 \times 10^7}$ as **3.2** EXP **7** or **3.2** EE **7**	On some calculators, the display 3.2 07 or 3.2 07 should be interpreted as $\mathbf{3.2 \times 10^7}$

Apply 2

1 Write these numbers in standard form.

a 3700
b 23 000 000
c 200 200
d 8 500 000 000
e 35
f 0.005
g 0.13
h 0.000000178
i 0.00000000002
j 0.5

2 Write these standard form numbers as ordinary numbers.

a 7×10^3
b 7.6×10^5
c 4.2×10^4
d 6.085×10^2
e 7.6635×10^1
f 5.1×10^8
g 3×10^{-1}
h 1.25×10^{-3}
i 3.086×10^{-4}
j 6.6×10^{-10}

3 Get Real!
The distance from the Earth to the Moon is approximately 384 000 000 m. Write this number in standard form.

4 Get Real!

The mass of an electron is approximately 0.000000000000000000000000000000910939 kilograms (there are 30 zeros). Write this number in standard form.

5 Get Real!

The table shows the diameters of the planets of the solar system.

Planet	Diameter (km)
Mercury	4.9×10^3
Venus	1.2×10^4
Earth	1.3×10^4
Mars	6.8×10^3
Jupiter	1.4×10^5
Saturn	1.2×10^5
Uranus	5.2×10^4
Neptune	4.9×10^4
Pluto	2.3×10^3

Place the planets in order of size, starting with the smallest.

6 Write the number 60^3 in standard form.

7 Calculate the following, giving your answers in standard form.

a $(4 \times 10^4) \times (2 \times 10^7)$

b $(5 \times 10^5) \times (3 \times 10^9)$

c $(5 \times 10^{-4})^2$

d $\dfrac{2.2 \times 10^1}{5.5 \times 10^{-6}}$

e $(3.3 \times 10^6) \times (3 \times 10^4)$

f $(2.5 \times 10^8) \times (5 \times 10^{-3})$

g $(4.5 \times 10^5) \times (2 \times 10^{11})$

h $\dfrac{4 \times 10^4}{2 \times 10^3}$

i $(1.5 \times 10^7)^2$

j $(2.2 \times 10^6) \div (4.4 \times 10^4)$

k $\dfrac{3.9 \times 10^5}{1.3 \times 10^8}$

8 Use your calculator to work out the following. Give your answers in standard form.

a $(3 \times 10^5) \times (3 \times 10^7)$

b $(6 \times 10^{-4})^2$

c $1 \div (2.5 \times 10^8)$

d $(4.55 \times 10^5) \times (6.2 \times 10^7)$

e $\dfrac{8 \times 10^{11}}{4 \times 10^3}$

f $(1.5 \times 10^7)^2$

g $(5 \times 10^5) + (3 \times 10^6)$

h $(2.4 \times 10^5) \times (3.5 \times 10^7)$

i $\dfrac{3.9 \times 10^8}{1.3 \times 10^{-5}}$

j $(8 \times 10^2) + (8 \times 10^4)$

k $(5 \times 10^5) \times (3.2 \times 10^9)$

l $(2.2 \times 10^2) \div (3.5 \times 10^{11})$

m $(5.2 \times 10^4) - (5.2 \times 10^3)$

n $(1.1 \times 10^3) - (1.11 \times 10^4)$

9 Get Real!

Some large numbers are:

One million = 10^6
One billion = 10^9
One trillion = 10^{12}

a Write the number one billion in ordinary form.

b Write the number 50 million in standard form.

c How many millions are there in one trillion?

d Multiply three billion by four trillion. Give your answer in standard form.

10 A rectangle has length 1.4×10^4 metres and width 2.2×10^3 metres.
Calculate the area and perimeter of the rectangle.
Give your answers in standard form.

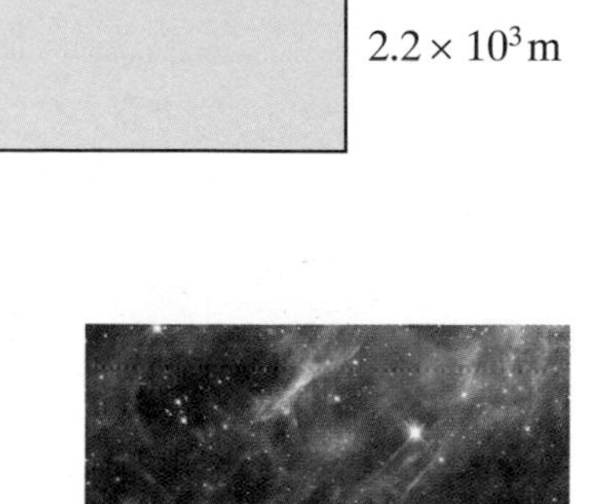

11 Alan says that $(4 \times 10^4) + (2 \times 10^4) = (6 \times 10^4)$.
Brian says that $(4 \times 10^4) + (2 \times 10^4) = (6 \times 10^8)$.
Who is correct?
Give a reason for your answer.

12 Given that $p = 4 \times 10^2$ and $q = 2 \times 10^{-1}$, calculate:

a $p \times q$ **b** $p \div q$ **c** $p + q$ **d** $p - q$ **e** p^2

13 Get Real!
The distance to the edge of the observable universe is approximately 4.6×10^{26} metres.
Express this distance in kilometres, giving your answer in standard form.

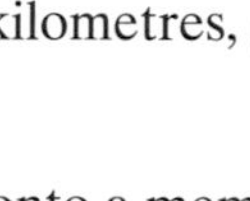

14 Get Real!
Anil saves some images onto a memory stick. Each image requires 32 000 bytes of memory. How many images can he save if the memory stick has a memory of 1.36×10^8 bytes?
Give your answer in standard form.

15 Get Real!
The speed of light is approximately 3.0×10^8 m/s.
How far will light travel in one week?
Give your answer in standard form.

Explore

- A **googol** is the number 10^{100}, that is, one followed by one hundred zeros
- What can you find out about the googol?

Investigate further

Indices and standard form

ASSESS

The following exercise tests your understanding of this chapter, with the questions appearing in order of increasing difficulty.

1 a Sam says numbers have two square roots.
George says some numbers have no square roots.
Who is right? Give a reason for your answer.

b Amelia joins in the conversation and says that all numbers have two cube roots.
Is she right? Give a reason for your answer.

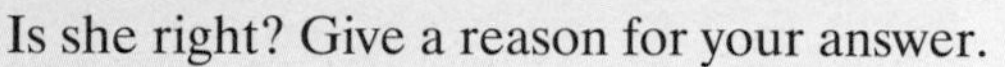

2 **a** Work out the following, giving your answers in index form.

i $4^6 \times 4^2$ **v** $6^4 \times 6^2 \times 6^3$ **ix** $5^8 \div 5^7$

ii $11^5 \times 11^3$ **vi** $10^4 \div 10^2$ **x** $2^3 \div 2^3$

iii $(5^3)^2$ **vii** $21^7 \div 21^5$

iv $7^5 \times 7$ **viii** $16^{10} \div 16^9$

b Find the value of:

i $3^2 \times 4^2$ **ii** $3^4 \div 5^2$ **iii** $6^5 \times 6^3 \div 6^4$ **iv** $\dfrac{(10^8 \times 10^7)}{(10^7 \times 10^6)}$

c Which is larger:

i 3^5 or 5^3 **ii** 11^2 or 2^{11} **iii** 2^4 or 4^2?

3 **a** The areas, in square kilometres, of some oceans and seas are shown below. Write them in increasing order of size and convert them to ordinary numbers.

Arctic Ocean:	1.4×10^7 km^2
Atlantic Ocean:	8.24×10^7 km^2
Pacific Ocean:	1.65×10^8 km^2
Mediterranean Sea:	2.50×10^6 km^2
Gulf of Mexico:	1.54×10^6 km^2

b Which is bigger: 1.1×10^8 or 99 999 999?

c Find the value of n in each of these.

i $3.5 \times 10^n = 350\,000$ **ii** $5.69 \times 10^n = 56.9$ **iii** $4.006 \times 10^n = 400\,600$

4 **a** Write these numbers in standard form.

i 0.003 **ii** 0.00000655 **iii** 0.1

b Write these standard form numbers as ordinary numbers.

i 1×10^{-9} **ii** 4.22×10^{-6} **iii** 3.9958×10^{-5}

c Calculate the value of $3.52 \times 10^4 \times 2.2 \times 10^{-3}$.

d Calculate the value of $3.52 \times 10^4 \div 2.2 \times 10^{-3}$.

e A company employs 4.7×10^3 workers and the workers use, on average, 2.3×10^2 litres of water per year.
How many litres of water does the company use in a year?

f The mass of a hydrogen atom is 1.7×10^{-24} g.
One litre of air contains 2.5×10^{22} atoms of hydrogen.
What is the mass of the hydrogen atoms in one litre of air?

5 Find the values of the following, leaving your answers as fractions where appropriate.

i 5^{-1} **iii** $12^3 \div 12^4$ **v** $2^6 \div 2^8$ **vii** $(\frac{1}{2})^{-3}$ **ix** $27^{\frac{2}{3}}$

ii 23^0 **iv** 3^{-2} **vi** $(\frac{1}{4})^0$ **viii** $8^{\frac{1}{3}}$ **x** $16^{-\frac{3}{4}}$

13 Sequences

OBJECTIVES

D **Examiners would normally expect students who get a D grade to be able to:**

Write the terms of a sequence or a series of diagrams given the nth term

C **Examiners would normally expect students who get a C grade also to be able to:**

Write the nth term of a sequence or a series of diagrams

What you should already know ...

- Continue a sequence of numbers or diagrams
- Write terms in a sequence of numbers or diagrams
- Write the term-to-term rule in a sequence of numbers or diagrams

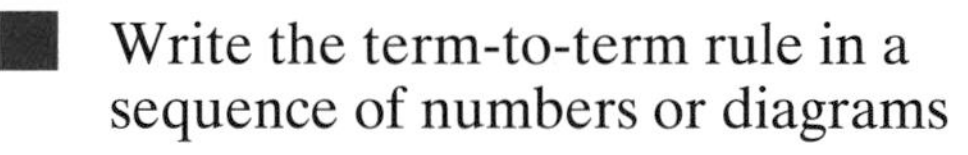

VOCABULARY

Sequence – a list of numbers or diagrams that are connected in some way

In this sequence of diagrams, the number of squares is increased by one each time:

The dots are included to show that the sequence continues

...

Term – a number, variable or the product of a number and a variable(s) such as 3, x or $3x$

***n*th term** – this phrase is often used to describe a 'general' term in a sequence; if you are given the nth term, you can use this to find the terms of a sequence

Square numbers – 1, 4, 9, 16, 25, ...

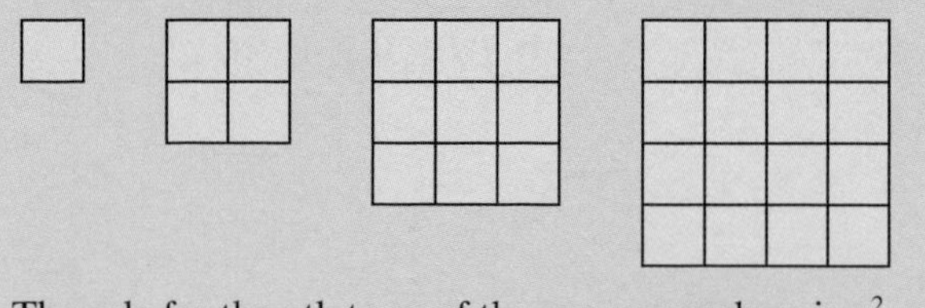

The rule for the nth term of the square numbers is n^2

Cube numbers – 1, 8, 27, 64, 125, ...

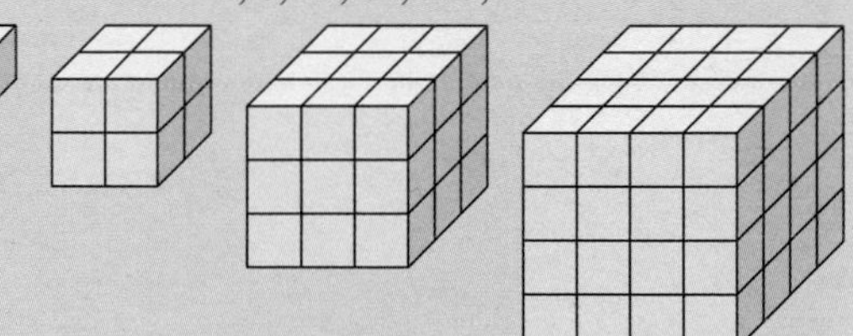

The rule for the nth term of the cube numbers is n^3

Triangle numbers – 1, 3, 6, 10, 15, ...

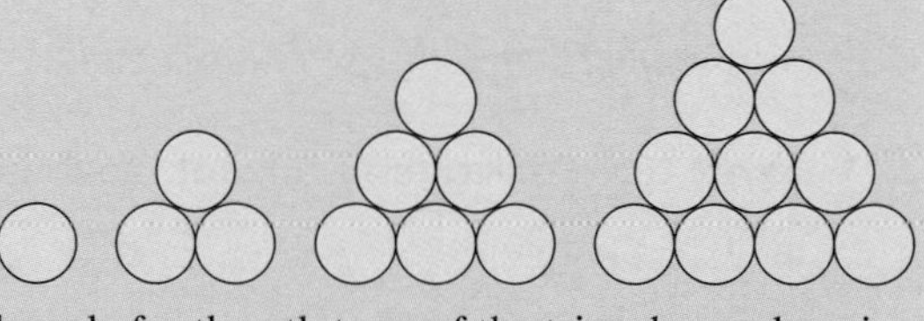

The rule for the nth term of the triangle numbers is $\frac{1}{2}n(n + 1)$

Fibonacci numbers – a sequence where each term is found by adding together the two previous terms

1, 1, 2, 3, 5, 8, 13, 21, ...

(2 = 1+1, 3 = 1+2, 5 = 2+3, 8 = 3+5, 13 = 5+8, 21 = 8+13)

Learn 1 The *n*th term of a sequence

Example: Given the nth term of a sequence is $2n + 3$, use this to find the first four terms of the sequence.

If the nth term is $2n + 3$, you can use this to find the first term by replacing n with 1 in the formula.
Similarly, you can find the second term by replacing n with 2 in the formula and the 100th term by replacing n with 100 in the formula.
So the sequence whose nth term is $2n + 3$ is:

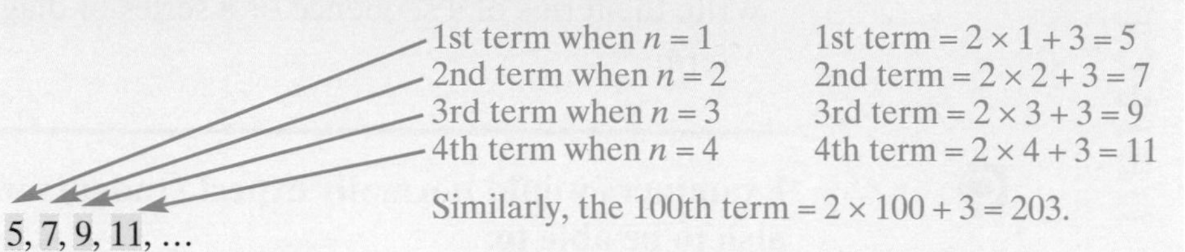

The above sequence is called a linear sequence because the differences between terms are all the same. In this example, the differences are all +2.
We say that the term-to-term rule is +2.

5, 7, 9, 11, …
(+2, +2, +2)

In general, to find the nth term of a linear sequence, you can use the formula:

$$n\text{th term} = \textbf{difference} \times n + (\textbf{first term} - \textbf{difference})$$
$$= dn + (a - d)$$

a is the first term = 5
d is the difference between terms = 2

So for 5, 7, 9, 11, …
(+2, +2, +2)

$$n\text{th term} = \text{difference} \times n + (\text{first term} - \text{difference})$$
$$= 2 \times n + (5 - 2)$$
$$= 2n + 3$$

Apply 1

Apart from question 9, this is a non-calculator exercise.

1 Write the term-to-term rule for the following sequences.

a 3, 7, 11, 15, ...
b 0, 5, 10, 15, ...
c 1, 2, 4, 8, 16, ...
d 3, 4.5, 6, 7.5, ...
e 20, 16, 12, 8, ...
f 2, 3, 4.5, 6.25, ...
g 54, 18, 6, 2, ...
h 0.01, 0.1, 1, 10, ...

2 The term-to-term rule is +6.
Write five different sequences that fit this rule.

3 Write the first five terms of the sequence whose nth term is:

a $n+3$

b $5n-3$

c n^2+3

d n^2-5

e $3n^2$

f $\dfrac{n}{n+2}$

4 Jenny writes the sequence 3, 7, 11, 15, ...
She says that the nth term is $n + 4$.
Is she correct?
Give a reason for your answer.

5 Copy and complete this table.

Pattern (n)	Diagram	Number of matchsticks (m)
1		3 matchsticks
2		5 matchsticks
3		7 matchsticks
4		
5		

a What do you notice about the pattern of matchsticks?

b Write the formula for the number of matchsticks (m) in the nth pattern.

c How many matchsticks will there be in the tenth pattern? Check your answer by drawing the tenth pattern and counting the number of matchsticks.

d There are 41 matchsticks in the 20th pattern. How many matchsticks are there in the 21st pattern? Give a reason for your answer.

6 Write the nth term in these linear sequences.

a 3, 7, 11, 15, ...

b 0, 5, 10, 15, ...

c 8, 14, 20, 26, ...

d 23, 21, 19, 17, ...

e 100, 95, 90, ...

f $-5, -1, 3, 7, ...$

g 4, 6.5, 9, 11.5, ...

h $-5, 3, 11, 19, ...$

7 **Get Real!**

Jackie builds fencing from pieces of wood as shown below.

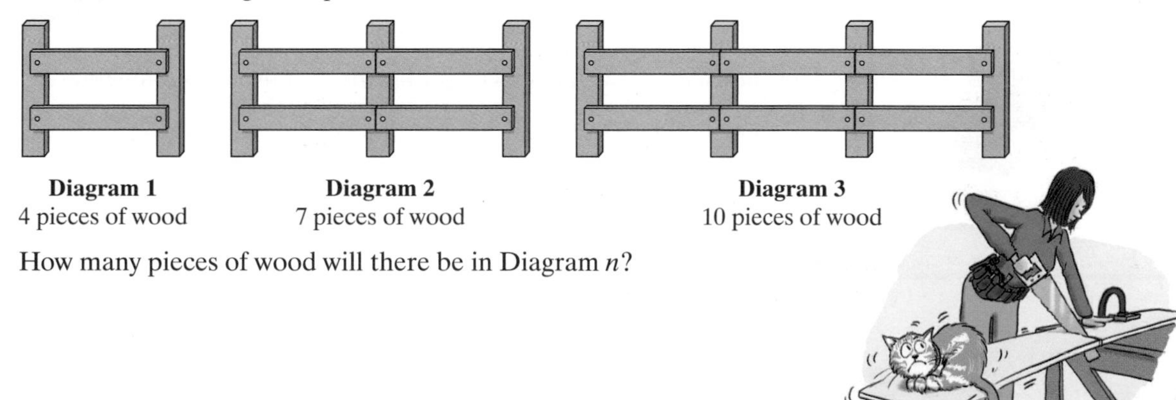

Diagram 1
4 pieces of wood

Diagram 2
7 pieces of wood

Diagram 3
10 pieces of wood

How many pieces of wood will there be in Diagram n?

8

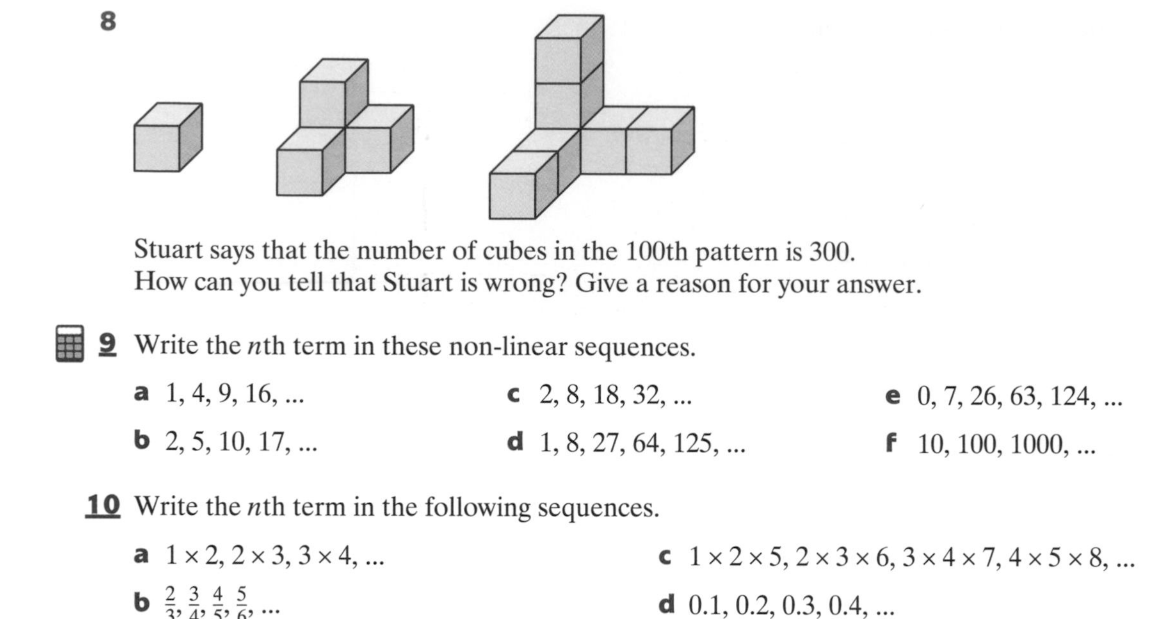

Stuart says that the number of cubes in the 100th pattern is 300.
How can you tell that Stuart is wrong? Give a reason for your answer.

9 Write the nth term in these non-linear sequences.

a 1, 4, 9, 16, ...
b 2, 5, 10, 17, ...
c 2, 8, 18, 32, ...
d 1, 8, 27, 64, 125, ...
e 0, 7, 26, 63, 124, ...
f 10, 100, 1000, ...

10 Write the nth term in the following sequences.

a $1 \times 2, 2 \times 3, 3 \times 4, \ldots$
b $\frac{2}{3}, \frac{3}{4}, \frac{4}{5}, \frac{5}{6}, \ldots$
c $1 \times 2 \times 5, 2 \times 3 \times 6, 3 \times 4 \times 7, 4 \times 5 \times 8, \ldots$
d 0.1, 0.2, 0.3, 0.4, ...

11 Farukh is exploring number patterns.
He writes down the following products in a table.

1×1	1
11×11	121
111×111	12 321
1111×1111	1 234 321
$11\,111 \times 11\,111$	
$111\,111 \times 111\,111$	

Copy and complete the last two rows in the table.

Farukh says he can use the table to work out $1\,111\,111\,111 \times 1\,111\,111\,111$
Is he correct? Give a reason for your answer.

Explore

- Workers are paid 1p on their first day, 2p on the second day, 4p on the third day, 8p on the fourth day and so on

- How much will a worker get paid on the fifth day?
- How much money will a worker get for the first five days altogether?
- How much money will the worker get for the first ten days altogether?

Investigate further

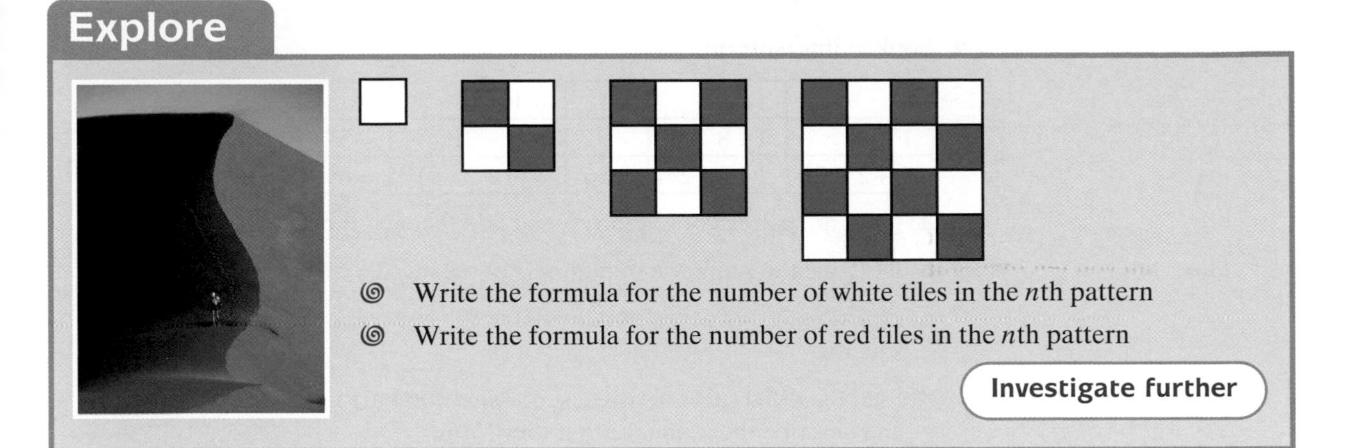

Explore

- Write the formula for the number of white tiles in the nth pattern
- Write the formula for the number of red tiles in the nth pattern

Investigate further

Sequences

The following exercise tests your understanding of this chapter, with the questions appearing in order of increasing difficulty.

1 Write the first three terms and the 5th, 20th and 50th terms of the sequences with nth term:

a $2n + 1$ **b** $5n - 2$ **c** $n^2 + 1$.

2 Find the nth terms of the following sequences.

a 6, 8, 10, 12, ... **c** 8, 6, 4, 2, ... **e** 4, 7, 12, 19, ...

b 3, 13, 23, 33, ... **d** −2, 5, 12, 19, ...

3 The diagrams show a quadrilateral, a pentagon and a hexagon with all possible diagonals drawn.

a Draw polygons with seven and eight sides and put in all the possible diagonals.

b Copy and complete this table.

Number of sides	4	5	6	7	8
Number of diagonals	2				

c Use your table to predict the number of diagonals in polygons having:

i 9 sides **ii** 10 sides **iii** 11 sides **iv** 12 sides.

d Use your answers to find the formula that gives the number of diagonals in a polygon with n sides.

e Use your formula to predict the number of diagonals in polygons having:

i 15 sides **ii** 20 sides **iii** 50 sides **iv** 100 sides.

4 Look at this pattern:

1^3	1	= 1	$= 1^2$	$= 1^2$
$1^3 + 2^3$	1 + 8	= 9	$= (1 + 2)^2$	$= 3^2$
$1^3 + 2^3 + 3^3$	1 + 8 + 27	= 36	$= (1 + 2 + 3)^2$	$= 6^2$

a Copy the pattern of sequences and add the next three lines.

b Describe the sequence of numbers in the middle column.

c Describe the sequence of numbers in the last column that create the squares.

d Find the **differences** between successive numbers in the middle column and describe the sequence that they form.

e Use the ideas in the table (without adding any more lines) to write down the value of $(1 + 2 + 3 + 4 + 5 + 6 + 7 + 8 + 9 + 10)^2$.

Try a real past exam question to test your knowledge:

5 The nth term of a sequence is $3n - 1$.

a Write down the first and second terms of the sequence.

b Which term of the sequence is equal to 32?

c Explain why 85 is not a term in this sequence.

Spec A, Int Paper 2, Nov 04

14 Coordinates

OBJECTIVES

D **Examiners would normally expect students who get a D grade to be able to:**

Solve problems involving straight lines

Draw lines such as $y = 2x + 3$

C **Examiners would normally expect students who get a C grade also to be able to:**

Find the midpoint of a line segment

Use and understand coordinates in three dimensions

What you should already know ...

- Negative numbers and the number line
- How to use coordinates in the first quadrant, such as how to plot the point (3, 2)
- How to use coordinates in all four quadrants, such as how to plot the points (3, –2) and (–4, –3)
- How to draw lines such as $x = 3$ and $y = x$

VOCABULARY

Coordinates – a system used to identify a point; an x-coordinate and a y-coordinate give the horizontal and vertical positions

Origin – the point (0, 0) on a coordinate grid

Axis (pl. **axes**) – the lines used to locate a point in the coordinates system; in two dimensions, the x-axis is horizontal, and the y-axis is vertical. This system of Cartesian coordinates was devised by the French mathematician and philosopher, René Descartes

In three dimensions, the x- and y-axes are horizontal and at right angles to each other and the z-axis is vertical

Horizontal – from left to right; parallel to the horizon

Vertical – directly up and down; perpendicular to the horizontal

Gradient – a measure of how steep a line is

$$\text{Gradient} = \frac{\text{change in vertical distance}}{\text{change in horizontal distance}} = \frac{y}{x}$$

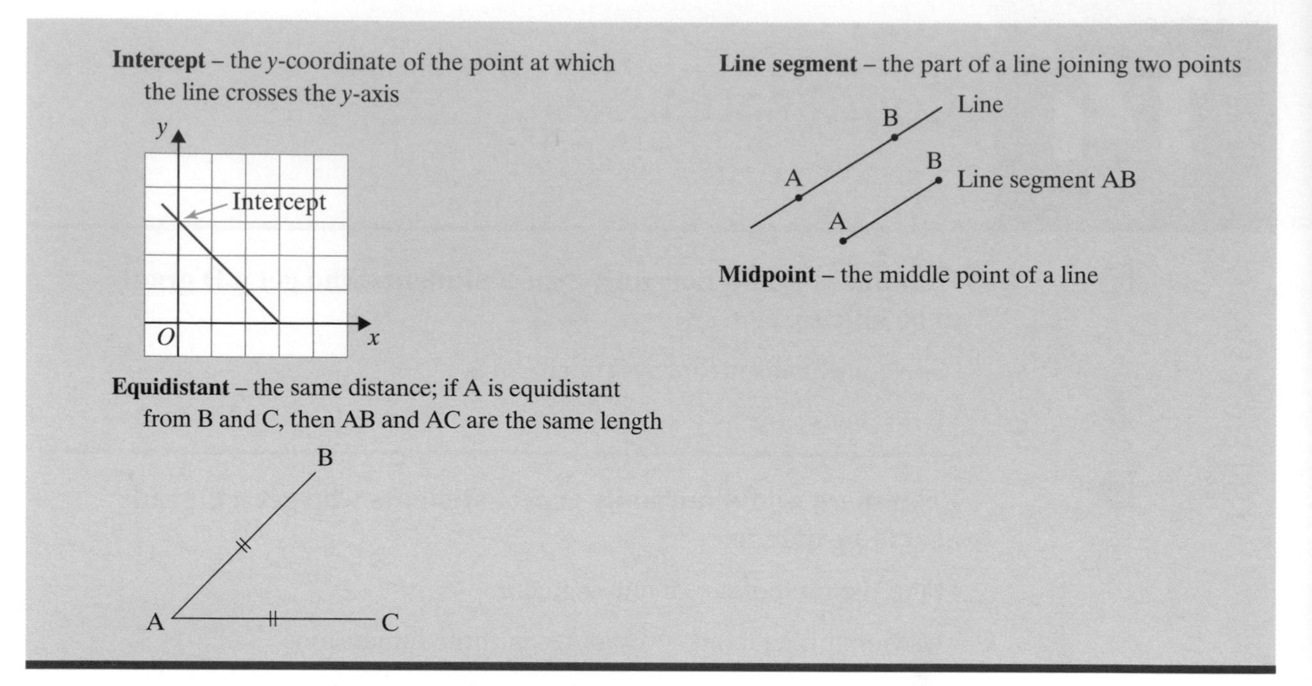

Intercept – the y-coordinate of the point at which the line crosses the y-axis

Line segment – the part of a line joining two points

Midpoint – the middle point of a line

Equidistant – the same distance; if A is equidistant from B and C, then AB and AC are the same length

Learn 1 Coordinates and lines

Example: Write the equation of the line in the diagram.

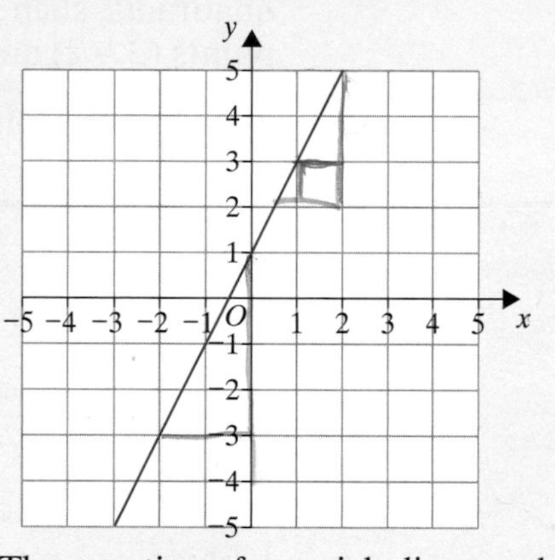

The equation of a straight line graph is

$y = mx + c$ ← c is the intercept on the y-axis

m is the gradient

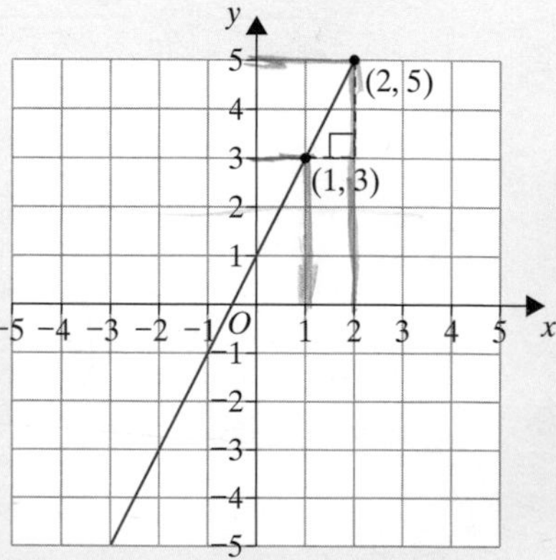

The gradient tells you how steep the line is.

You can work out the gradient if you know the coordinates of two points on the line.

$$\text{Gradient} = \frac{\text{change in vertical distance}}{\text{change in horizontal distance}}$$

$$= \frac{y_2 - y_1}{x_2 - x_1}$$

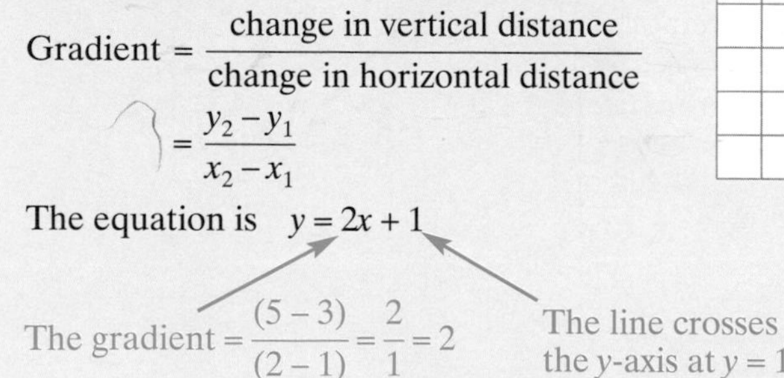

The equation is $y = 2x + 1$

The gradient $= \frac{(5-3)}{(2-1)} = \frac{2}{1} = 2$

The line crosses the y-axis at $y = 1$

Apply 1

1 Toby says the line $y = x + 2$ passes through the point (3, 1).
Colin says the line $y = x + 2$ passes through the point (1, 3).
Who is right, Toby or Colin?

2 Write the coordinates of the points where these lines cross.

a $x = 2$ and $y = 4$

b $x = 5$ and $y = -4$

c $x = -3$ and $y = -2$

d $x = 0$ and $y = 1$

e $x = 2.5$ and $y = -3.5$

f $x = 2$ and $y = x$

3 Javindra says that the line $y = x + 3$ is the same line as $x = y - 3$. Is she right?
Give a reason for your answer.

4 For each grid, write the equation of the line.

a

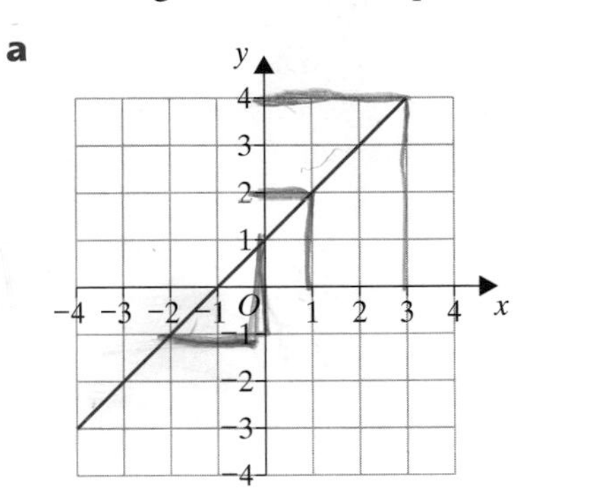

b

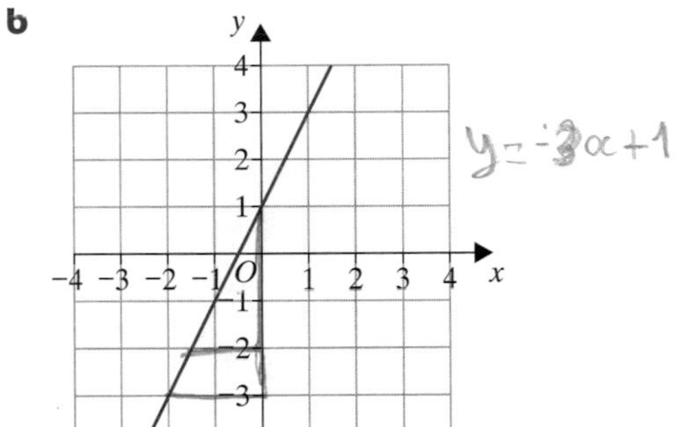

c

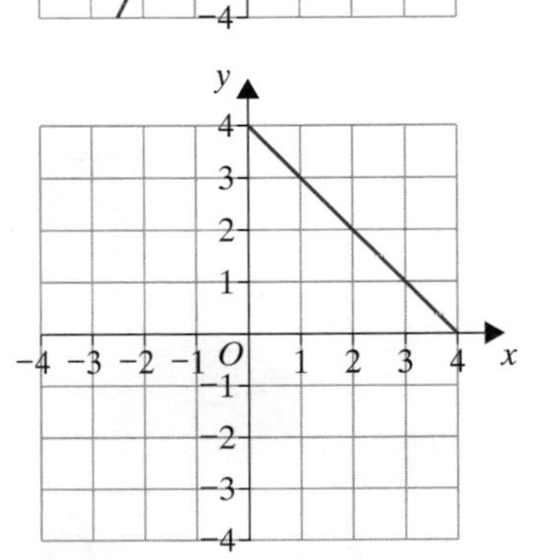

d

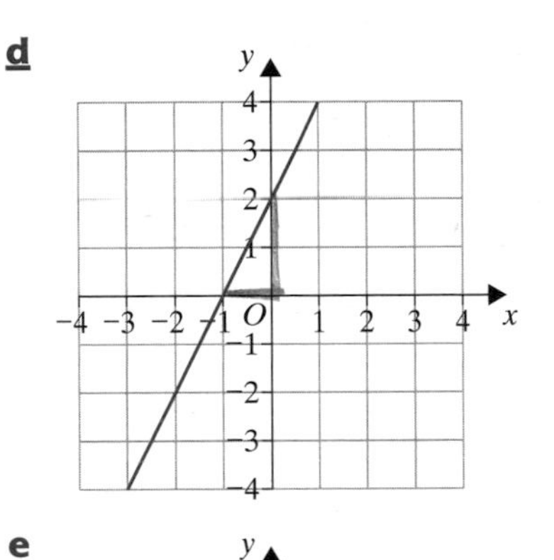

e

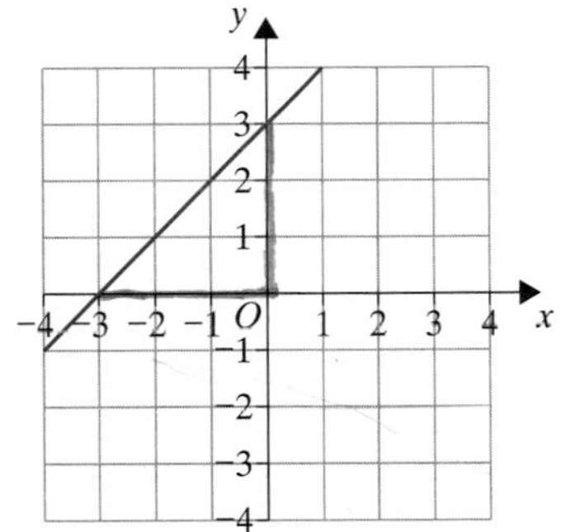

f

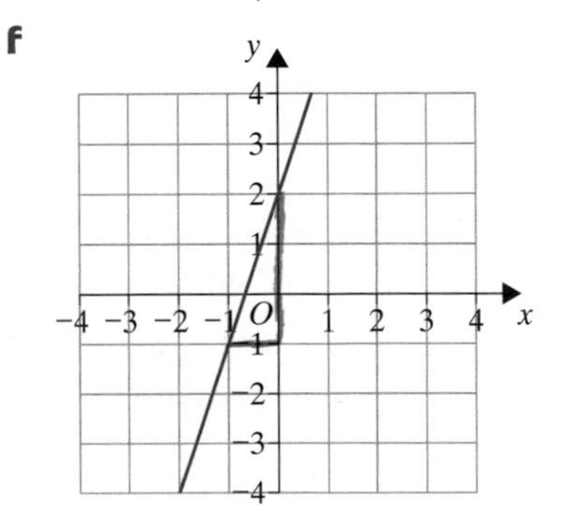

5 Write the equations of the lines A and B in the diagram.

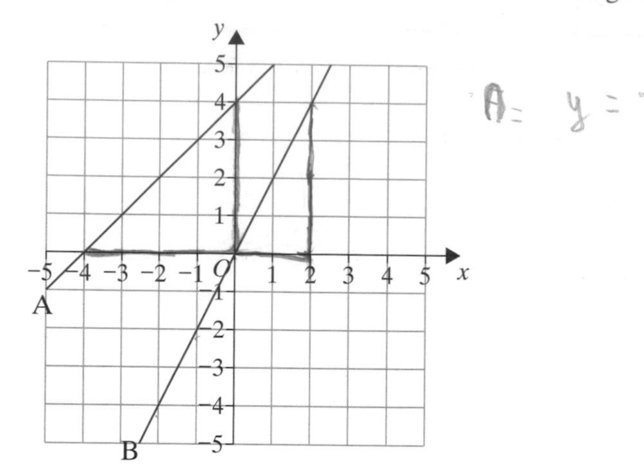

6 Get Real!

A farmer owns a field, and wants to put a fence around it.
He describes the shape as follows:

'It's a four-sided field, but it isn't a rectangle.
You can draw a plan of it like this:
Draw a coordinate grid with the x-axis and y-axis labelled from -5 to 5.
Draw in the four lines $x = 4$, $y = 5$, $y = 2x + 3$ and $x + y = -2$.'

Draw a plan of the farmer's field.

7 Where do these pairs of lines cross?

a $x = 4$ and $y = x + 2$

b $x = -2$ and $y = x - 1$

c $x = 3$ and $y = 2x$

d $x = -2$ and $y = 3x - 1$

e $y = 4$ and $y = x + 3$

f $y = -2$ and $y = 3x + 1$

8 Draw a coordinate grid with the x-axis and y-axis labelled from -6 to 6.
Draw and label these lines.

a $y = 2x + 1$

b $y = 3x - 2$

c $y = 4 - 2x$

d $y + 2x = 7$

Explore

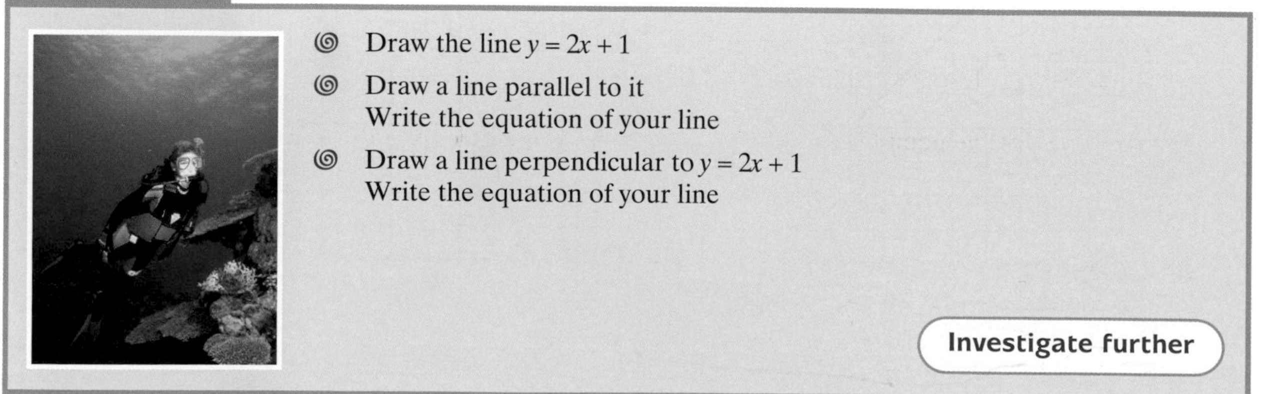

- Draw the line $y = 2x + 1$
- Draw a line parallel to it
  Write the equation of your line
- Draw a line perpendicular to $y = 2x + 1$
  Write the equation of your line

Investigate further

Learn 2 The midpoint of a line segment

Example: Write the coordinates of the point halfway between A and B.

The midpoint of the line segment joining from (a, b) to (c, d) is $\left(\frac{a+c}{2}, \frac{b+d}{2}\right)$.

The mean of the x-coordinates — The mean of the y-coordinates

A is the point $(-4, 3)$; B is $(1, 2)$.

The midpoint is at

$$\left(\frac{-4+1}{2}, \frac{3+2}{2}\right)$$

$$= \left(\frac{-3}{2}, \frac{5}{2}\right)$$

$$= (-1.5, 2.5)$$

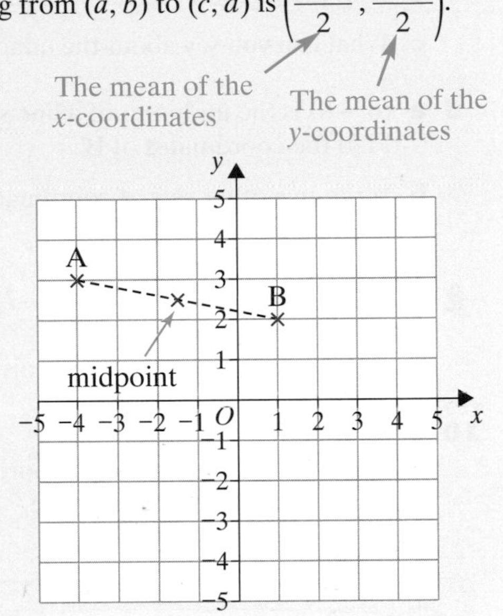

Apply 2

1 Write the coordinates of the point halfway between (1, 6) and (7, 2).

2 A triangle ABC has vertices at A(8, −4), B(2, 6) and C(−4, 2).

a Find the coordinates of X, the midpoint of AB.

b Find the coordinates of Y, the midpoint of AC.

c Draw a grid with the x-axis and y-axis labelled from −4 to 8.
Plot the points A, B, C, X and Y.

d Draw the lines XY and BC. What do you notice about them?

3 Work out the coordinates of the point halfway between (4, −5) and (1, 3).

4 If A is the point (5, −1) and B is the point (−7, −4), what are the coordinates of the midpoint of the line AB?

5 Liam says that the point (2, 1.5) is halfway between (−4, 2) and (8, −5).
Is he correct?
Give a reason for your answer.

6 C is the midpoint of the line AB.
The coordinates of C are (4, −1) and
B is the point (2, 5).
What are the coordinates of A?

7 A quadrilateral ABCD has coordinates:

A(4, −2); B(7, −4); C(−2, 4); D(−5, 6).

a Find the midpoint of the diagonal AC.

b Find the midpoint of the diagonal BD.

c What can you say about the quadrilateral ABCD?

8 **a** (6, −8) is the midpoint of a line segment AB. A is the point (3, −6). Find the coordinates of B.

b Write five other sets of coordinates that have a midpoint at (6, −8) on line segment AB.

9 Brian says that (3, 2), (9, −3), (3, −7) and (−3, −3) are the corners of a square. Tony says that he is wrong.
Find the midpoints of the diagonals to find out who is right.

10 **Get Real!**

A computer programmer uses coordinates to plot points in the screen when designing a computer game.

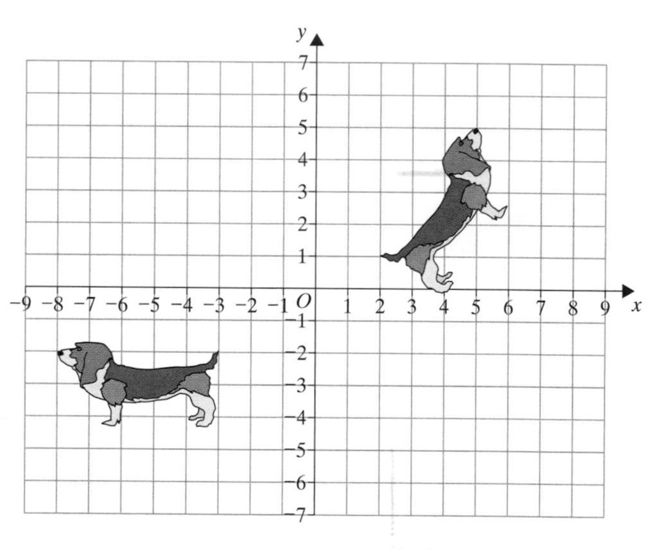

She draws a picture of a dog, with its nose at (−8, −2) and its tail at (−3, −2). She draws a reflection of the dog with its nose at (5, 5) and its tail at (2, 1). She wants to draw the mirror line for the reflection. She knows the mirror line must go halfway between the dog and its reflection.

a What are the coordinates of the point halfway between the dog's nose and its reflection?

b What are the coordinates of the point halfway between the dog's tail and its reflection?

11 Lori draws a circle with its centre at (−3, −2). She draws a diameter from the point (2, −8). Find the coordinates of the other end of the diameter.

12 Find your way through the maze by only occupying boxes where M is the midpoint of AB. Diagonal moves are not allowed.

1 **START**	2 A(3, 2) B(3, 7) M(6, 4.5)	3 A(−2, 11) B(4,−4) M(−1, 3.5)	4 A(6, −9) B(3, 9) M(4.5, 9)	5 A(−11, 29) B(−15, 22) M(−13, 25.5)	6 **END**
7 A(4, 5) B(2, 9) M(3, 7)	8 A(1, 2) B(2, 3) M(3, 5)	9 A(4, 2) B(−4, −2) M(0, 2)	10 A(−1.4, 0.2) B(1.3, 0.6) M(−0.05, 0.4)	11 A(2.3, −0.8) B(−1.3, 0.3) M(0.5, −0.25)	12 A(3, 1) B(4, 2) M(5, 3)
13 A(4, −4) B(−2, 6) M(1, 1)	14 A(8, −2) B(3, −4) M(5.5, −3)	15 A(−13, 24) B(−12, 9) M(−12.5, 17)	16 A(0, 7) B(−7, 0) M(−3.5, 3.5)	17 A(−1, 9) B(9, −1) M(4, −4)	18 A(−12, 19) B(−12, 2) M(0, 10.5)
19 A(0, 0) B(2, 3) M(2, 3)	20 A(−4, −4) B(−2, 7) M(−3, 1.5)	21 A(4, −11) B(−4, −11) M(0, 0)	22 A(−8.5, 3.4) B(4.1, 2.2) M(−2.2, 2.8)	23 A(23, −13) B(13, −12) M(18, −12.5)	24 A(4.2, −3.7) B(−2.4, 1.3) M(0.9,−1.2)
25 A(4, −12) B(−11, 8) M(−4.5, −2)	26 A(−4, 7) B(2, 5) M(−1, 6)	27 A(−3, 5) B(2, −6) M(−0.5, −0.5)	28 A(0, 9.2) B(−2.4, 3.2) M(−1.2, 6.4)	29 A(1, 5) B(3, 1) M(4, 3)	30 A(23, 17) B(−14, 11) M(4.5, 14)
31 A(4.2, 3) B(1.9, 8) M(3.1, 5.5)	32 A(−9, −8) B(−3, 9) M(6, 0.5)	33 A(−1, 4) B(−3, −4) M(−2, 0)	34 A(−3, 4) B(−2, −3) M(−2.5, 0.5)	35 A(4.2, −3.3) B(2.4, 1.3) M(3.3, −1)	36 A(−3, 2) B(2, −3) M(−0.5, −0.5)

Explore

- Write the coordinates of five pairs of points which have a midpoint at (4, 1)
- What can you say about the coordinates of any two points which have a midpoint at (x, y)?

Investigate further

Learn 3 Coordinates in three dimensions

Remember to write coordinates alphabetically: (x, y, z)

Example:

In the diagram of the cuboid
$OA = 2$ units, $AB = 5$ units and $AD = 3$ units.
O is the origin.
Write down the coordinates of A, B, C and D.

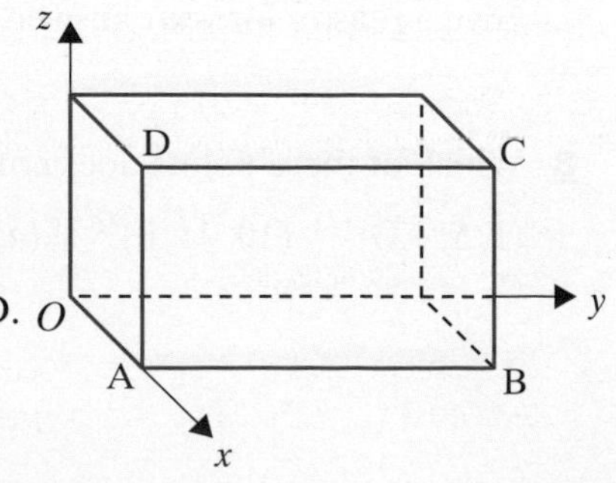

A = (2, 0, 0)   C = (2, 5, 3)
B = (2, 5, 0)   D = (2, 0, 3)

x-coordinate first, then y, then z
Draw and label diagrams to make the work easier

Apply 3

Questions 1 and 2 are about the cuboid in the diagram.

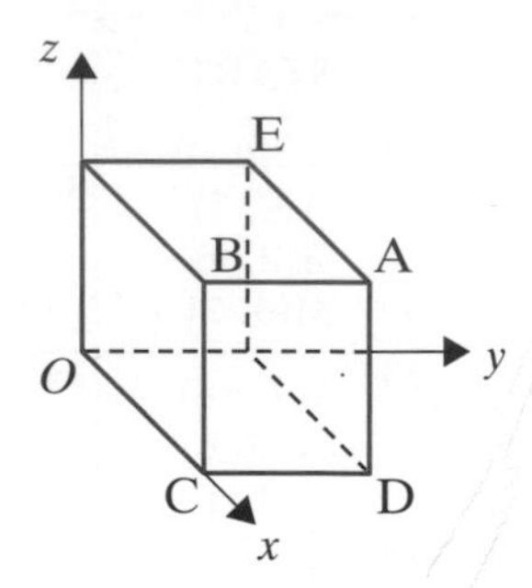

1 In the diagram, A is the point (6, 4, 5).
Write the coordinates of the points B, C, D and E.

2 An identical cuboid is placed on top of the one in the diagram.
Write the coordinates of the top corner directly above A.

3 The diagram shows a model of a shed. The point A has coordinates (2, 0, 8), and B is the point (0, 5, 6). The pentagonal end is symmetrical.

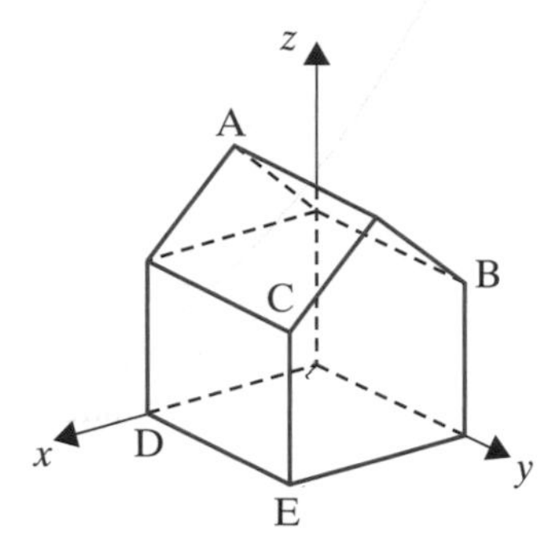

a Work out the coordinates of points C, D and E.

The model is rotated, so that vertex D is at the origin, and E lies on the y-axis.

b Work out the new coordinates of the vertices A, B, C and E.

4 A square-based pyramid is placed on top of a cube as shown. The cube has sides of length 8 units, and the pyramid has a square base of sides 8 units and a vertical height of 6 units.
If the solid is placed with a vertex of the cube at the origin, and edges running along the x-, y- and z-axes, what are the coordinates of the top of the pyramid?

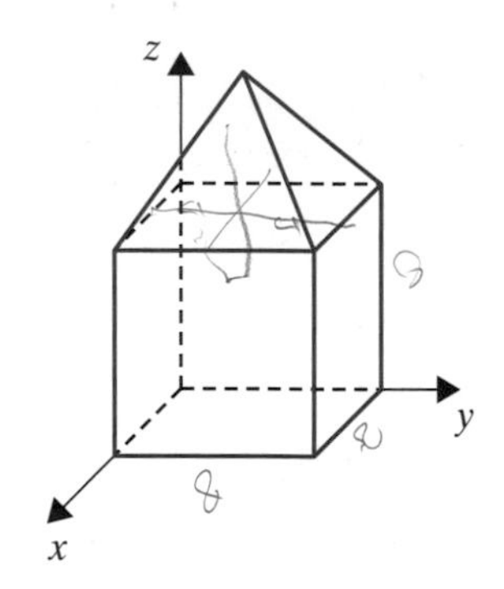

5 What is the midpoint of the line joining (6, 3, 7) to (2, 5, 9)?

6 What is the midpoint of the line joining (−2, 4, −5) to (6, 2, −6)?

7 Karen says that the points (2, 4, 5), (5, 7, 3) and (8, 10, 1) all lie on the same straight line. Is she right?
Give a reason for your answer.

8 Which of these points does not lie on the same straight line as the other three?

(1, 5, −3) (10, −1, 0) (4, 3, −2) (7, 1, −1)

9 Alex is designing a toy street. A plan of it looks like this:

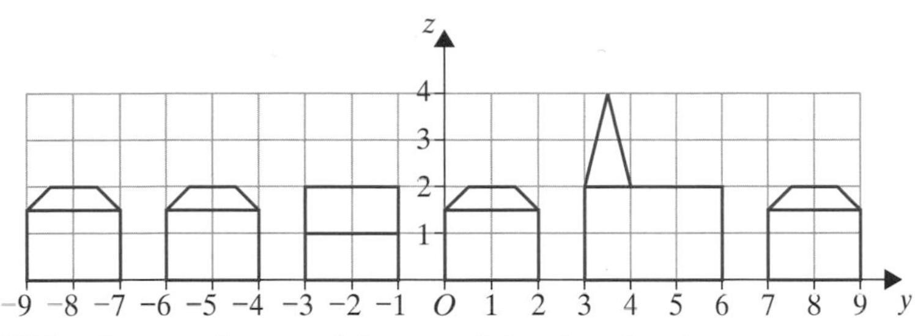

The front elevation looks like this:

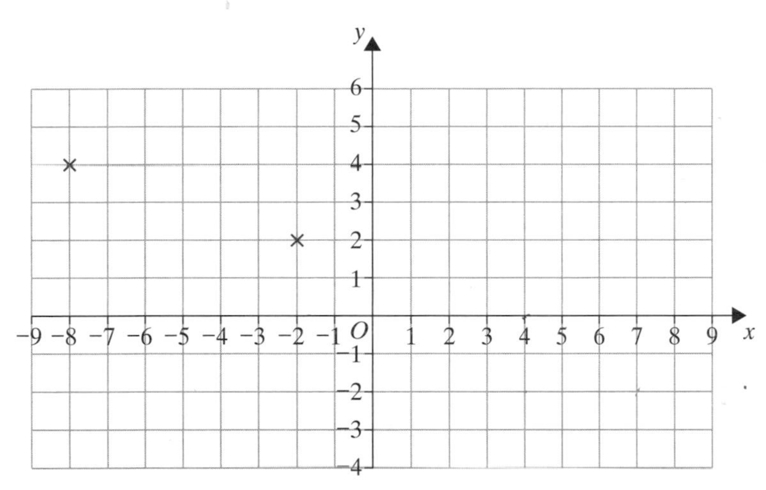

Write the coordinates of the top of the church spire.

10 Get Real!

An electricity company plans on a map where to put its electricity pylons. The results are shown below.

They plan to put the first one at $(-8, 4)$, and the second at $(-2, 2)$.

a If they space them equally and in a straight line, where will they place the third?

b The top of the first one is $(-8, 4, 1)$. Assuming they are all the same height and on level ground, what are the three-dimensional coordinates of the top of the fourth pylon?

c Halfway between the pylons, the cable sags to a height of 0.8 units. What are the coordinates of the cable halfway between the first and second pylons?

11 Get Real!

The plan below shows a bathroom. All measurements are in millimetres.

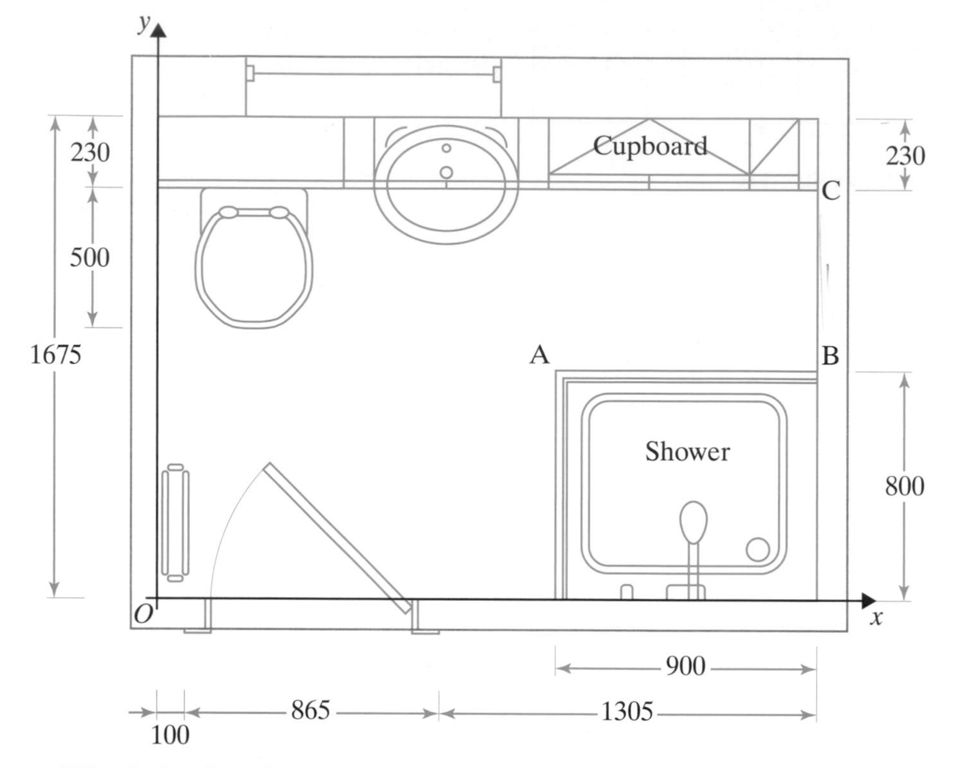

a What is the size of the gap between the shower and the cupboard (the distance BC on the plan)?

b The shower tray is 80 mm tall. The shower cubicle, which sits on top of the tray, is 2000 mm tall.
What is the height of the top of the shower cubicle from the floor?

c What are the (three-dimensional) coordinates of the top of the shower cubicle at A?

Explore

- A $2 \times 2 \times 2$ cube is placed with a corner at O, the origin, and other corners at A(2, 0, 0), B(0, 2, 0), C(2, 2, 0), D(0, 0, 2), E(0, 2, 2), F(2, 0, 2) and G(2, 2, 2)
- How many right angles can you find by joining up corners of the cube?
- How many angles of 45° can you find?
- How many angles of 60° can you find?

Investigate further

Coordinates

ASSESS

The following exercise tests your understanding of this chapter, with the questions appearing in order of increasing difficulty.

1 For each grid write the equation of the straight line.

a

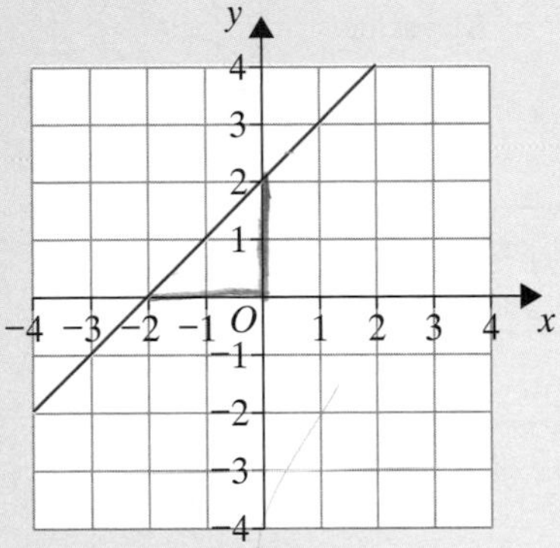

b

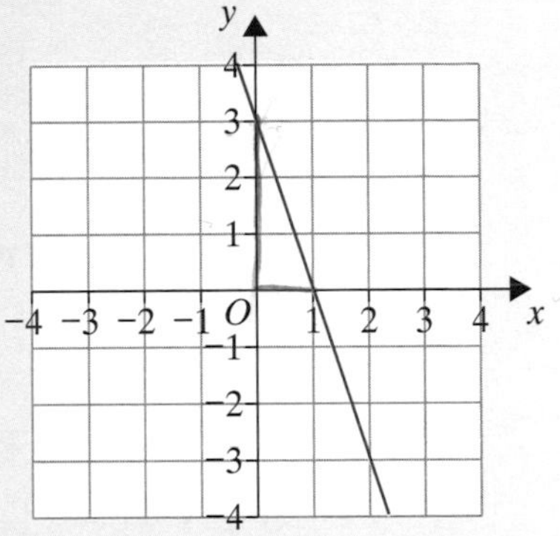

c

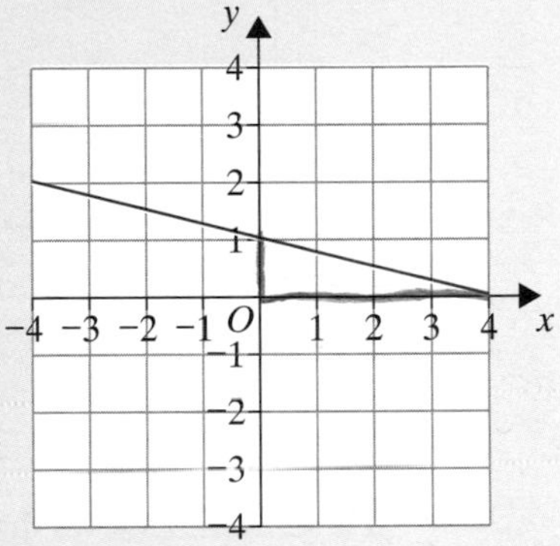

2 Draw a coordinate grid on x- and y-axes labelled from -6 to $+6$.
Draw and label these lines:

a $y = 3x + 2$

b $y = 5 - 3x$

c $y + 2x = 4$

3 Find the midpoints of the line segments.
Draw sketch diagrams to illustrate your answers.

a $(2, 3)$ and $(6, 7)$

b $(9, -9)$ and $(3, 4)$

c $(-4, 4)$ and $(6, 10)$

d $(2, 9)$ and $(-2, -9)$

e $(-4, 10)$ and $(6, -8)$

4 **a** On graph paper, plot the coordinates of the quadrilateral ABCD, given by the points $(-6, -7)$, $(-4, 8)$, $(2, 3)$ and $(5, -5)$.

b Calculate the coordinates of P, Q, R and S, the midpoints of AB, BC, CD and DA.

c Plot P, Q, R and S and draw the quadrilateral PQRS.

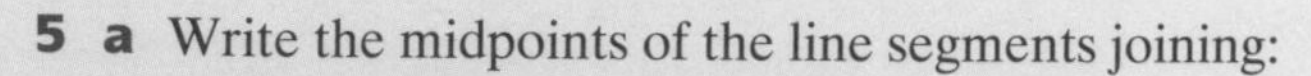

5 a Write the midpoints of the line segments joining:

i $(3, -1, 2)$ and $(4, 5, -5)$

ii $(-1, -5, -8)$ and $(2, -3, 7)$

iii $(4, -7, 2)$ and $(5, 9, -3)$

b

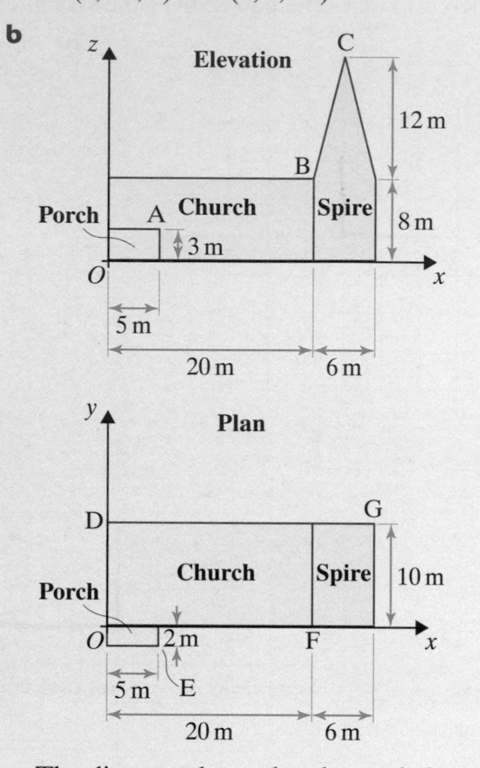

The diagram shows the plan and elevation of a church with a spire.
The x-, y- and z-axes are as shown on the diagram.
Write the coordinates of the points A to G marked on the diagram.

15 Collecting data

OBJECTIVES

D **Examiners would normally expect students who get a D grade to be able to:**

Classify and know the difference between various types of data

Use a variety of different sampling methods

Design and use data collection sheets and questionnaires

C **Examiners would normally expect students who get a C grade also to be able to:**

Identify possible sources of bias in the design and use of data collection sheets and questionnaires

A **Examiners would normally expect students who get an A grade also to be able to:**

Use stratified sampling methods

What you should already know ...

- Counting and, in particular, counting in fives (for tally charts)
- Design and use tally charts for discrete and grouped data
- Design and use two-way tables for discrete and grouped data

VOCABULARY

Tally chart – a useful way to organise the raw data; the chart can be used to answer questions about the data, for example,

Number of pets	Tally
0	~~IIII~~ IIII
1	~~IIII~~ ~~IIII~~ II
2	~~IIII~~ II
3	III
4	II

The tallies are grouped into five so that

IIII = 4
~~IIII~~ = 5
~~IIII~~ I = 6

This makes the tallies easier to read

Two-way table – a combination of two sets of data presented in a table form, for example,

	Men	Women
Left-handed	7	6
Right-handed	20	17

Quantitative data – data that can be counted or measured using numbers, for example, number of pets, height, weight, temperature, age, shoe size, etc.

Qualitative or **categorical data** – data that cannot be measured using numbers, for example, type of pet, car colour, taste, people's opinions/feelings, etc.

Discrete data – data that can only be counted and take certain values, for example, number of cars (you can have 3 cars or 4 cars but nothing in between, so $3\frac{1}{2}$ cars is not possible)

Continuous data – data that can be measured and take any value; length, weight and temperature are all examples of continuous data

Survey – a way of collecting data; there are a variety of ways of doing this, including face-to-face, or via telephone, e-mail or post using questionnaires

Respondent – the person who answers the questionnaire

Direct observation – collecting data first-hand, for example, counting cars at a motorway junction or observing someone shopping

Primary data – data that you collect yourself; this is new data and is usually gathered for the purpose of a task or project

Secondary data – data that someone else has collected; this might include data in books, newspapers, magazines, etc. or data that has been loaded onto a database

Data collection sheets – these are used to record the responses to the different questions on a questionnaire; they can also be used with computers to load data onto a database

Pilot survey – a small-scale survey to check for any unforeseen problems with the main survey

Convenience or **opportunity sampling** – a survey that is conducted using the first people who come along, or those who are convenient to sample (such as friends and family)

Random sampling – this requires each member of the population to be assigned a number; the numbers are then chosen at random

Systematic sampling – this is similar to random sampling except that it involves every nth member of the population; the number n is chosen by dividing the population size by the sample size

Quota sampling – this method involves choosing a sample with certain characteristics, for example, select 20 adult men, 20 adult women, 10 teenage girls and 10 teenage boys to conduct a survey about shopping habits

Cluster sampling – this is useful where the population is large and it is possible to split the population into smaller groups or clusters

Stratified sampling – this involves dividing the population into a series of groups or 'strata' and ensuring that the sample is representative of the population as a whole, for example, if the population has twice as many boys as girls, then the sample should have twice as many boys as girls

Learn 1 Collecting data

Examples:

a Categorise each of the following:

i Weight of whales in an aquarium
ii Favourite make of car of students in the sixth form
iii Number of bottles of lemonade sold at a store each day
iv Best pop group in the charts.

	Quantitative	Qualitative	Continuous	Discrete
i Weight of whales in an aquarium	✓		✓	
ii Favourite make of car of students in the sixth form		✓		
iii Number of bottles of lemonade sold at a store each day	✓			✓
iv Best pop group in the charts		✓		

b Write down one advantage and one disadvantage of the following methods of collecting data.

i Personal surveys.
ii Postal surveys.
iii Direct observations.

i This is the most common method of collecting data and involves an interviewer asking questions of the interviewee. This method is sometimes called a face-to-face interview.

Advantages
- The interviewer can ask more complex questions and explain them if necessary.
- The interviewer is likely to be more consistent when they record the responses.
- The interviewee is more likely to answer the questions than with postal or e-mail surveys.

Disadvantages
- This method of interviewing takes a lot of time and can be expensive.
- The interviewer can influence the answers and this may cause bias.
- The interviewee is more likely to lie or to refuse to answer a question.

The telephone survey is a special case of the personal survey with similar advantages and disadvantages

ii Postal surveys make use of mailing lists (or the electoral register) and involve people being selected and sent a questionnaire.

Advantages
- The interviewees can take their time answering and give more thought to the answer.
- The possibility of interviewer bias is avoided.
- The cost of a postal survey is usually lower.

Disadvantages
- Postal surveys suffer from low response rates which may cause bias.
- The process can take a long time to get questionnaires out and await their return.
- Different people might interpret questions in different ways when giving their answers.

The e-mail survey is a special case of the postal survey and an increasingly popular method for collecting data

iii Direct observation, as the name implies, means observing the situation directly. Direct observation might include, for example, counting cars at a motorway junction or observing someone to see what shopping they buy. It can take place over a short or a long period of time.

Advantages
- Direct observation is a reliable method which allows observations to take place in the interviewees 'own environment'.

Disadvantages
- For some experiments, the interviewee may react differently because they are being observed.
- This method takes a lot of time and can be expensive.

c Write down three requirements of a good questionnaire.

- **Appropriate** to the survey being carried out and not asking unnecessary questions. — For example, asking someone for their address may not be appropriate for most questionnaires ... unless it is a survey on where people live
- **Unbiased** so that they do not lead the respondent to give a particular answer. — For example, asking the question 'Do you believe we should have a new shopping centre?' is a biased (leading) question
- **Unambiguous** so that they are clear and straightforward to the respondent. — For example, asking the question 'Do you agree or disagree that we should have a new shopping centre?' is not very clear

Apply 1

1 For each of the following say whether the data is quantitative or qualitative.

a The number of people at a cricket test match.

b The weights of newborn babies.

c How many cars a garage sells.

d Peoples' opinions of the latest Hollywood blockbuster.

e The best dog at Crufts.

f The time it takes to run the London Marathon.

g The colour of baked beans.

h How well your favourite football team played in their last match.

i The number of text messages received in a day.

2 For each of the following say whether the data is discrete or continuous.

a The number of votes for a party at a general election.

b The ages of students in your class.

c The number of beans in a tin.

d The weight of rubbish each household produces each week.

e How many people watch the 9 o'clock news.

f How long it takes to walk to school.

g The number of sheep Farmer Angus has.

h The weights of Farmer Angus' sheep.

i The heights of Year 10 students in your school.

3 Connect the following to their proper description.
The first one has been done for you.

Lengths of fish caught in a competition
Number of goals scored by a football team
Ages of teachers at a school
Favourite colours in the tutor group
Number of sweets in a bag
Favourite type of music
Person's foot length
Person's shoe size
Cost of stamps
Best player on the Wales rugby team

Quantitative and discrete
Qualitative
Quantitative and continuous

4 For each of the following questions, identify whether the information requested is:

- quantitative and discrete
- quantitative and continuous
- qualitative.

Questionnaire

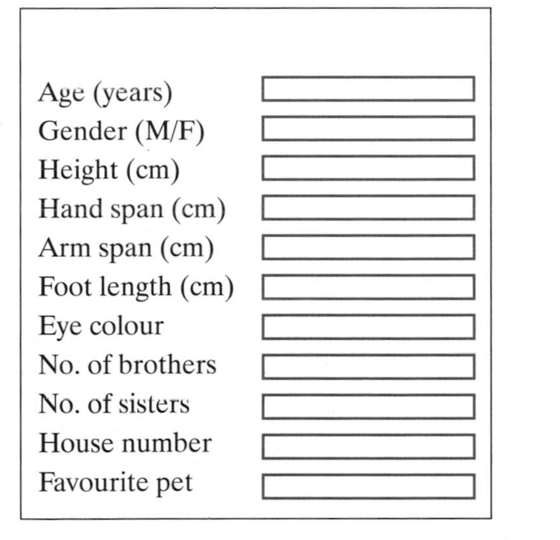

Age (years)
Gender (M/F)
Height (cm)
Hand span (cm)
Arm span (cm)
Foot length (cm)
Eye colour
No. of brothers
No. of sisters
House number
Favourite pet

5 The following questions are taken from different surveys.
Write down one criticism of each question.
Rewrite the question in a more suitable form.

a How many hours of TV do you watch each week?

Less than 1 hour ☐ More than 1 hour ☐

b What is your favourite football team?

Real Madrid ☐ Luton Town ☐

c How do you spend your leisure time? (You can only tick one box.)

Doing homework ☐ Playing sport ☐ Reading ☐

Computer games ☐ On the Internet ☐ Sleeping ☐

d You do like football, don't you?

Yes ☐ No ☐

e How much do you earn each year?

Less than £10 000 ☐ £10 000 to £20 000 ☐ More than £20 000 ☐

f How often do you go to the cinema?

Rarely ☐ Sometimes ☐ Often ☐

g Do you or do you not travel by taxi?

Yes ☐ No ☐

6 Peter and Paul are writing questions for a research task.

a This is one question from Peter's questionnaire:

> Skateboarding is an excellent pastime. Don't you agree?
> Tick one of the boxes.
>
> Strongly agree ☐ Agree ☐ Don't know ☐

Write down two criticisms of Peter's question.

b These questions are from Paul's questionnaire:

> Do you buy CDs? ☐ Yes ☐ No
>
> If yes, how many CDs do you buy on average each month?
>
> ☐ 2 or less ☐ 3 or 4 ☐ 5 or 6 ☐ more than 6

Write down two reasons why these are good questions.

7 Write down a definition for:

a quantitative data

b qualitative data

c continuous data

d discrete data.

8 Write down two advantages of undertaking a pilot survey.

9 Write down five things that make a good questionnaire.

Explore

Here are some investigations for you to consider.

- A company slogan states 'In tests 9 out of every 10 cats prefer it'
- Investigate ways in which the company could have arrived at this answer
- You might consider carrying out your own survey
- How might you get such results?

- In a supermarket poll of 100 people 85 said they had at least two cars!
- Is this correct?
- Investigate ways in which the company could have arrived at this answer
- How might you get such results?

Investigate further

Learn 2 Sampling methods

Examples:

You are given a list of 500 students (200 boys and 300 girls) and wish to choose a sample of 50.

Explain how you would use the following sampling methods.

a Convenience sampling
b Random sampling
c Systematic sampling
d Quota sampling
e Cluster sampling
f Stratified sampling

a Convenience sampling or opportunity sampling means that you just take the first people who come along or those who are convenient to sample. The likelihood is that you choose the first 50 students that you meet or otherwise choose 50 students from among your friends.

b Random numbers can be taken from random number tables or generated by a calculator using the RAN or RND button. Assign each student a number between 0 and 499 and generate random numbers to choose 50 students.

c Systematic sampling is similar to random sampling except that you take every nth member of the population. The value n is found by dividing the population size by the sample size giving $\frac{500}{50} = 10$ so that every 10th student is chosen from the population. A random number is used to start so that the number 4 would suggest taking the 4th, 14th, 24th, 34th... students.

d Quota sampling is popular in market research and involves choosing a sample with certain characteristics (the choice of who to ask is left to the interviewer). The likely requirement is that you are asked to sample 20 boys and 30 girls.

e Cluster sampling is useful where the population is large and it is possible to split the population into smaller groups or clusters. The most obvious choice is to consider tutor groups as clusters and sample the whole of two tutor groups, although this is not likely to result in exactly 50 students.

f Stratified sampling involves dividing the population into a series of groups or 'strata' and ensuring that the sample is representative of the population as a whole. For example, if the population has twice as many boys as girls then the sample should have twice as many boys as girls.

You choose 20 boys and 30 girls randomly using one of the above sampling methods. Stratified sampling would also take account of different year groups and these would need to be divided in a similar ratio.

g The table shows the number of students in each year group of a school.

Year	7	8	9	10	11
Number	200	200	240	220	140

A stratified sample ensures that the sample is representative of the population as a whole. To take a stratified sample, you need to appreciate that 200 students out of 1000 students are from Year 7, so that $\frac{200}{1000}$ of the sample should be from Year 7

Completing this information for each year group:

Year	7	8	9	10	11
Number	200	200	240	220	140
Fraction	$\frac{200}{1000}$	$\frac{200}{1000}$	$\frac{240}{1000}$	$\frac{220}{1000}$	$\frac{140}{1000}$
For a sample size of 50	$\frac{200}{1000} \times 50$ = 10 students	$\frac{200}{1000} \times 50$ = 10 students	$\frac{240}{1000} \times 50$ = 12 students	$\frac{220}{1000} \times 50$ = 11 students	$\frac{140}{1000} \times 50$ = 7 students

If the required sample size is 50, then $\frac{200}{1000} \times 50$ will be from Year 7, that is 10 students from Year 7

When completing this information for each section, remember to check that the totals for each year group add up to the sample size of 50

Apply 2

1 The following table shows the ordered ages for 100 people in a London shopping centre.

22	22	22	22	22	22	23	23	24	24
24	24	24	24	24	24	24	25	25	25
25	25	25	25	25	26	26	26	26	26
26	26	26	26	26	27	27	27	27	27
28	28	28	28	28	28	28	28	29	29
29	29	29	30	31	31	31	31	31	31
32	32	33	33	33	33	33	33	34	34
34	34	34	34	34	34	34	34	35	35
35	35	35	36	36	36	36	36	36	36
37	38	40	40	42	42	44	47	48	50

a Take a random sample of 20 people using the RAN or RND button on your calculator.
Find the mean of this sample.

b Take a systematic sample of 20 people using every fifth number.
Find the mean of this sample.

c Take a cluster sample of the first 20 people in the table.
Find the mean of this sample.

d Which sample do you think is most representative of the 100 people?
Give a reason for your answer.

2 Which sampling method is most appropriate for each of the following surveys? Give a reason for your answer.

a The amount of pocket money students receive in different year groups at your school.

b The favourite programmes of 15 to 19 year olds.

c The average life expectancy of people around the world.

d The favourite holiday destinations of people in the sixth form at your school.

e The opinions of two hundred 25 to 39 year olds on their favourite soap opera.

f Information on voting intentions at a general election.

5-4

3 Consider each of the following surveys and say whether the sample is representative. Give a reason for your answer.

a Aiden is trying to find out what students in the school think about school dinners. He decides to use cluster sampling, asking the first 30 Year 7s as they leave the dining room.

b Betty wants to find out the most popular names for newborn babies – she goes to the Internet, finds the relevant website and takes the first 100 names on the list.

c Cameron wants to find out if all premiership footballers think that having overseas players is a good thing. He decides to take a 10% sample and ask all the players at Chelsea and Arsenal.

d Davina wishes to check how many people travel on the underground. She telephones 100 people at home in the evening and asks them if they have travelled on the underground that week.

e Eric undertakes a convenience sample of 20 friends to see if they wear glasses. He says that 45% of students at his college wear glasses.

4 Explain the difference between a random sample and a systematic sample.

5 Find two articles containing around 200 words each.
Find which article contains longer words.

a Choose an appropriate sampling method and give a reason for your choice.

b Choose an appropriate sample size and give a reason for your choice.

6 A college wishes to undertake a survey on the eating habits of its students.
Explain how you would take:

a a random sample of 100 students

b a systematic sample of 100 students

c a stratified sample of 100 students.

7 The table shows the number of students in each year group of a college.

Year	10	11	12	13
Number	300	300	220	180

Explain how you would take a stratified sample of size 50.

8 The table shows the number of people employed in a department store.

Occupation	Management	Sales	Security	Office
Number	10	130	25	35

Explain how you would take a stratified sample of size:

a 40 **b** 50

9 Write one advantage and one disadvantage of each of these sampling methods.

a Convenience sampling

b Random sampling

c Systematic sampling

d Quota sampling

e Stratified sampling

Explore

- Design your own questionnaire and undertake a survey
- You should consider a hypothesis and write some suitable questions
- Undertake a pilot survey
- Amend your questionnaire and explain any improvements
- Consider a suitable sampling method – explain why you have chosen this method
- Carry out the survey and comment on your results

Investigate further

Collecting data

ASSESS

The following exercise tests your understanding of this chapter, with the questions appearing in order of increasing difficulty.

1 For each of the following, state whether the data is quantitative or qualitative.

a The heights of the people watching a tennis match.

b The colours of cars in a car park.

c The sweetness of different orange juices.

d The lengths of pencils in a pencil case.

e The numbers of students in different classes.

f The numbers of leaves on different types of trees.

g The ages of people at a night club.

h The musical abilities of students in a class.

2 For each of the following, state whether the data is continuous or discrete.

a The heights of the people watching a tennis match.

b The times taken by athletes to complete a race.

c The numbers of sweets in a sample of packets.

d The lengths of pencils in a pencil case.

e The numbers of students in different classes.

f The numbers of leaves on different types of trees.

g The shoe sizes of people at a party.

h The ages of people at a night club.

3 a Criticise each of the following questionnaire questions.

i How many hours of television have you watched in the last 2 months?

ii Do you or do you not watch news programmes?

iii How often do you have a shower?

b Criticise each of the following questionnaire questions and suggest alternatives to find out the required information.

i What do you think about our new improved fruit juice?

ii How much do you earn?

iii Do you or do you not agree with the new bypass?

iv Would you prefer to sit in a non-smoking area?

4 a Explain why each of the following may not produce a truly representative sample.

i Selecting people at random outside a supermarket.

ii Selecting every 10th name from an electoral register starting with M.

iii Selecting names from a telephone directory.

b In a telephone poll conducted one morning, 20 people were asked whether they regularly used a bus to get to work. Give three reasons why this sample might not be truly representative.

5 a Explain how you could use the electoral register to obtain a systematic sample.

b A college wishes to undertake a survey on the part-time employment of its students. Explain how you would take:

i a random sample of 100 students

ii a stratified sample of 100 students.

6 The table shows the number of people employed in a factory.

Occupation	Management	Office	Sales	Shop floor
Number	10	15	30	145

a Explain why a random sample of the employees might not be suitable.

b Explain how you would take a stratified sample of size 40.

c Explain how you would take a stratified sample of size 50.

16 Percentages

OBJECTIVES

D **Examiners would normally expect students who get a D grade to be able to:**

Increase or decrease a quantity by a given percentage

Express one quantity as a percentage of another

C **Examiners would normally expect students who get a C grade also to be able to:**

Work out a percentage increase or decrease

B **Examiners would normally expect students who get a B grade also to be able to:**

Understand how to use successive percentages

Work out compound interest

Work out reverse percentage problems

What you should already know ...

- Place values in decimals and putting decimals in order of size
- How to express fractions in their lowest terms (or simplest form)
- How to change between fractions, decimals and percentages
- Work out a percentage of a given quantity

VOCABULARY

Percentage – a number of parts per hundred, for example, 15% means $\frac{15}{100}$

Numerator – the number on the top of a fraction

Numerator → $\frac{3}{8}$ ← Denominator

Denominator – the number on the bottom of a fraction

Multiplier – a number used to multiply an amount

Interest – money paid to you by a bank, building society or other financial institution if you put your money in an account or the money you pay for borrowing from a bank

Simple interest – pays interest only on the sum of money originally invested

Compound interest – pays interest on both the original sum and the interest already earned

Principal – the money put into the bank or borrowed from the bank

Rate – the percentage at which interest is added, usually expressed as per cent per annum (year)

Time – usually measured in years for the purpose of working out interest

Amount – the total you will have in the bank or the total you will owe the bank, at the end of the period of time

Balance – the amount of money you have in your bank account or the amount of money you owe after you have paid a deposit

Deposit – an amount of money you pay towards the cost of an item, with the rest of the cost to be paid later

Discount – a reduction in the price, perhaps for paying in cash or paying early

VAT (Value Added Tax) – a tax that has to be added on to the price of goods or services

Depreciation – a reduction in value, for example, due to age or condition

Credit – when you buy goods 'on credit' you do not pay all the cost at once; instead you make a number of payments at regular intervals, often once a month

Learn 1 Increasing or decreasing by a given percentage

Examples:

a Parveen's bus fare to town is 80p. The bus fares go up by 5%. How much is the new fare?

Non-calculator	Calculator
10% of 80p = 8p 5% of 80p = 4p New fare = 84p	Original fare = 100% New fare = (100 + 5)% = 105% = 1.05 New fare = 1.05 × 80p = 84p

b Find the new price of a £350 TV after a 4% reduction.

Non-calculator	Calculator
1% of £350 = £3.50 4% of £350 = £3.50 × 4 = £14 New price = £350 – £14 = £336	Original price = 100% New price = (100 – 4) % = 96% = 0.96 New price = 0.96 × £350 = £336

Apply 1

1 Increase 25 cm by 10%.

2 Decrease 700 g by 5%.

3 Decrease £450 by 20%.

4 Increase £3 by 8%.

5 **Get Real!**
Todd is paid £300 per week.
He gets a 4% pay rise.
What is his new weekly pay?

6 Get Real!
A package holiday is priced at £660.
Gary gets a 10% discount for booking before the end of January.
How much does he pay?

7 Get Real!
Emma gets a 15% discount on purchases from Aqamart.
How much does she pay for a TV priced at £500?

8 Get Real!
A music centre costs £280.
VAT at $17\frac{1}{2}$% has to be added to the bill.
What is the total cost of the music centre?

9 Get Real!
Jared buys a jacket in a sale.
The price ticket says £70.
There is a label on the rack saying 'Take 25% off all marked prices'.
How much will Jared pay for the jacket?

10 Increase 125 cm by 16%.

11 Increase 340 g by 9%.

12 Decrease £560 by 22%.

13 Decrease £9.55 by 12%.

14 Get Real!
The population of Baytown was 65 970 in 1990.
By the year 2000, Baytown's population had gone up by 27%.
What was the population in 2000?

15 Get Real!
Becky buys a new car for £12 499.
Over 2 years, it depreciates by 45%.
What is the value of the car after 2 years?

16 Get Real!
A garden shed is for sale at '£550 + VAT'.
If VAT is $17\frac{1}{2}$%, what is the total cost of the shed?

<u>17</u> Get Real!
The bill for a repair is £57.35
VAT at $17\frac{1}{2}$% has to be added to the bill.
What is the total cost of the repair?

18 Paul needs to increase 45 kg by 5%.
He writes down $45 \times 1.5 = 67.5$ kg.
Is he correct?
Give a reason for your answer.

Explore

- Jo wants to buy a music centre priced at £650
- She has to put down £100 as a deposit
- There are two ways she can pay the rest of the price (the balance)
  1 The EasyPay Option:
    - 14% credit charge on the balance
    - 12 equal monthly payments

  2 The PayLess Option:
    - 3% added each month to the amount owing at the beginning of the month
    - pay £50 per month until the balance is paid off

  (Note: in the last month Jo will only pay the remaining balance, not a full £50)
- Using a calculator, investigate these two options to advise Jo which one is best
- Would your advice be different if EasyPay charged 11% or PayLess charged $3\frac{1}{2}$% each month?

Investigate further

Learn 2 Using successive percentage changes

Examples:

a There are 200 fish in a pond. 75% of them are goldfish.
36% of these goldfish are less than 6 cm long.
How many fish in the pond are goldfish and less than 6 cm long?

First find 75% of 200: 50% of 200 = 100
25% of 200 = 50
75% of 200 = 150 fish are goldfish

Then find 36% of 150: 36% of 100 = 36
36% of 50 = 18
36% of 150 = 54 goldfish are less than 6 cm long

For successive percentages, work out the first percentage and use your answer to work out the second percentage (usually the best method if you are not allowed to use a calculator)

b Paul buys a car for £25 000. It depreciates by 35% in the first year and 20% in the second year. What is it worth after two years?

The original price is 100%
In the first year, it becomes (100 – 35)% which is 65%
£25 000 × 0.65 = £16 250

At the start of the second year, the price is £16 250 (The new 100%)
This becomes (100 – 20)% which is 80%
£16 250 × 0.80 = £13 000

After two years it is worth £13 000

Using multipliers this can be worked out on a calculator as:
£25 000 × 0.65 × 0.80 = £13 000

Apply 2

1 There are 800 students at Uptown College. 20% of them have a driving licence. 35% of those with driving licences drive to college.
How many students drive to college?

2 In London, 250 teachers attend a meeting.
60% of them are women.
18% of these women are under 30 years old.
How many of the women are under 30?

3 Kate earns £900 per month. On average, she spends 25% of this at the supermarket. 60% of her supermarket spending is on food.
How much does Kate spend on food each month?

4 Jack bought shares worth £4000 in January 2002. By January 2003, the value of his shares had risen by 15%. Between January 2003 and January 2004, their value rose by 5%.
What was the value of the shares in January 2004?

5 Alice is answering question **4** above. She writes:
15% + 5% equals 20%
20% is the same as one fifth
One fifth of £4000 is £800
What mistake has Alice made?

6 Ellie was paid £8.00 per hour in 2003. In 2004, she got a pay rise of 5%, and in 2005, her pay rise was $2\frac{1}{2}$%.
How much was Ellie paid at the end of 2005?

7 Which of these is bigger:

a a 25% increase followed by a 10% decrease

b a 10% decrease followed by a 25% increase?

8 Assad says that a decrease of 20% followed by a decrease of 10% is the same as a decrease of 30%.
Is Assad correct? Give a reason for your answer.

9 The population of Mellowby village is 1250.
44% of them are pensioners.
26% of the pensioners live alone.
How many pensioners live alone in Mellowby?

10 There were 160 people at a concert.
55% of them bought a programme.
$12\frac{1}{2}$% of these programmes were left behind after the concert.
How many programmes were left behind?

11 Marie bought a painting for £750. In the first year, the value of the painting went up by 16% and in the second year it went down by 35%.
Find the value of the painting after two years.

12 Brad bought shares worth £15 000 in April 2001. By April 2002, the value of his shares had dropped by 8.2%. In the following year, the value of the shares dropped by 11.5%.
What were Brad's shares worth in April 2003?

13 A puppy weighed 3.2 kg at 8 weeks old. His weight went up by 6% in the next week and by 7% in the following week.
What did he weigh when he was 10 weeks old?

Explore

- Jack's beanstalk grows 35% each month
- In January it is 12 m high
- How many months does it take to reach 50 m?
- When does it reach 100 m?

Investigate further

Learn 3 Compound interest

Examples:

a Find the compound interest on £3000 invested for 2 years at 5% per annum.

	Principal	Interest
Year 1	£3000	5% of £3000 = $\frac{5}{100} \times$ £3000 = 150
Year 2	£3000 + £150 = £3150	5% of £3150 = $\frac{5}{100} \times$ £3150 = 157.5

This is £157.50

Total interest = £150 + £157.50 = £307.50

b Find the compound interest on £8000 invested for 3 years at 4.6% per annum.

The compound interest formula is:

$$\text{Amount} = \text{Principal} \times \left[1 + \frac{\text{Rate}}{100}\right]^n \quad \text{where } n \text{ is the number of years}$$

	Principal	Interest by 4.6%
Year 1	£8000	£8000 × 1.046 = £8368
Year 2	£8368	£8368 × 1.046 = £8752.928
Year 3	£8752.928	£8752.928 × 1.046 = £9155.5627

Amount = £9155.56
Interest = £9155.56 − £8000 = £1155.56

This has to be rounded to £9155.56

In this example, successive principals were multiplied by 1.046 so the final amount = £8000 × 1.046 × 1.046 × 1.046 or £8000 × 1.046^3 = £9155.5627

Apply 3

1 Find the compound interest on £500 invested for 2 years at 4% per annum.

2 Find the compound interest on £2000 invested for 2 years at 7% per annum.

3 £4000 is invested at 3% per annum compound interest. Find the amount at the end of 2 years.

4 £1000 is invested at 10% per annum compound interest. Find the amount at the end of 3 years.

5 **Get Real!**
The value of machinery in a factory depreciates by 20% each year.
The machinery was bought for £74 500.
What was its value after 2 years?

6 Find the compound interest on £5850 invested for 3 years at 3.4% per annum.

7 Find the compound interest on £2000 invested for 5 years at $7\frac{1}{2}$% per annum.

8 £14 000 is invested at $9\frac{1}{4}$% per annum compound interest. Find the amount at the end of 6 years.

9 Jo borrows £4500 for 3 years at 3% per annum simple interest. Kevin borrows the same amount at 3% per annum compound interest. How much more does Kevin have to pay back?

10 **Get Real!**
Liam has inherited some money from his grandmother.
He wants to invest it for 5 years.
He could put it in AqaBank or in SmartMoney.
AqaBank pays 5.27% per annum simple interest.
SmartMoney pays 4.81% per annum compound interest.
Where should Liam invest his inheritance to earn the most interest?

11 £6000 is invested at $4\frac{3}{4}$% per annum compound interest. After how many years will this amount to more than £7000?

12 **Get Real!**
A bird colony is decreasing at 16% per annum.
If the original population was 600 birds, after how many years will there be fewer than 200 birds left?

Learn 4 Expressing one quantity as a percentage of another and finding a percentage increase or decrease

Examples:

a Express 84p as a percentage of £20.

Make sure both quantities are in the **same units**
Write them as a fraction and multiply by 100 to change to a percentage

Working in pounds, change 84p to £0.84

0.84 as a fraction of 20 is $\frac{0.84}{20}$

To convert a fraction to a percentage you multiply by 100:

$$\frac{0.84}{20} = \frac{0.84}{20} \times 100\%$$

$$= \frac{0.84}{\cancel{20}_1} \times \cancel{100}^5\%$$

$$= 4.2\%$$

Working in pence, change £20 to 2000p

84 as a fraction of 2000 is $\frac{84}{2000}$

To convert a fraction to a percentage you multiply by 100:

$$\frac{84}{2000} = \frac{84}{2000} \times 100\%$$

$$= \frac{\cancel{84}^{42}}{\cancel{2000}_{\cancel{1000}_{10}}} \times \cancel{100}^1\%$$

$$= 4.2\%$$

b Find the percentage increase when the temperature goes up from 20°C to 26°C.

Temperature increase = 6°
6 as a fraction of 20 is $\frac{6}{20}$
To convert a fraction to a percentage you multiply by 100:

$$\frac{6}{20} = \frac{6}{20} \times 100\%$$

$$= \frac{6}{\cancel{20}_1} \times \cancel{100}^5\%$$

$$= 30\% \text{ increase}$$

Make sure both quantities are in the **same units**

Write this as a fraction and multiply by 100 to change to a percentage **(The original quantity has to be on the bottom of the fraction)**

c Find the percentage decrease when the price of a toy falls from £12.50 to £11.75

Price decrease = £0.75
£0.75 as a fraction of £12.50 is $\frac{0.75}{12.50}$
To convert a fraction to a percentage you multiply by 100:

$$\frac{0.75}{12.50} = \frac{0.75}{12.50} \times 100\%$$

$$= \frac{0.75}{\cancel{12.50}_{\cancel{25}_1}} \times \cancel{100}^{\cancel{200}^8}\%$$

$$= 6\% \text{ decrease}$$

Apply 4

1 Express 12p as a percentage of 16p.

2 Express 20 kg as a percentage of 50 kg.

3 Express 85 mm as a percentage of 10 cm.

4 Express 42p as a percentage of £7.

5 Express 385 g as a percentage of 35 kg.

6 **Get Real!**
There are 800 students in Uptown College.
96 of these students walk to college each day.
What percentage of the students walk to college?

7 **Get Real!**
Chris has 50 books on his shelves.
29 of these books are science fiction.
What percentage of his books are science fiction?

8 **Get Real!**
The price of a packet of biscuits goes up from 30p to 36p.
Find the percentage increase.

9 **Get Real!**
Becky's curtains were 60 cm long before she washed them.
After the wash they were only 51 cm long.
Find the percentage decrease in length.

10 **Get Real!**
Sam buys a guitar for £125 and sells it for £160.
Find his percentage profit.

11 Express 7 cm as a percentage of 12 cm.

12 Express £17.64 as a percentage of £72.

13 Express 65 g as a percentage of 5 kg.

14 Express 90 cm as a percentage of 8 m.

15 Express 38p as a percentage of £12.

16 Anna's answer to question **10** is 21.9%.
What mistake has she made?

17 **Get Real!**
Mel wanted to buy a sofa priced at £1450.
The salesman asked for a deposit of £348.
What percentage of the price was this?

18 Get Real!
Out of 3600 claims on household insurance, 522 were for broken windows.
What percentage of claims were for broken windows?

19 Grace has £6.50 in her purse.
She puts 50p in a charity box.
What percentage of her money has gone to charity?

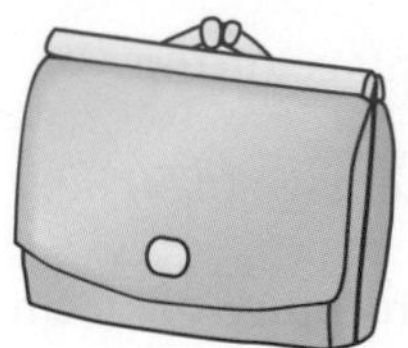

20 Get Real!
The value of a house goes down from £166 000 to £141 100.
Find the percentage decrease.

21 Get Real!
A landlord puts the rent on a flat up from £280 per month to £301 per month.
Find the percentage increase.

22 Get Real!
Jo buys a painting for £975 and sells it a year later for £700.
Find her percentage loss.

Learn 5 Reverse percentage problems

Examples:

a A pair of jeans are priced at £35 in a sale. They have been reduced by 30%.
What was their original price?

Work out what percentage the new price is of the original price.
Divide by this value to find 1%.
Multiply by 100 to find 100%.

100% – 30%

70% of the original price = £35
1% of the original price = £35 ÷ 70 = £0.50
100% of the original price = £0.50 × 100 = £50

b The cost of a holiday has gone up by 15%. It is now £483.
What was the price before the increase?
Using the method above:

100% + 15%

115% of the original price = £483
1% of the original price = £483 ÷ 115 = £4.20
100% of the original price = £4.20 × 100 = £420

Apply 5

1 Jared's car has depreciated by 40% since he bought it.
It is now valued at £7200.
How much did he pay for it?

2 The cost of Suzi's car insurance has gone up by 10% this year.
She now pays £473.
What did she pay last year?

3 Household goods have been reduced by 25% in a sale.
A washing machine is now priced at £465.
What was the price of the washing machine before the sale?

4 Dean bought a guitar in a sale for £54.
He knew it had been reduced by 10%.
Dean said 'I saved £5.40 by getting the guitar in the sale.'
Explain why Dean is wrong.

5 The train fare from Ansaville to Bisterton has gone up by 8%.
It is now £8.10
What was the fare before the increase?

6 The number of patients on a doctor's register has increased to 2055.
This is an increase of 37% over 5 years.
How many patients were on the register 5 years ago?

7 Chris took out a loan from a finance company.
He had to pay it back with interest after one year.
The finance company charged interest at 28% per annum.
Chris paid back £4480. How much was the original loan?

8 The machinery in a factory is valued at £765 000.
Depreciation was 18% over the past year.
What was the value of the machinery a year ago?

9 A motorist is charged £246.44 to have her car serviced.
This charge includes VAT at $17\frac{1}{2}$%.
What was the charge before the VAT was added?

10 A computer is priced at £795, which includes VAT at $17\frac{1}{2}$%.
The computer is bought by a charity, who can reclaim the VAT.
How much can the charity reclaim?

HINT Some of the last four questions will require reverse percentages and some will not.

11 On the first night of the school play, there were 250 people in the audience.
On the second night, the audience numbers went up by 18%.
How many people were in the audience on the second night?

12 The price of admission to the cinema has gone up by 25%.
Tim pays £4.80 for his ticket.
Andy says the old price must have been £3.60
Explain why Andy is wrong.

13 A house in Upville was valued at £255 000 in January 2005.
Over the next year, houses in Upville depreciated by 12%.
What was the value of the house in January 2006?

14 The number of employees in a factory went up by 22%.
The factory now employs 366 people.
How many were there before the increase?

Percentages

ASSESS

The following exercise tests your understanding of this chapter, with the questions appearing in order of increasing difficulty.

1 a Write 60p as a percentage of £1.20

b What is 150 cm as a percentage of 3 km?

c David has bought Victoria an 18 carat gold bracelet.
Pure gold is 24 carat.
What percentage of Victoria's bracelet is gold?

d A bag of sand is labelled as 50 kg.
It actually contains 2.5% more.
How much sand does it contain?

e Ms Berry has picked 1.2 kg of blackberries for making jam.
She needs 15% more to make her recipe.
What weight of blackberries does the recipe require?

2 a A book is designed to have 650 pages.
When the author finished the manuscript he found he had written 754 pages.
What percentage increase is this?

b A box of 144 pens is bought for £10 and individual pens are sold at 10p each.
What is the percentage profit?

c Toad of Toad Hall bought his latest car for £18 000.
A week later he crashed it and, after repair, sold it for £11 700.
What was his percentage loss?

d 100 apples are bought for £17 but 5% are found to be damaged and not saleable. The rest are sold at 20p each.
What is the percentage profit?

e Pythagoras makes a calculator error while using his famous theorem!
He wants to find the value of $\sqrt{112}$ but instead finds $\sqrt{121}$.
What is the percentage error in his calculation?

3 a The population in a village of 3600 people grows by 7% per year.
What is the population after:
i 1 year **ii** 2 years?

b A firm employs 6500 workers.
The work force depreciates by 14% each year.
How many workers are there after:
i 1 year **ii** 2 years?

c Use the compound interest formula to find the interest on £25 000 at 4% over 5 years.

d Farmer Barleymow has sold 20% of his land to a builder.
She now owns 60 acres of land.
How much land did Farmer Barleymow originally own?

e After a discount of 12%, a kitchen is priced at £4224.
What was the price before the discount?

f Mr Moneybags invests a sum of money in a dubious share venture.
The investment initially increases by 10% but then decreases by 16% to £23 100.
How much was his original investment?

17 Equations

OBJECTIVES

D **Examiners would normally expect students who get a D grade to be able to:**

Solve equations such as $3x - 4 = 5 + x$ or $2(5x + 1) = 28$

C **Examiners would normally expect students who get a C grade also to be able to:**

Solve equations such as $3x - 12 = 2(x - 5)$, $\frac{2x}{3} - \frac{x}{4} = 5$ or $\frac{7 - x}{3} = 2$

B **Examiners would normally expect students who get a B grade also to be able to:**

Solve equations such as $\frac{2x - 1}{6} + \frac{x + 3}{3} = \frac{5}{2}$

What you should already know ...

- Collecting like terms
- Multiplying out brackets (by a number only, which may be negative)
- Cancelling fractions
- Adding and subtracting fractions
- Solving equations where the unknown appears once only

VOCABULARY

Linear expression – a combination of terms where the highest power of the variable is 1

Term – a number, variable or the product of a number and a variable(s) such as 3, x or $3x$

Equation – a statement showing that two expressions are equal, for example, $2y - 7 = 15$

Linear equation – an equation where the highest power of the variable is 1; for example, $3x + 2 = 7$ is a linear equation but $3x^2 + 2 = 7$ is not

Unknown – the letter in an equation such as x or y

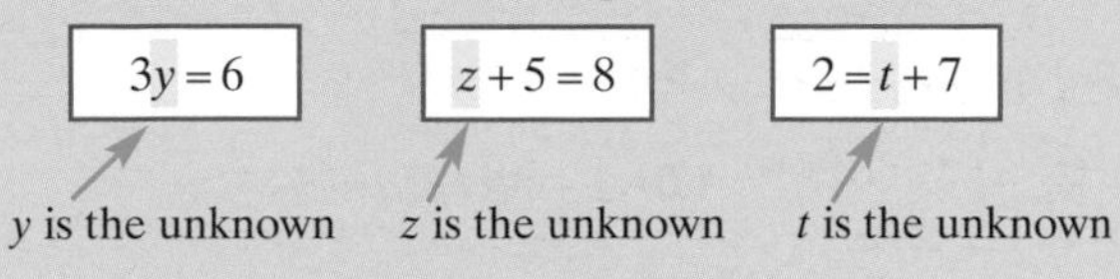

Coefficient – the number (with its sign) in front of the letter representing the unknown, for example:

$4p - 5$ — 4 is the coefficient of p

$2 - 3p^2$ — −3 is the coefficient of p^2

Solution – the value of the unknown in an equation, for example the solution of the equation $3y = 6$ is $y = 2$

Solve – when you solve an equation you find the solution

Operation– a rule for combining two numbers or variables, such as add, subtract, multiply or divide

Inverse operation – the operation that undoes or reverses a previous operation, for example, subtract is the inverse of add:

$15 + 8 = 23$ Add 8
$23 - 8 = 15$ Subtract 8 to return to the starting number 15

Integer – any positive or negative whole number or zero, for example, $-2, -1, 0, 1, 2 \ldots$

Brackets – these show that the terms inside should be treated alike, for example,

$2(3x + 5) = 2 \times 3x + 2 \times 5 = 6x + 10$

Simplify – to make simpler by collecting like terms

Substitute – find the value of an expression when the variable is given a value, for example when $x = 4$, the expression $3x + 2 = 3 \times 4 + 2 = 14$

Variable – a symbol representing a quantity that can take different values such as x, y or z

Learn 1 Equations where the unknown appears on both sides

Examples: Solve these equations:

a $4x + 1 = 9 - x$ **b** $3y + 5 = 5y - 4$

a

$4x + 1 = 9 - x$
$4x + x + 1 = 9 - x + x$ Add x to both sides
$5x + 1 = 9$
$5x + 1 - 1 = 9 - 1$ Subtract 1 from both sides
$5x = 8$
$\frac{5x}{5} = \frac{8}{5}$ Divide both sides by 5
$x = 1.6$

b

$3y + 5 = 5y - 4$
$3y - 3y + 5 = 5y - 3y - 4$ Subtract $3y$ from both sides
(If you took $5y$ from both sides, you would get $-2y$ on the left-hand side)
$5 = 2y - 4$
$5 + 4 = 2y - 4 + 4$ Add 4 to both sides
$9 = 2y$
$\frac{9}{2} = \frac{2y}{2}$ Divide both sides by 2
$4.5 = y$
$y = 4.5$ Writing the equation with y as the subject

Collect together on one side all the terms that contain the unknown.
Collect together on the other side all the terms that do not contain the unknown.
Remember that a sign belongs to the term *after* it.

Apply 1

Solve these equations:

1. $4x - 5 = 2x + 7$
2. $5y - 1 = y + 9$
3. $7z - 3 = 11 + 3z$
4. $t + 8 = 2 - 3t$
5. $3p + 2 = 5 - p$
6. $5 + 2q = 12 - 3q$
7. $8 + a = 5 - 2a$
8. $3b - 10 = 5b - 1$
9. $3c - 5 = 6 - c$
10. $2 + 5d = 7d + 3$
11. $2 - 3e = 8 - 6e$
12. $4f + 11 = 5 + f$

13 Andy is solving the equation $5x - 3 = 2 - 3x$
He writes down $2x - 3 = 2$
Is this correct?
Give a reason for your answer.

14 Sara is solving the equation $4y + 2 = 3 - y$
She writes down $3y = 5$
Is this correct?
Give a reason for your answer.

15 Tom solves the equation $6x - 7 = 9 + 2x$ and gets the answer $x = 2$
Can you find Tom's mistake?

16 Jan solves the equation $7y + 4 = 9 - 3y$ and gets the answer $y = 2$
Can you find Jan's mistake?

17 $4z - 1 =$ ● $- z$
The answer to this equation is $z = 6$
What is the number under the ink blob?

18 $3a +$ ● $= 1 - a$
The answer to this equation is $a = -1$
What is the number under the ink blob?

19 If $b = 4$, find the value of $2b - 5$
Hence explain why $b = 4$ is not the solution of the equation $2b - 5 = 9 - 3b$

20 If $c = -3$, find the value of $8 - 2c$
Hence explain why $c = -3$ is not the solution of the equation $3c + 11 = 8 - 2c$

21 Aidan thinks of a number, doubles it and subtracts 7.
The answer is 15.
Write this as an equation.
Solve the equation to find Aidan's number.

22 Bindia thinks of a number, multiplies it by 5 and adds 3.
The answer is 38.
Write this as an equation.
Solve the equation to find Bindia's number.

23

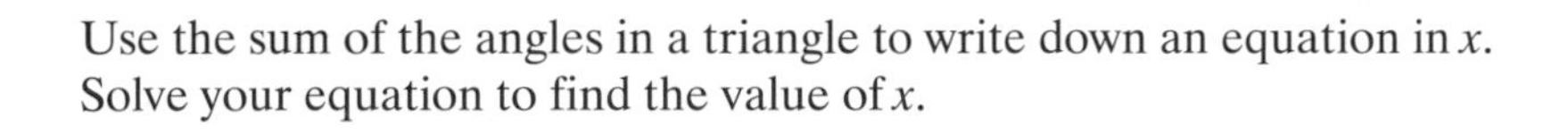

Use the sum of the angles in a triangle to write down an equation in x.
Solve your equation to find the value of x.

24

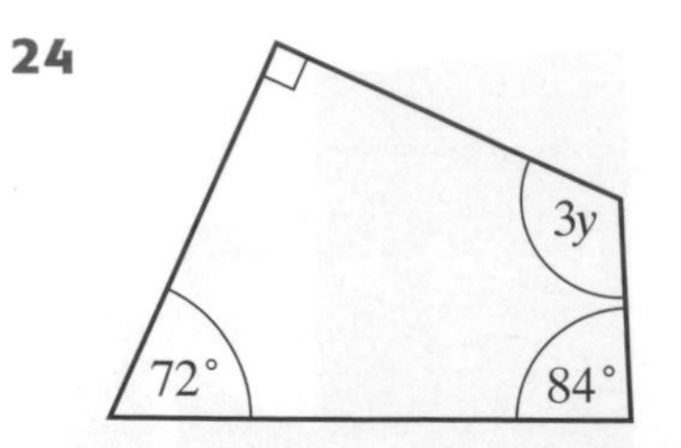

Use the sum of the angles in a quadrilateral to write down an equation in y.
Solve your equation to find the value of y.

25 Four angles around a point are $x°$, $(x + 25)°$, $(2x + 72)°$ and $(3x - 38)°$.
Use this information to write an equation and solve it to find the value of the largest angle.

26 Zoë is z years old. Liam is twice as old as Zoë.
Mandy is 6 years younger than Liam. Neil is 10 years older than Zoë.
The total of their ages is 100 years.
Use this information to write an equation and solve it to find Zoë's age.

Explore

- Make up five different equations that have the solution $x = 8$
- Use a different style for each equation:
  - one which requires division
  - one which requires multiplication
  - one which requires addition
  - one which requires subtraction
  - one which requires a combination of reverse operations
- Explain how you did this
- Make up three different equations that have the solution $y = -12$
- Make up three different equations that have the solution $z = \frac{3}{5}$

Investigate further

Learn 2 Equations with brackets

Examples: Solve these equations:

a $3(2x - 3) = 18$ **b** $5 - 3(y + 1) = 7 - 4y$

a $3(2x - 3) = 18$ Multiply out the brackets first, then follow the rules for solving equations

$6x - 9 = 18$ Remember to multiply both terms in the brackets by 3

$6x = 18 + 9$ Add 9 to both sides

$6x = 27$

$x = 4.5$ Divide both sides by 6

Alternative method: $3(2x - 3) = 18$

$2x - 3 = 6$ Divide both sides by 3

$2x = 9$ Add 3 to both sides

$x = 4.5$ Divide both sides by 2

This alternative method cannot be used for all equations with brackets, as the next example shows

b $5 - 3(y + 1) = 7 - 4y$ Multiply out the brackets first, then follow the rules for solving equations

$5 - 3y - 3 = 7 - 4y$ Multiplying by -3 changes the sign in the brackets

$2 - 3y = 7 - 4y$ The numbers on the left-hand side have been collected

$-3y = 5 - 4y$ Subtract 2 from both sides

$-3y + 4y = 5$ Add $4y$ to both sides

$y = 5$

Apply 2

Solve these equations:

1 $5(x - 3) = 20$

2 $4(y + 2) = 3y + 9$

3 $8 = 2(z - 1)$

4 $5(p + 1) = 3p + 11$

5 $3(q - 11) = 7 - q$

6 $2(3t - 10) = 13$

7 $5a - 2 = 2(a - 4)$

8 $3(3b - 4) = 2 + 7b$

9 $3 + 8c = 6(c - 1)$

10 $14d - 1 = 3(2d + 1)$

11 $2(5 - 2e) = 8 - 3e$

12 $7 - f = 3(5 - f)$

13 $3(5 + 2x) = x - 5$

14 $2(y - 3) + 4(2y - 7) = 6$

15 $8 - 2(z + 3) = 5 - 3z$

16 $19 = 4 - 2(t - 1)$

17 $6(2p - 5) - 4(p - 2) = 14$

18 $5(q - 7) - 3(2q - 4) + 25 = 0$

19 Sadhia thinks of a number, adds 5 and then doubles the result.
Her answer is 64.
Write this as an equation.
Solve the equation to find Sadhia's number.

20 Todd thinks of a number, subtracts 8 and then multiplies the result by 3.
His answer is 42.
Write this as an equation.
Solve the equation to find Todd's number.

21

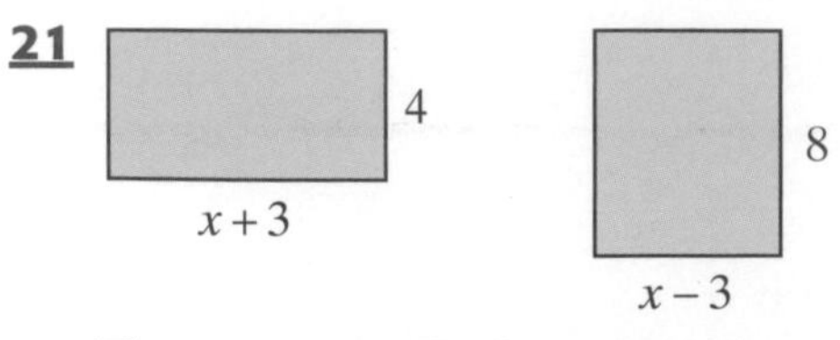

The two rectangles have the same area (all measurements are in cm).
Use this information to write down an equation.
Solve your equation to find the value of x.

22 The side of an equilateral triangle is $(3y - 1)$ cm.
The length of a rectangle is $(2y + 5)$ cm and its width is $(2y - 3)$ cm.
The perimeter of the triangle is equal to the perimeter of the rectangle.
Use this information to write an equation.
Solve your equation to find the value of y.

Learn 3 Equations with fractions

Examples: Solve these equations:

a $\frac{x}{4} - 6 = 3$ **b** $\frac{2x}{3} - \frac{x}{4} = 5$ **c** $\frac{3x+5}{2} = 7$

a $\frac{x}{4} - 6 = 3$

$\frac{x}{4} = 3 + 6$ Add 6 to both sides

$\frac{x}{4} = 9$

$x = 9 \times 4$ Multiply both sides by 4

$x = 36$

b $\frac{2x}{3} - \frac{x}{4} = 5$ The lowest common denominator is 12

${}^{4}\cancel{12} \times \frac{2x}{\cancel{3}_1} - {}^{3}\cancel{12} \times \frac{x}{\cancel{4}_1} = 5 \times 12$ Multiply **each term** by 12

$4 \times 2x - 3 \times x = 5 \times 12$ Cancel (this should clear all the fractions)

$8x - 3x = 60$

$5x = 60$

$x = 12$

c $\frac{3x+5}{2} = 7$ ← This is the same as $\frac{1}{2}(3x + 5) = 7$

${}^{1}\cancel{2} \times \frac{3x+5}{\cancel{2}_1} = 7 \times 2$ Multiply **both** sides by 2 to clear the fraction

$3x + 5 = 14$

$3x = 14 - 5$ Subtract 5 from both sides

$3x = 9$

$x = 3$ Divide both sides by 3

Remove the fraction by multiplying both sides by the denominator.
If there is more than one fraction, multiply by the lowest common denominator.
There are 'invisible brackets' around the terms on top of an algebraic fraction.

Apply 3

Solve these equations:

1 $\frac{x}{2} + 7 = 9$

2 $\frac{y}{3} - 1 = 5$

3 $6 = 2 + \frac{z}{5}$

4 $3 + \frac{a}{2} = 7$

5 $4 - \frac{b}{3} = 1$

6 $\frac{c}{6} + 4 = 3$

7 $\frac{3x + 2}{5} = 4$

8 $\frac{2y - 3}{4} = 3$

9 $1 = \frac{9 - z}{3}$

10 $\frac{1}{4}(3p + 5) = 2$

11 $\frac{1}{2}(5q + 3) = 14$

12 $\frac{1}{3}(2t - 7) = 4$

13 $\frac{1}{2}(5a - 2) = a - 4$

14 $\frac{1}{8}(b - 1) = 10 - b$

15 $2c - 11 = \frac{1}{3}(2 - c)$

16 $\frac{x}{3} + \frac{x}{4} = 7$

17 $\frac{y}{2} - \frac{y}{5} = 6$

18 $\frac{3z}{4} - \frac{7z}{10} = 2$

19 $\frac{2p}{3} = 7 - \frac{p}{2}$

20 $\frac{q}{4} - \frac{1}{2} = \frac{q}{12}$

21 $\frac{3t}{5} + \frac{1}{6} = \frac{2t}{3}$

22 Jayne says the answer to the equation $\frac{x + 5}{2} = 4 - x$ is $x = -1$

Use substitution to check whether Jayne is correct.

23 Tom and Jared are solving the equation $\frac{3y - 7}{4} = y + 2$

Tom gets the answer $y = 1$ and Jared gets $y = -15$

Check their answers to see if either of them is correct.

24 Explain why you cannot solve the equation $\frac{4z - 1}{2} = 1 + 2z$

Learn 4 More complex equations with fractions

Example: Solve $\frac{2x - 1}{6} + \frac{x + 3}{3} = \frac{5}{2}$

$\frac{2x - 1}{6} + \frac{x + 3}{3} = \frac{5}{2}$ The lowest common denominator of 6, 3 and 2 is 6

$\not{6} \times \frac{(2x - 1)}{\not{6}_1} + {}^{2}\not{6} \times \frac{(x + 3)}{\not{3}_1} = {}^{3}\not{6} \times \frac{5}{\not{2}_1}$ Multiply both sides of the equation by 6 and cancel

This is where you have to remember the 'invisible brackets'

$(2x - 1) + 2(x + 3) = 15$

$2x - 1 + 2x + 6 = 15$

$4x + 5 = 15$

$4x = 10$ Subtract 5 from both sides

$x = 2.5$ Divide both sides by 4

Apply 4

Solve these equations.

1 $\frac{x+3}{2}+\frac{x+4}{3}=2$

2 $\frac{y+5}{6}+\frac{y-1}{2}=10$

3 $\frac{z+1}{2}-\frac{z-1}{4}=4$

4 $\frac{a-4}{5}-\frac{a-2}{6}=0$

5 $\frac{3b+5}{4}-\frac{5b-13}{3}=1$

6 $\frac{c-2}{7}-\frac{3-c}{14}=1$

7 $\frac{d+12}{3}+\frac{d+3}{6}=\frac{3d+11}{2}$

8 $\frac{2e+1}{6}-\frac{e-1}{3}=\frac{e-3}{2}$

9 $\frac{p-1}{3}-\frac{p+2}{6}=\frac{p}{10}$

10 $\frac{2q-5}{3}-\frac{2q-3}{4}=\frac{q-6}{12}$

Equations

ASSESS

The following exercise tests your understanding of this chapter, with the questions appearing in order of increasing difficulty.

1 Solve these equations:

a $5x+1=15-2x$

b $4x+8=6x-5$

c $4w+7=37-2w$

d $\frac{1}{2}z-1=11-\frac{1}{4}z$

2 Solve these equations:

a $2(3p-1)=28$ **b** $3(a-1)=2(a+1)$ **c** $\frac{2s+14}{5}=1$

3 Write down an equation and use it to solve the following problems.

a Demelza's dad is three times as old as Demelza. The sum of their ages is 52. How old is Demelza?

b Three consecutive odd numbers add up to 93. Find these numbers.

c Jack and Jill went up the hill to fetch a pail of water. The water weighed 15 kg more than the pail. The total weight was 18 kg. How heavy was the pail?

d Tweedledum and Tweedledee bought a cake and cut it into 3 pieces, one piece for each of them and one piece for Alice. Tweedledum's piece was 50 g heavier than Tweedledee's piece. Tweedledee's piece was 30 g heavier than Alice's piece. The total mass was 710 g. What was the mass of Alice's piece?

4 Solve these equations.

a $\frac{a-2}{3}+\frac{a+6}{4}=16$

b $\frac{3x+5}{2}-\frac{2x-1}{3}=-3$

18 Reflections and rotations

OBJECTIVES

D **Examiners would normally expect students who get a D grade to be able to:**

Reflect shapes in lines such as $x = 2$ or $y = -1$

Rotate shapes about the origin

Describe fully reflections and rotations about the origin

Identify reflection symmetry in 3-D solids

C **Examiners would normally expect students who get a C grade also to be able to:**

Reflect shapes in the lines $y = x$ and $y = -x$

Rotate shapes about any point

Describe fully reflections and rotations about any point

Find the centre of a rotation and describe it fully

Combine reflections and rotations

What you should already know ...

- Coordinates and equations of lines, such as $x = 3$, $y = -2$, $y = x$, $y = -x$, ...
- Names of 2-D shapes and 3-D shapes
- Reflect a shape in a mirror line, including axes
- Describe the symmetry of 2-D shapes
- Sketch 3-D shapes and draw 3-D solids on an isometric grid

VOCABULARY

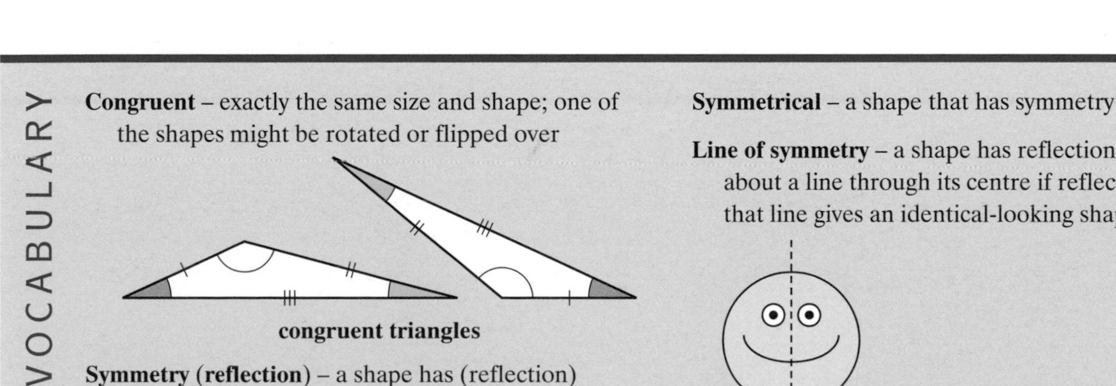

Congruent – exactly the same size and shape; one of the shapes might be rotated or flipped over

Symmetry (reflection) – a shape has (reflection) symmetry if a reflection through a line passing through its centre produces an identical-looking shape. The shape is said to be symmetrical

Symmetrical – a shape that has symmetry

Line of symmetry – a shape has reflection symmetry about a line through its centre if reflecting it in that line gives an identical-looking shape

Symmetry (rotation) – a shape has (rotation) symmetry if a rotation about its centre through an angle greater than 0° and less than 360° gives an identical-looking shape

Order of rotation symmetry – the number of ways a shape would fit on top of itself as it is rotated through 360°

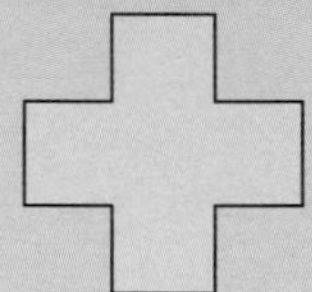

rotation symmetry order 4

(Shapes that are not symmetrical have rotation symmetry of order 1 because a rotation of 360° always produces an identical-looking shape)

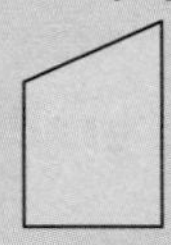

rotation symmetry order 1 (i.e. not symmetrical)

Transformation – reflections and rotations are examples of transformations as they transform one shape onto another

Reflection – a transformation involving a mirror line (or axis of symmetry), in which the line from the shape to its image is perpendicular to the mirror line. To describe a reflection fully, you must describe the position or give the equation of its mirror line, for example, the triangle A is reflected in the mirror line $y = 1$ to give the image B

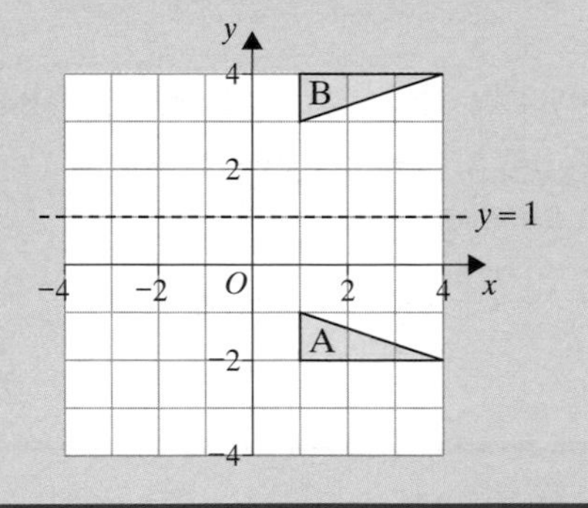

Axis of symmetry – the mirror line in a reflection

Image – the shape after it undergoes a transformation, for example, reflection or rotation

Rotation – a transformation in which the shape is turned about a fixed point called the centre of rotation. To describe a rotation fully, you must give the centre, angle and direction (a *positive angle* is *anticlockwise* and a *negative angle* is *clockwise*), for example, the triangle A is rotated about the origin through 90° anticlockwise to give the image C

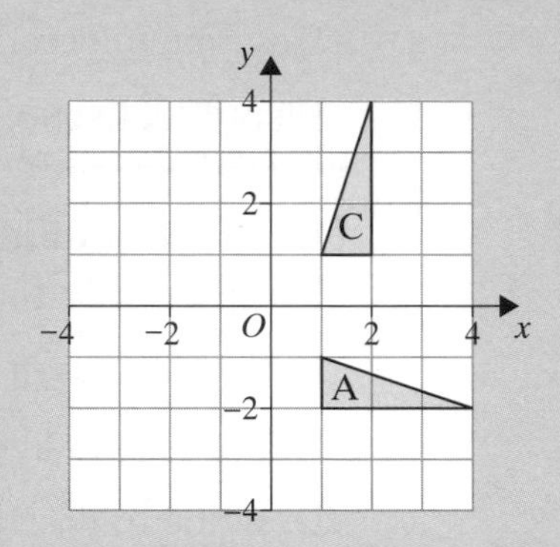

Centre of rotation – the fixed point around which the object is rotated

Learn 1 Drawing reflections and rotations

Examples:

a Reflect triangle T:
- **i** in the y-axis and label the image U
- **ii** in the line $y = 1$ and label the image V
- **iii** in the line $y = x$ and label the image W.

b Write down the image when:
- **i** triangle U is rotated through 90° clockwise (−90°) about the origin O
- **ii** triangle U is rotated through 180° (clockwise or anticlockwise) about the point (0, 1)
- **iii** triangle V is rotated through 90° anticlockwise (+90°) about the point (1, 1).

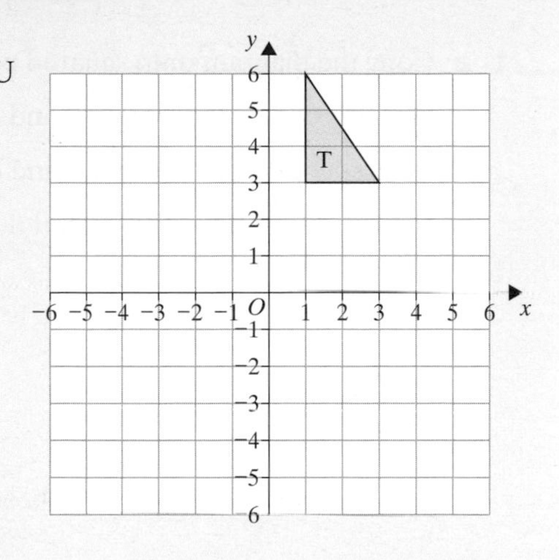

Reflection and rotation are transformations that can be used to move a shape from one position (the object) to another position (the image). In each case the object and image are congruent (the same size and shape).

a Reflecting triangle T:
- **i** in the y-axis gives the image U
- **ii** in the line $y = 1$ gives the image V
- **iii** in the line $y = x$ gives the image W.

b The image when:
- **i** triangle U is rotated through 90° clockwise (−90°) about the origin O is triangle W
- **ii** triangle U is rotated through 180° (clockwise or anticlockwise) about the point (0, 1) is triangle V
- **iii** triangle V is rotated through 90° anticlockwise (+90°) about the point (1, 1) is triangle W.

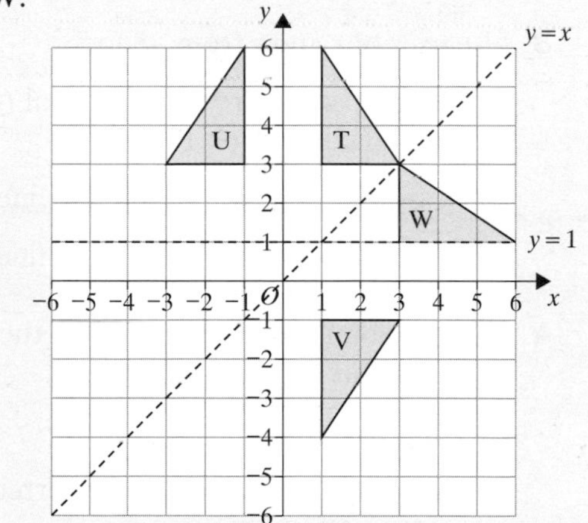

You can use tracing paper to check these

Reflection
For each point on the object, there is a point on the image at an equal distance from the mirror line, but on the other side of it. The mirror line is a line of symmetry.

Rotation
The object is turned through an angle about the centre of rotation. The angle of rotation can be given in degrees or as a fraction of a turn and can be clockwise (negative) or anticlockwise (positive).

Apply 1

1 **a** Copy the diagram onto squared paper.

i Reflect T in the line $y = 1$ and label its image U.

ii Reflect P in the line $y = 1$ and label its image Q.

b Repeat part **a** on new axes, but now reflect each shape in the line $x = 1$.

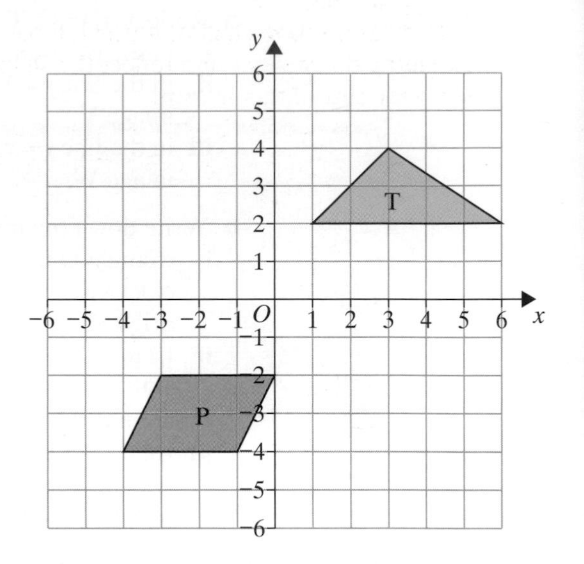

2 Use axes of x and y from -8 to 8.

a Draw a rhombus, R, with vertices (0, 1), (2, 4), (0, 7) and $(-2, 4)$.

b Reflect R in the line $x = 3$ and label the image S.

c Reflect R in the line $x = -3$ and label the image T.

3 Use axes of x and y from -8 to 8.

a Draw a pentagon with vertices at $(-2, -3)$, $(2, -3)$, $(4, 0)$, $(0, 2)$, $(-4, 0)$. Label the pentagon P.

b Draw the reflection of P in the line $y = 2$. Label the image Q.

c Draw the reflection of P in the line $y = -3$. Label the image R.

4 When Naomi was asked to reflect the shape S in the line $x = 1$, she drew this diagram.

a What did she do wrong?

b Draw a diagram to show the correct image.

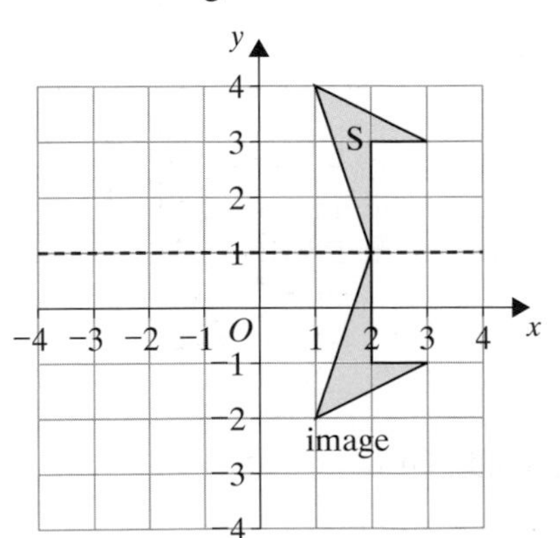

5 Get Real!

A Rangoli is a floor decoration used to welcome visitors to an Indian home.

This diagram shows half of a Rangoli pattern. Copy it onto isometric dotty paper and reflect the shapes in the line AB to show the rest of the pattern.

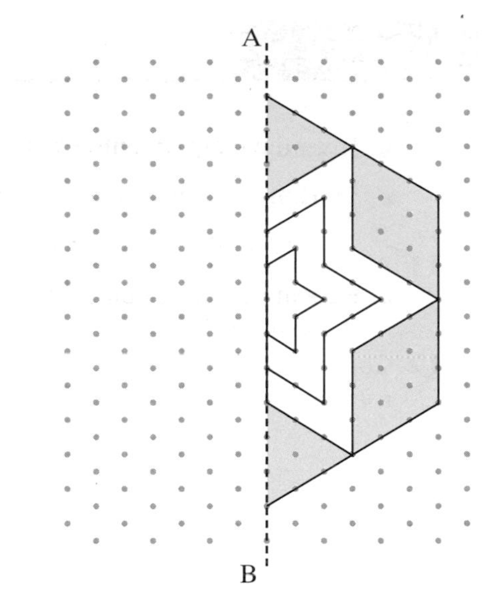

6 **a** Draw the mirror line $y = x$ on axes x and y from -6 to 6.

b **i** Draw the triangle, T, by joining the points (1, 6), (3, 6) and (3, 3).

ii Reflect T in the line $y = x$ and label its image U.

c **i** Draw the kite, K, by joining (0, -2), (2, -3), (0, -6) and (-2, -3).

ii Reflect K in the line $y = x$ and label it L.

7 Use axes of x and y from -8 to 8.

a Draw:

i triangle T with vertices at (-4, 7), (1, 6) and (-2, 3)

ii quadrilateral Q with vertices at (-6, -7), (2, -6), (1, -2) and (-3, -3).

b Draw the image of each shape after reflection in the line $y = -x$.

c Each shape and its image are congruent. Mark the sides and angles to show which are equal to each other.

8 **a** Copy the diagram onto squared paper.

b Draw the image of the T-shape after a rotation of 90° clockwise about the origin O. Label it A.

c Draw the image of the T-shape after a rotation of 180° about the origin O. Label it B.

d Draw the image of the T-shape after a rotation of 90° anticlockwise about the origin O. Label it C.

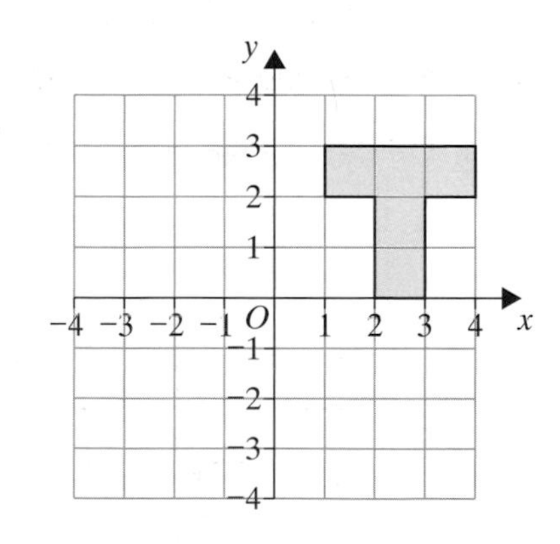

9 **a** On squared paper draw axes of x and y from -6 to 6.
Join the points (1, 1), (3, 1) and (1, 5) to form a triangle and label it A.

b Rotate A through 180° about the origin O and label the image B.

10 a Draw axes of x and y from –6 to 6 and a quadrilateral with vertices (–6, 1), (–2, 1), (–1, 3) and (–6, 6).
Label the quadrilateral Q.

b Rotate Q through a half turn about the origin O and label the image R.

c Mark the corresponding sides and angles in Q and R.

11 Angela says that when point P(2, –3) is rotated 90° clockwise about the origin the image is Q(3, 2). What has Angela done wrong?

12 Copy each shape onto isometric paper and show its image after the rotation described.

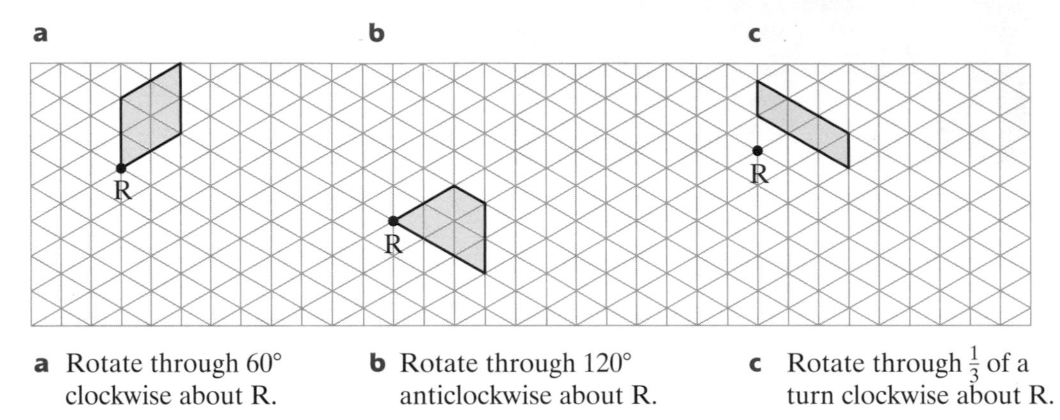

a Rotate through 60° clockwise about R.

b Rotate through 120° anticlockwise about R.

c Rotate through $\frac{1}{3}$ of a turn clockwise about R.

13 On squared paper draw axes of x and y from –8 to 8.

a Draw a trapezium with vertices (2, 1), (4, 0), (4, 3) and (2, 2).
Label the trapezium T.

b Draw the image of T after a half turn about the point (3, 4).
Label the image U.

c Draw the image of T after a half turn about the point (3, –2).
Label the image V.

d Draw the image of T after a half turn about the point (–2, –2).
Label the image W.

e What do you notice about the three images?

14 On squared paper draw axes of x and y from –6 to 6.

a Draw a triangle with vertices (–2, 4), (–2, 2) and (–6, 2).
Label the triangle A.

b Draw the image of A after a rotation of 90° anticlockwise about the point (0, 4). Label it B.

c Draw the image of A after a rotation of 90° anticlockwise about the point (–2, 0). Label it C.

d What can you say about the coordinates of the images?

15 **a** Using axes of x and y from −8 to 8, draw the pentagon, P, with vertices at (3, −5), (7, −6), (7, −4), (5, −2) and (3, −3).

b Draw the image of P after a quarter turn clockwise about (3, 0). Label it Q.

c Draw the image of P after a quarter turn clockwise about (0, 2). Label it R.

d Draw the image of P after a quarter turn clockwise about (6, 4). Label it S.

Explore

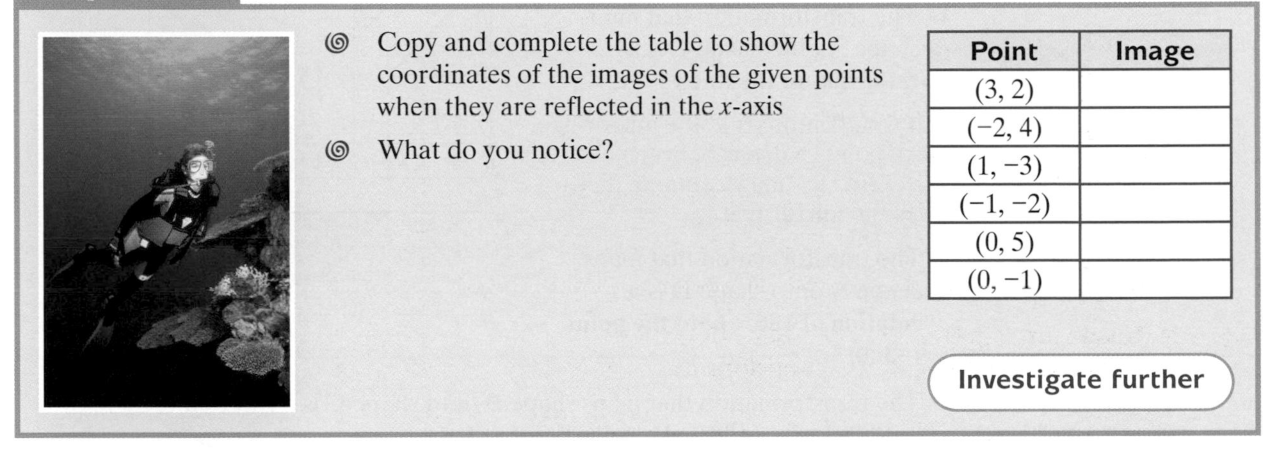

- Copy and complete the table to show the coordinates of the images of the given points when they are reflected in the x-axis
- What do you notice?

Point	Image
(3, 2)	
(−2, 4)	
(1, −3)	
(−1, −2)	
(0, 5)	
(0, −1)	

Investigate further

Explore

- Copy and complete the table to show the coordinates of the images of the given points when they are rotated through 180° about the origin, O
- What do you notice?

Point	Image
(3, 2)	
(−2, 4)	
(1, −3)	
(−1, −2)	
(0, 5)	
(0, −1)	

Investigate further

Learn 2 Describing reflections and rotations

Examples: Describe fully the transformation that maps (moves):

a shape A onto shape B
b shape A onto shape C
c shape A onto shape D
d shape B onto shape C.

a The transformation that maps shape A onto shape B is a **reflection in the line $x = -1$.**

b The transformation that maps shape A onto shape C is a **reflection in the line $y = -x$.**

If it is difficult to find the mirror line, mark points halfway between each vertex and its image. Joining these gives the mirror line

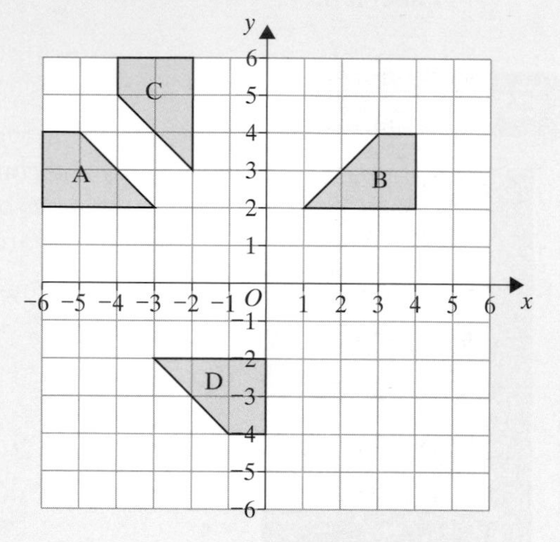

c The transformation that maps shape A onto shape D is a **rotation of 180° about the point (−3, 0).**

d The transformation that maps shape B onto shape C is a **rotation of 90° anticlockwise (+90°) about the point (−1, 1).**

You must say anticlockwise (+) or clockwise (−) for every angle except 180°

You can use tracing paper to check these

To describe a **reflection** fully you must give the **equation of the mirror line.**

To describe a **rotation** fully you must give the **centre** and the **angle** and say whether it is **clockwise** or **anticlockwise.**

Apply 2

1 Write the equation of the mirror line that maps:

a A onto B
b A onto C
c B onto D
d F onto G
e A onto E
f B onto H
g C onto F
h G onto D
i C onto D
j D onto I

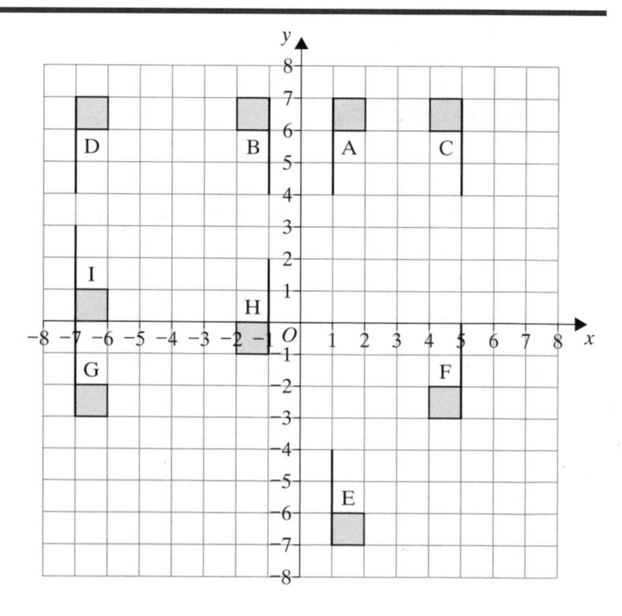

2 **a** Using axes of x and y from −8 to 8, draw trapezium A with vertices (3, 1), (7, 1), (6, 3) and (4, 3).

b Describe fully the transformation that maps trapezium A onto the trapezium B, with vertices (1, 3), (1, 7), (3, 6) and (3, 4).

c Describe fully the transformation that maps trapezium A onto the trapezium C, with vertices (−1, −3), (−1, −7), (−3, −6) and (−3, −4).

3 Describe fully the rotation that maps:

a A onto B

b A onto C

c A onto D

d D onto B

e B onto F

f A onto E

g C onto G

h H onto A

i B onto E

j D onto C

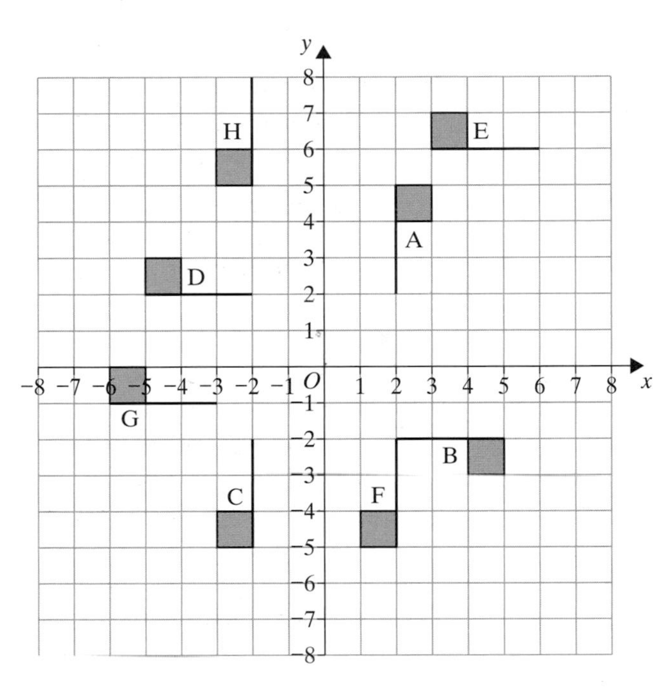

4 The diagram shows a quadrilateral ABCD and its image PQRS after a transformation.

a Give a full description of the transformation.

b Find and name the length that is equal to:

i AB **iii** BD

ii CD **iv** AC

c Find and name the angle that is equal to:

i ∠CDA **iii** ∠ACD

ii ∠CAD

d Find and name a triangle that is congruent to:

i DAB **ii** CAD

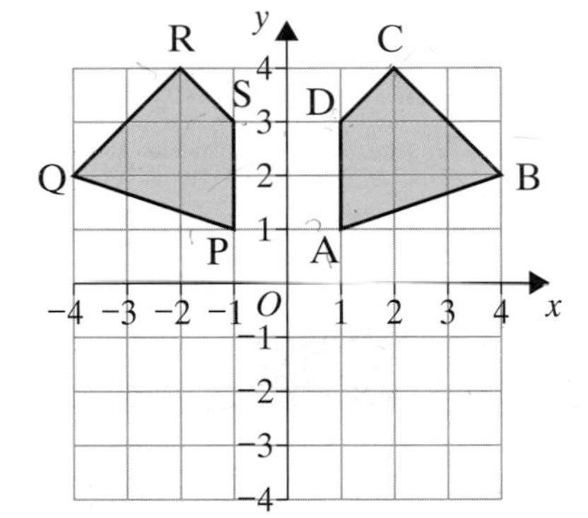

5 Describe fully the transformation that maps:

a P onto Q

b P onto R

c P onto S

d Q onto R

e Q onto S

f R onto S

g S onto P

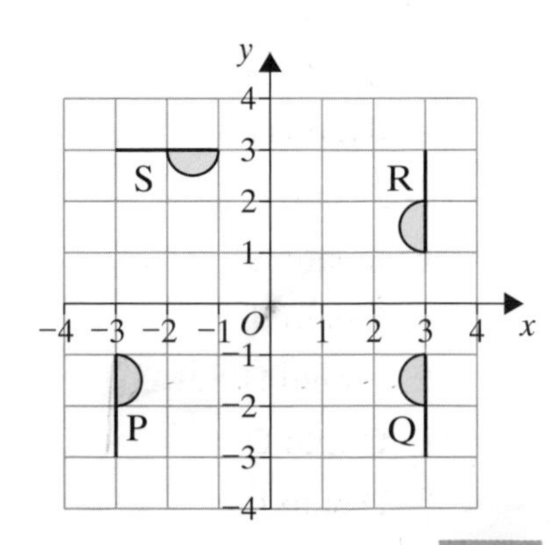

6 The diagram shows a pentagon ABCDE and its image PQRST after a transformation.

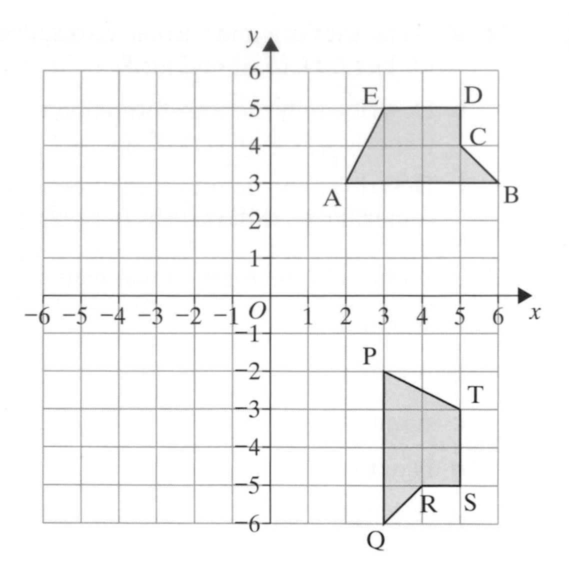

a Give a full description of the transformation.

b Find and name the length that is equal to:

i PQ **iii** PS

ii RS **iv** PR

c Find and name the angle that is equal to:

i ∠RST **iii** ∠PRT

ii ∠RQP **iv** ∠QOP

d Find and name a triangle that is congruent to:

i PQR **iii** PRS

ii PTS

7 **Get Real!**

a Describe fully the transformation that maps the minute hand on a clock from its position at one o'clock to its position at ten past one.

b Describe fully the transformation that maps the hour hand on a clock from its position at half past four to its position at five o'clock.

8 **a** Using axes of x and y from -6 to 6, draw the square with vertices (2, 2), (5, 2), (5, 5) and (2, 5).
Label this square S.

b **i** Draw and label square T with vertices (2, −2), (5, −2), (5, −5) and (2, −5).

ii Find as many transformations as you can that map S onto T.

Describe each transformation fully.

c **i** Draw and label square U with vertices (−2, −2), (−5, −2), (−5, −5) and (−2, −5).

ii Find as many transformations as you can that map S onto U.
Describe each transformation fully.

d Find as many transformations as you can that map U onto T.
Describe each transformation fully.

9 Using axes of x and y from −8 to 8, draw three triangles:
A(2, 2), B(7, 2), C(5, 4)
K(1, −4), L(6, −4), M(3, −6)
X(−4, −5), Y(−2, −2), Z(−2, −7).

a Find and write down two line segments that are equal in length to:

i AB **ii** BC **iii** AC

b Describe fully the transformation that maps triangle ABC onto triangle KLM.

c Describe fully the transformation that maps triangle ABC onto triangle XYZ.

d Find and name a triangle that is congruent to triangle OAB.

10 PQR and XYZ are congruent equilateral triangles.
They intersect at A, B, C, D, E and F, which are the vertices of a regular hexagon with centre O.
Describe fully the transformation that maps triangle PQR onto triangle XYZ in which:

a P is mapped onto X, Q onto Y and R onto Z

b P is mapped onto Z, Q onto X and R onto Y

c P is mapped onto Z, Q onto Y and R onto X

d P is mapped onto Y, Q onto X and R onto Z.

Explore

- What do you notice about the coordinates of the pairs of object and image points given in the table?
- Draw the triangle ABC and its image DEF
- Describe fully the transformation that maps ABC onto DEF
- On new axes draw the triangle PQR and its image STU
- Describe fully the transformation that maps PQR onto STU

Object	Image
A(1, 2)	D(2, −1)
B(5, 4)	E(4, −5)
C(3, −1)	F(−1, −3)
P(−5, 2)	S(2, 5)
Q(1, −1)	T(−1, −1)
R(−2, −1)	U(−1, 2)

Investigate further

Learn 3 Combining reflections and rotations

Example:

The shape A is reflected in the line $x = 2$ to give shape B.
The shape B is rotated through 180° about the point (2, 0) to give shape C.

What single transformation maps shape A onto shape C?

When shape A is reflected in the line $x = 2$, the image is shape B.
When shape B is then rotated through 180° about the point (2, 0), the image is shape C.

The single transformation that maps shape A onto shape C is a reflection in the x-axis.

The x-axis is the same as the line $y = 0$

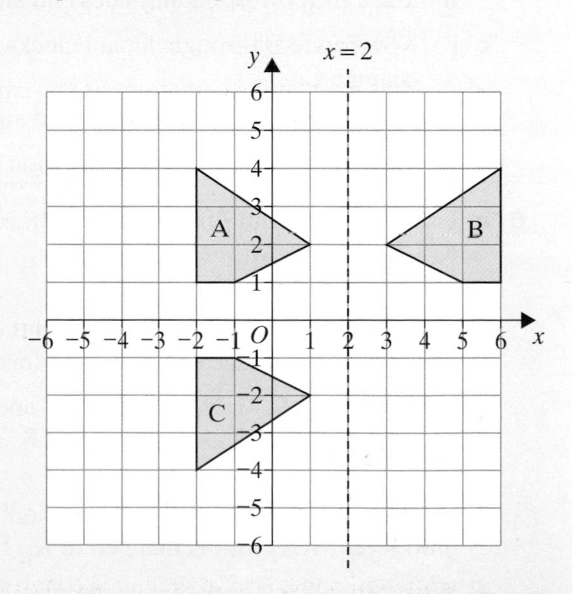

Apply 3

1 **a** Using axes of x and y from -6 to 6, draw the triangle, T, with vertices at (3, 4), (1, 2) and (6, 2).

b Draw the reflection of T in the x-axis and label it U.

c Now draw the reflection of U in the y-axis and label it V.

d Describe fully the single transformation that would transform T onto V.

2 **a** Using axes of x and y from -6 to 6, draw the quadrilateral, Q, with vertices at (3, 6), (1, 6), (3, 1) and (5, 4).

b Reflect Q in the y-axis and label the image R.

c Now rotate R through 180° about O and label the image S.

d Describe fully the single transformation that would map Q onto S.

3 **a** Using axes of x and y from -6 to 6, draw the trapezium, A, by joining points (2, 1), (5, 1), (5, 3) and (0, 3).

b Rotate A through 90° clockwise about O and label the image B.

c Now draw the reflection of B in the y-axis and label it C.

d What single transformation would move A onto C?

4 Use axes of x and y from -8 to 8.

a Draw the pentagon, P, with vertices at $(-1, 4)$, (4, 4), (3, 5), (3, 6) and (1, 6).

b Draw the reflection of P in the line $x = -2$ and label the image Q.

c Now rotate Q through 180° about $(-2, 3)$ and label the image R.

d Describe fully the single transformation that would map P onto R.

5 Use axes of x and y from -10 to 10.

a Draw the triangle, A, with vertices at $(-3, 1)$, $(-8, 4)$ and $(-5, 5)$.

b **i** Reflect A in the line $y = -x$ to give triangle B.

ii Mark the corresponding sides and angles in A and B.

c **i** Now rotate B through 90° anticlockwise about $(2, -2)$ to give triangle C.

ii Mark the corresponding sides and angles in C.

d Describe fully the single transformation that maps A onto C.

6 **a** Using axes of x and y from -6 to 6, draw the kite with vertices K(3, 5), L(1, 4), M(3, 1) and N(5, 4).
Label each vertex.

b Reflect KLMN in the y-axis and label the image $K_1L_1M_1N_1$ where K_1 is the image of K, L_1 is the image of L and so on.

c Now rotate $K_1L_1M_1N_1$ through 180° about O and label the image $K_2L_2M_2N_2$ where K_2 is the image of K_1, L_2 is the image of L_1 and so on.

d Describe fully the single transformation that would map KLMN onto $K_2L_2M_2N_2$ with K mapped to K_2, L to L_2 etc.

e What other transformation would map KLMN onto $K_2L_2M_2N_2$ but in which L would not be mapped to L_2?

7 What single transformation is equivalent to a reflection in the line $y = x$ followed by a reflection in the line $y = -x$?
Draw a diagram to illustrate your answer.

8 Gary says, 'If you rotate a shape 90° clockwise about O and then reflect it in the line $x = -2$, you always get the same image that you would get if you reflected the shape in the line $x = -2$ and then rotated it 90° clockwise about O'. Is this statement true?
Draw a diagram to illustrate your answer.

9 Use isometric dotty paper for this question.

a Draw two lines at an angle of 60° and a right-angled triangle, T, as shown.

b Reflect triangle T in mirror line M, then mirror line N.

c Give a full description of the single transformation that would have the same effect on the triangle.

d What happens if you reflect T in N, then M?

Explore

- Using x- and y-axes from 0 to 12 draw a triangle A, with vertices at (0, 1), (3, 2) and (2, 4)
- Reflect A in the line $x = 3$ and then reflect the result in the parallel line $x = 6$
- Describe fully the single transformation that would have the same effect
- Try this with other shapes

Investigate further

Learn 4 Reflection symmetry in 3-D solids

Example: How many planes of symmetry are there in a square-based pyramid?

There are 4 planes of symmetry in a square-based pyramid.

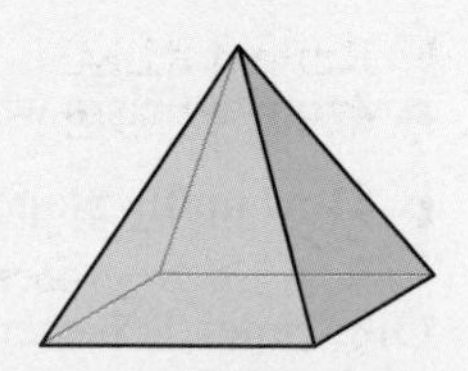

Square-based pyramid

Apply 4

1 Which of these 3-D solids have reflection symmetry?
In each case say how many different ways you can cut the solid into matching halves.

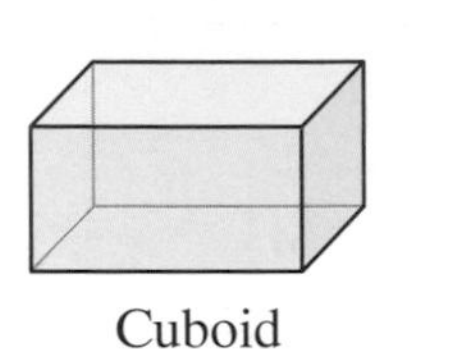
Cuboid

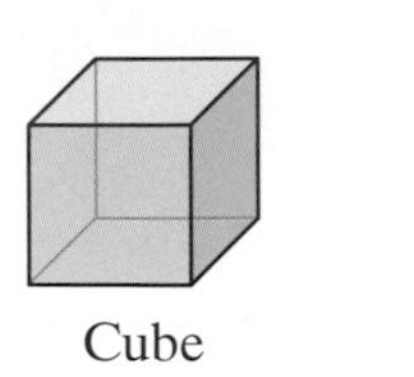
Cube

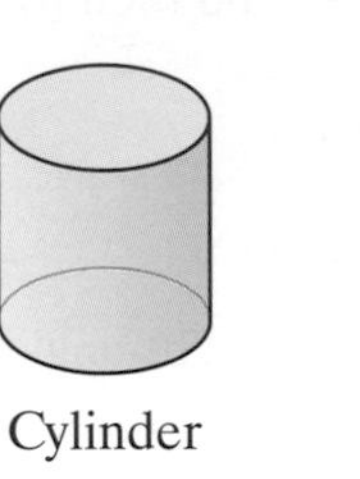
Cylinder

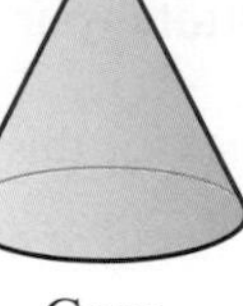
Cone

2 **Get Real!**
How many planes of symmetry are there in each of these objects?

a Coffee mug **b** Stool **c** Can **d** Salt cellar

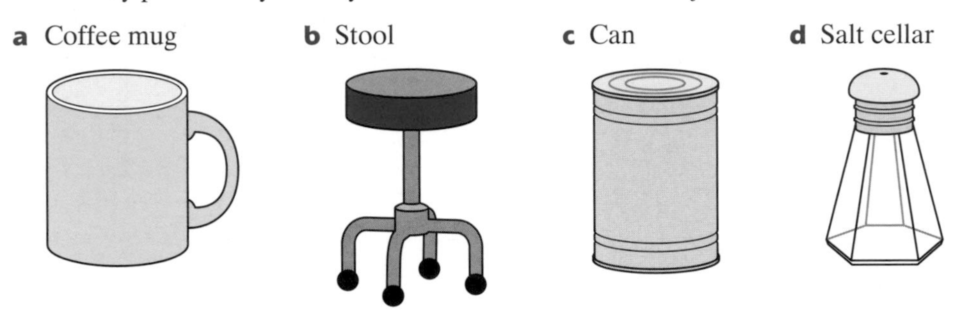

3 These solids are made from cubes.
How many planes of symmetry does each solid have?

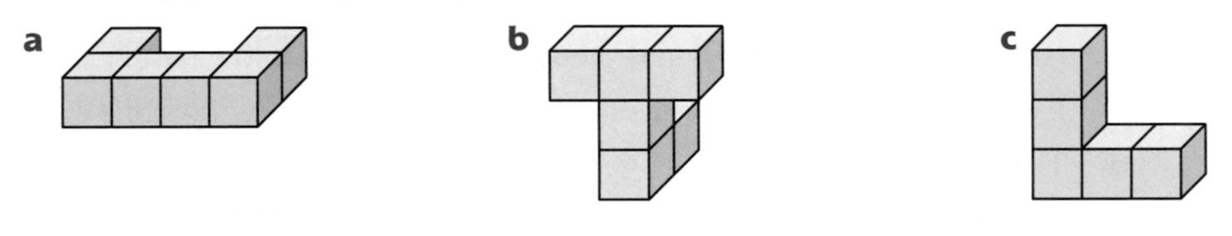

4 Each sketch shows half of a solid made from cubes and a plane of symmetry.
In each case copy and complete the solid.

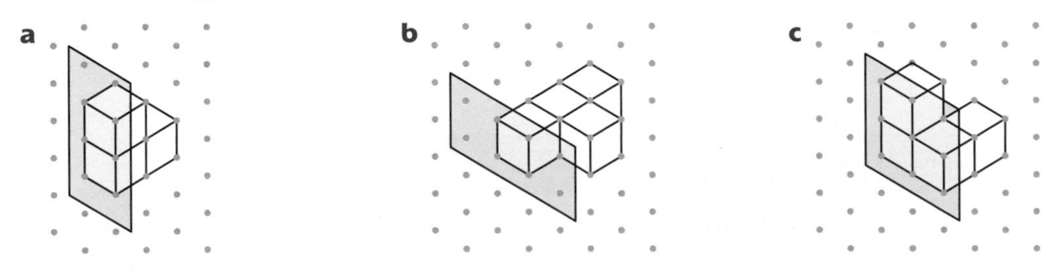

5 **a** Draw a prism whose cross-section is a right-angled isosceles triangle.

b How many planes of symmetry does this prism have?

6 On isometric paper draw:

a a 3-D solid that has no plane of symmetry

b a 3-D solid that has just one plane of symmetry

c a 3-D solid that has exactly two planes of symmetry.

Reflections and rotations

ASSESS

The following exercise tests your understanding of this chapter, with the questions appearing in order of increasing difficulty.

1 Reflect the object shown in the lines:

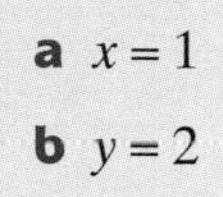

a $x = 1$

b $y = 2$

c $y = x$

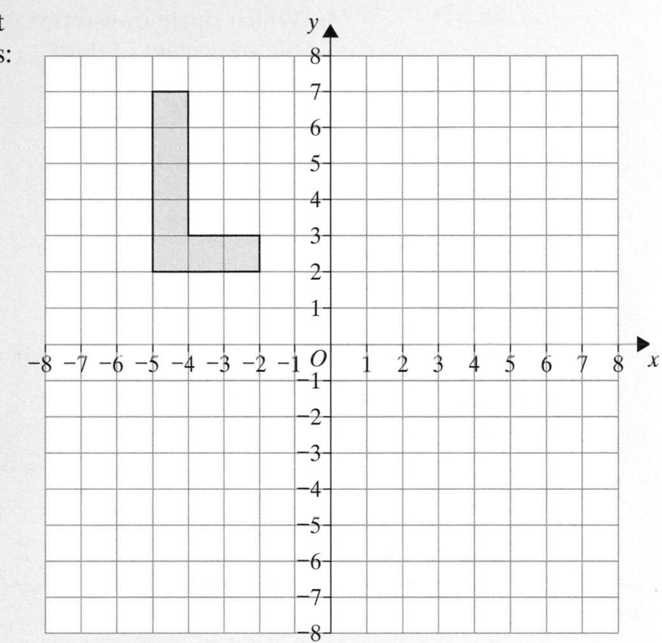

2 Copy the following diagrams and draw the image obtained by rotating the shape:

i 90° anticlockwise about the given point

ii 90° clockwise about the given point.

a

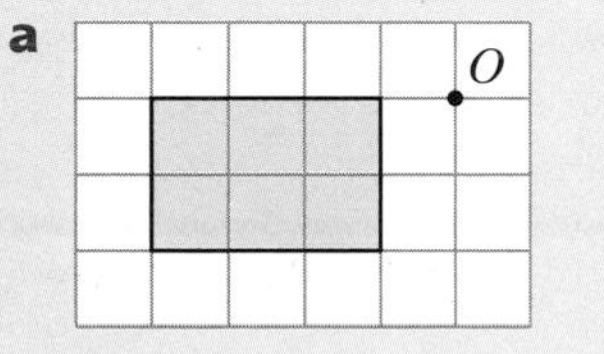

b

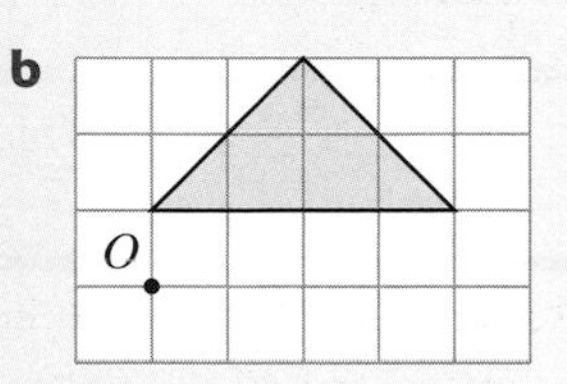

3 Find the images of the given shape after rotations about the origin of:

a 90° anticlockwise

b 90° clockwise

c $\frac{3}{4}$ of a turn anticlockwise.

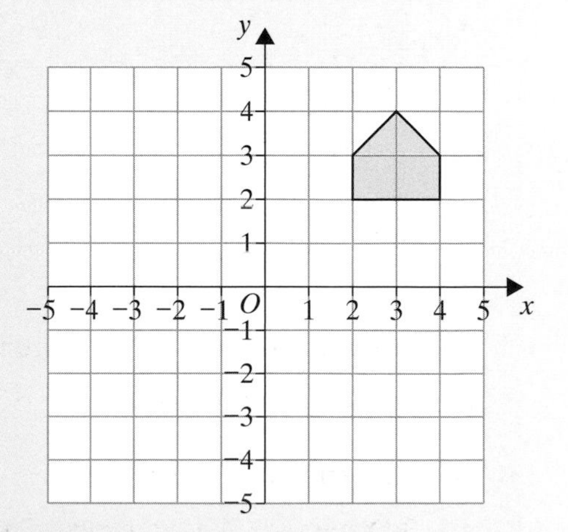

4 a Rotate the given figure by 90° clockwise about the origin followed by a reflection in the x-axis.

b Which single transformation is the equivalent of this?

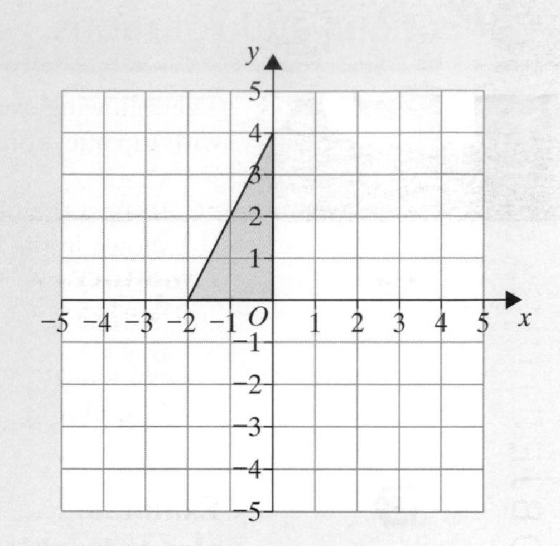

5 a Reflect the given figure in the line $y = x$ followed by a reflection in the line $y = -x$.

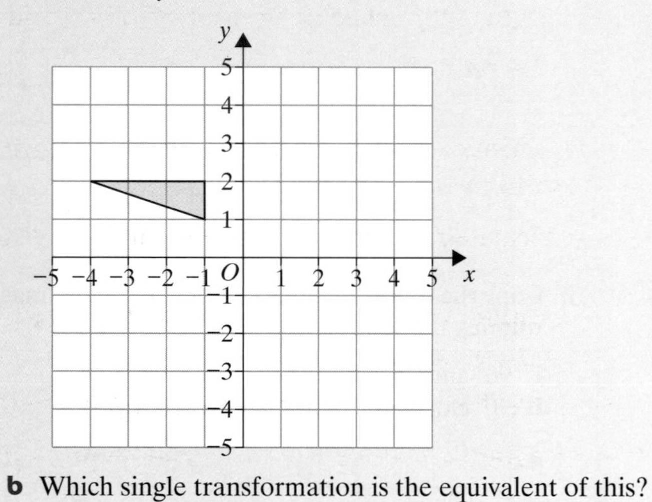

b Which single transformation is the equivalent of this?

19 Ratio and proportion

OBJECTIVES

D **Examiners would normally expect students who get a D grade to be able to:**

Solve simple ratio and proportion problems, such as finding the ratio of teachers to students in a school

C **Examiners would normally expect students who get a C grade also to be able to:**

Solve more complex ratio and proportion problems, such as sharing out money between two groups in the ratio of their numbers

Solve ratio and proportion problems using the unitary method

B **Examiners would normally expect students who get a B grade also to be able to:**

Calculate proportional changes using a multiplier

A **Examiners would normally expect students who get an A grade also to be able to:**

Solve direct and inverse proportion problems

Interpret graphs of direct and inverse proportion relationships

What you should already know ...

- How to add, subtract, multiply and divide numbers
- How to simplify fractions

VOCABULARY

Constant – a number that does not change, for example, the formula $P = 4l$ states that the perimeter of a square is always four times the length of one side; 4 is a constant and P and l are variables

Ratio – the ratio of two or more numbers or quantities is a way of comparing their sizes, for example, if a school has 25 teachers and 500 students, the ratio of teachers to students is 25 to 500, or 25 : 500 (read as 25 to 500)

Unitary ratio – a ratio in the form $1 : n$ or $n : 1$; for example, for every 100 female babies born, 105 male babies are born. The ratio of the number of females to the number of males is 100 : 105; as a unitary ratio, this is 1 : 1.05, which means that, for every female born, 1.05 males are born

Proportion – if a class has 12 boys and 18 girls, the proportion of boys in the class is $\frac{12}{30}$, which simplifies to $\frac{2}{5}$, and the proportion of girls is $\frac{18}{30}$, which simplifies to $\frac{3}{5}$ (the **ratio** of boys to girls is 12 : 18, which simplifies to 2 : 3) – a proportion compares one part with the whole; a ratio compares parts with one another

Unitary method – a way of calculating quantities that are in proportion, for example, if 6 items cost £30 and you want to know the cost of 10 items, you can first find the cost of one item by dividing by 6, then find the cost of 10 by multiplying by 10

6 items cost £30

1 item costs $\frac{£30}{6} = £5$

10 items cost $10 \times £5 = £50$

Direct proportion – if two variables are in direct proportion, one is equal to a constant multiple of the other, so that if one increases, the other increases and if one decreases then the other decreases

In general $x \propto y$ and $x = kx$

Inverse proportion – if two variables are in inverse proportion, their product is a constant; so that if one increases, the other decreases and vice versa

In general $x \propto \frac{1}{y}$ and $x = k\frac{1}{y}$ and $xy = k$

Learn 1 Finding and simplifying ratios

Examples:

a A school has 50 teachers and 900 students.
Write down the teacher : student ratio and express it in its simplest form.

First write the numbers in the correct order for the ratio and separate them with a colon symbol.

The colon symbol is used to express ratio

The teacher : student ratio is 50 : 900.

'50 : 900' is read as 'fifty to nine hundred'

Like cancelling a fraction, the ratio can be simplified.

Both numbers have been divided by 10

Ratio = 50 : 900 = 5 : 90 = 1 : 18

Both numbers have been divided by 5

$$\frac{50}{900} \overset{\div 10}{=} \frac{5}{90} \overset{\div 5}{=} \frac{1}{18}$$

This is like simplifying fractions

The ratio in its simplest form is 1 : 18.

This means that, for every teacher in this school, there are 18 students (or the number of teachers is $\frac{1}{18}$ of the number of students).

b A shopkeeper buys boxes of chocolates for £3.50 and sells them for £4.25
What is the ratio of the profit to the cost price?

The profit is the selling price minus the cost price = £4.25 – £3.50 = 75p.

Ratio of profit to cost price = 75p : £3.50

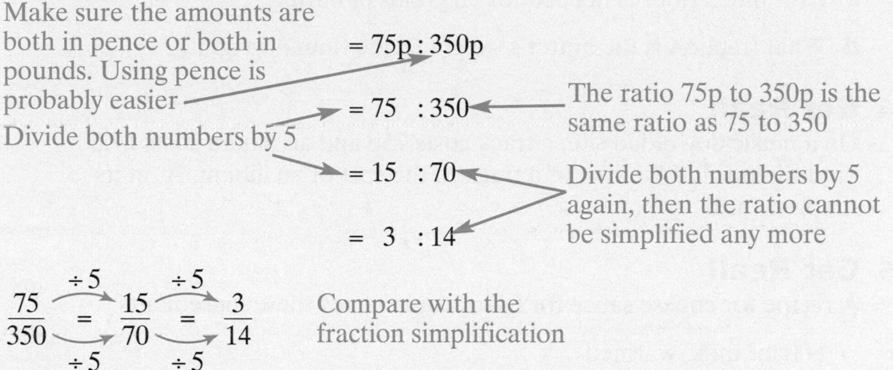

$$\frac{75}{350} = \frac{15}{70} = \frac{3}{14}$$

The ratio of 3 to 14 means that the shopkeeper makes a profit of £3 for every £14 she spends on boxes of chocolates (if she sells them all).

It also means that:

- the profit is $\frac{3}{17}$ of the selling price
- the cost price is $\frac{14}{17}$ of the selling price.

Check that you can see where the numerators and denominators have come from.

Also check that $\frac{3}{17}$ of £4.25 is 75p and $\frac{14}{17}$ of £4.25 is £3.50

Apply 1

1 Write each of these ratios as simply as possible.

a 2 : 4	**e** 2 : 12	**i** 24 : 36	**m** 0.3 : 0.8
b 2 : 6	**f** 2 : 14	**j** 25 : 100	**n** $2\frac{1}{2} : 7\frac{1}{2}$
c 2 : 8	**g** 12 : 36	**k** $\frac{2}{3} : \frac{4}{9}$	**o** 20% : 80%
d 2 : 10	**h** 18 : 24	**l** 1.5 : 2.5	**p** 25 : 200

2 **a** Write down three different pairs of numbers that are in the ratio 1 : 2.

b Write down three different pairs of numbers that are in the ratio 1 : 4.

c Explain how to find pairs of numbers that are in the ratio 1 : 4.

d Pippa writes the three pairs of numbers 6 and 9, 9 and 12, and 12 and 15.
She says these pairs of numbers are all in the same ratio.
What has Pippa done wrong?

3 Get Real!

A recipe for pastry needs 50 grams of butter and 100 grams of flour.

a What is the ratio of butter to flour? What is the ratio of flour to butter?

b How much butter is needed for 200 grams of flour?

c How much flour is needed for 30 grams of butter?

d What fraction is the butter's weight of the flour's weight?

4 Get Real!

On a music download site, a track costs 75p and an album costs £7.50
Find the ratio of the cost of a track to the cost of an album, A, in its simplest form.

5 Get Real!

A recipe for cheese sauce for four people needs these ingredients:

- 600 mℓ milk, warmed
- 100 g grated cheese
- 40 g flour
- 40 g butter
- seasoning

a List the ingredients needed to make enough cheese sauce for two people.

b Explain how to find the quantities to make enough cheese sauce for ten people.

6 Get Real!

a Find, in their simplest forms, the teacher : student ratios for these schools.

School	Number of teachers	Number of students
School 1	75	1500
School 2	15	240
School 3	22	374
School 4	120	1800
School 5	65	1365

The numbers have been made simple so that it is easy for you to work them out Real schools have harder numbers!

b **i** If a school with 50 teachers had the same teacher : student ratio as School 1, how many students would it have?

ii If a school with 2000 students had the same teacher : student ratio as School 1, how many teachers would it have?

c Which school has the 'best' teacher to student ratio? (That is, which school has the smallest number of students for each teacher?)

7 Get Real!

a Find the profit : cost price ratio for these items.

Item	Cost price	Selling price	Profit
Litre of petrol	85p		5p
Car	£4500	£5000	
Calculator	£15		£10
Book	£2.80	£3.50	
Magazine		£1.10	15p
Sandwich		£1.25	50p

b Use the profit : cost price ratio for the car to write fraction statements like those at the end of Learn **1**.

c In question **6**, all the ratios were in the form 1 : *something* (mathematically, $1 : n$) so they were easy to compare.
How could you compare the ratios in part **a**? Would a calculator help?

8 Get Real!

a In a salsa class, the ratio of women to men is 5 : 4.

i There are 10 women in the class. How many men are there?

ii The number of women and the number of men both double.
Does the ratio change? Explain your answer.

b In the jazz dance class, the ratio of men to women is 2 : 3 and there are 10 dancers altogether.

i How many men and how many women are there in the jazz dance class?

ii Two more men and two more women join the class.
Does the ratio of men to women increase, decrease or stay the same?
Explain your answer.

9 The ratio $x : y$ simplifies to 3 : 4.

a If x is 6, what is y?

b If y is 12, what is x?

c If y is 2, what is x?

d If x and y add to 35, what are x and y?

10 Make up another question like question **9** and give the answers.

11 Here is a pattern sequence.

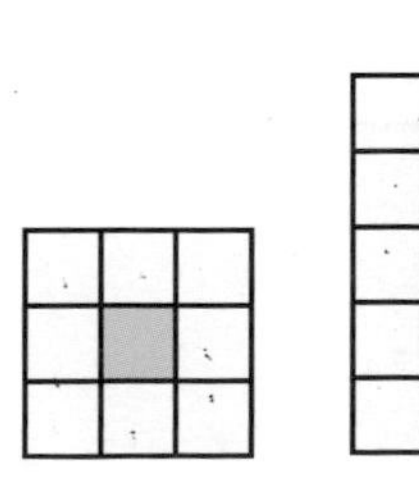

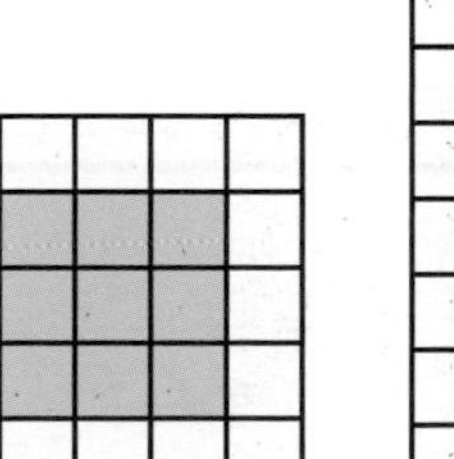

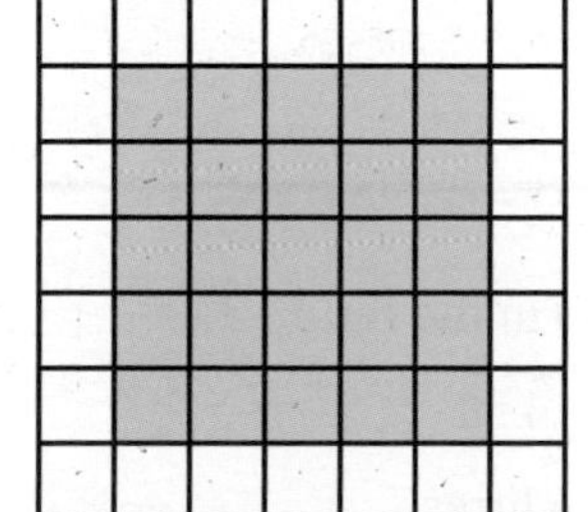

a Does the ratio 'number of green squares : number of yellow squares' increase, decrease or stay the same as the shapes get bigger?
Show how you worked out your answer.

b Draw your own sequence where the ratio of the number of green squares to the number of yellow squares stays the same as the shapes get bigger.

Explore

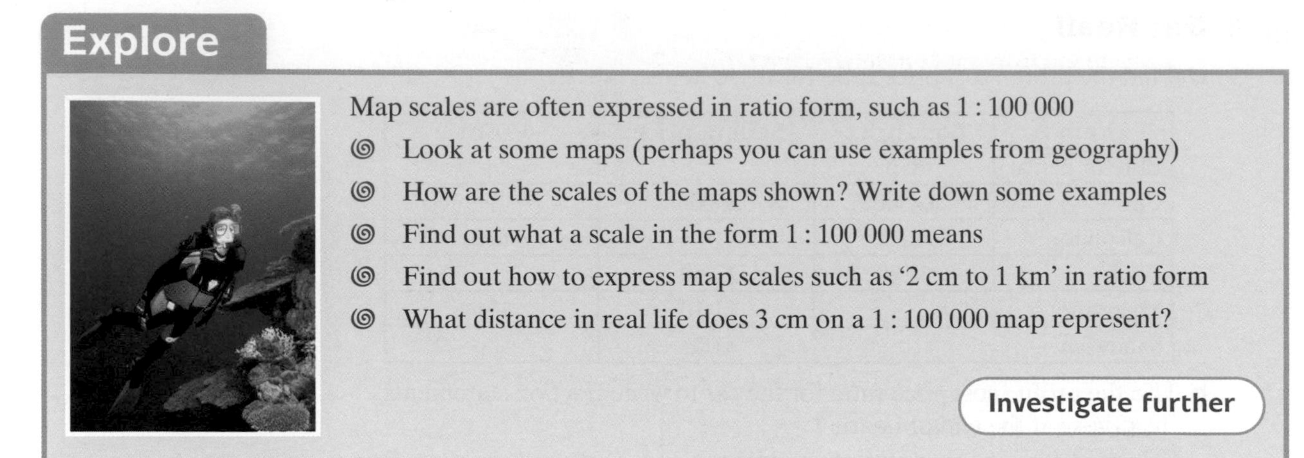

Map scales are often expressed in ratio form, such as 1 : 100 000

- Look at some maps (perhaps you can use examples from geography)
- How are the scales of the maps shown? Write down some examples
- Find out what a scale in the form 1 : 100 000 means
- Find out how to express map scales such as '2 cm to 1 km' in ratio form
- What distance in real life does 3 cm on a 1 : 100 000 map represent?

Investigate further

Learn 2 Using ratios to find quantities

Example:

In a school of 1000 students, the ratio of boys to girls is 9 : 11.
How many boys and how many girls are there in the school?

For this problem, you need to divide 1000 students in the ratio 9 : 11 to find the number of boys and the number of girls.

The ratio shows that *for every* 9 boys there are 11 girls. So *in every* 20 students, there are 9 boys and 11 girls, as $9 + 11 = 20$.

Out of every 20 students, 9 are boys

Out of every 20 students, 11 are girls

The fraction of boys in the school is $\frac{9}{20}$ and the fraction of girls is $\frac{11}{20}$

The number of boys in the school is $\frac{9}{20}$ of 1000.

$\frac{1}{20}$ of $1000 = \frac{1000}{20} = 50$

So $\frac{9}{20}$ of $1000 = 50 \times 9 = 450$

The number of girls in the school is $\frac{11}{20}$ of 1000, which is $11 \times 50 = 550$.

So the number of boys is 450 and the number of girls is 550.

Check that the number of boys and the number of girls add up to 1000, the total number of students in the school

Apply 2

1 Divide these numbers and quantities in the ratio 1 : 2.

a 150 **c** £4.50 **e** £1.50

b 300 **d** 6 litres **f** 1.5 litres

2 Divide the numbers and quantities in question **1** in the ratio 2 : 3.

3 Divide the numbers and quantities in question **1** in the ratio 3 : 7.

4 Divide the numbers and quantities in question **1** in the ratio 1 : 3 : 6.

5 Get Real!

Pastry is made from fat and flour in the ratio 1 : 2.

a How much flour is needed to make 150 g of pastry?

b How much fat is needed to make 6 ounces of pastry?

c How much pastry can you make if you have plenty of flour but only 60 g of fat?

6 *You should be able to do the first four of these schools without a calculator but you will need one for School E and for parts* ***b*** *and* ***c****.*

a Find the number of boys and the number of girls in these schools.

School	Total number of students	Boy : girl ratio
School A	750	1 : 1
School B	900	4 : 5
School C	1800	4 : 5
School D	1326	6 : 7
School E	1184	301 : 291

School E shows the most realistic ratio. What is the boy : girl ratio in your school or college?

b Find the boy : girl ratios in part **a** in the form 1 : n (in other words, find how many girls there are for every boy).

c Which school has the largest proportion of boys? Give a reason for your answer.

7 This table shows the ratio of carbohydrate to fat to protein in some foods.

a Find the amount of fat in 150 g of each of the foods.

Food	Carbohydrate : fat : protein
Chicken sandwich	1 : 1 : 1
Grilled salmon	0 : 1 : 1
Yoghurt (whole milk)	1 : 2 : 1
Taco chips	10 : 4 : 1
Bread	7 : 2 : 1
Milk	2 : 3 : 2

HINT Use a calculator for milk as the ratios do not work out easily. Round your answers to the nearest 5 grams.

b Which of these foods would you avoid if you were on a low-fat diet?

c How many grams of yoghurt would you need to eat to have 100 g of protein?

d Which of these foods would you avoid if you were on a low-carbohydrate diet?

8 Two people, Jamil and Jane, invested money in a business. Jamil invested £3450 and Jane invested £5500. At the end of the financial year, the profit is split between Jamil and Jane in the ratio of their investments. The profit is £7350. How much do Jamil and Jane each receive?

9 Bronze for coins can be made of copper, tin and zinc in the ratio 95 : 4 : 1.

a How much of each metal is needed to make 1 kilogram of bronze?

b How much of each metal is needed to make 10 kilograms of bronze?

c How much of each metal is needed to make half a kilogram of bronze?

d How much zinc would there be in a coin weighing 6 grams?

Learn 3 Ratio and proportion

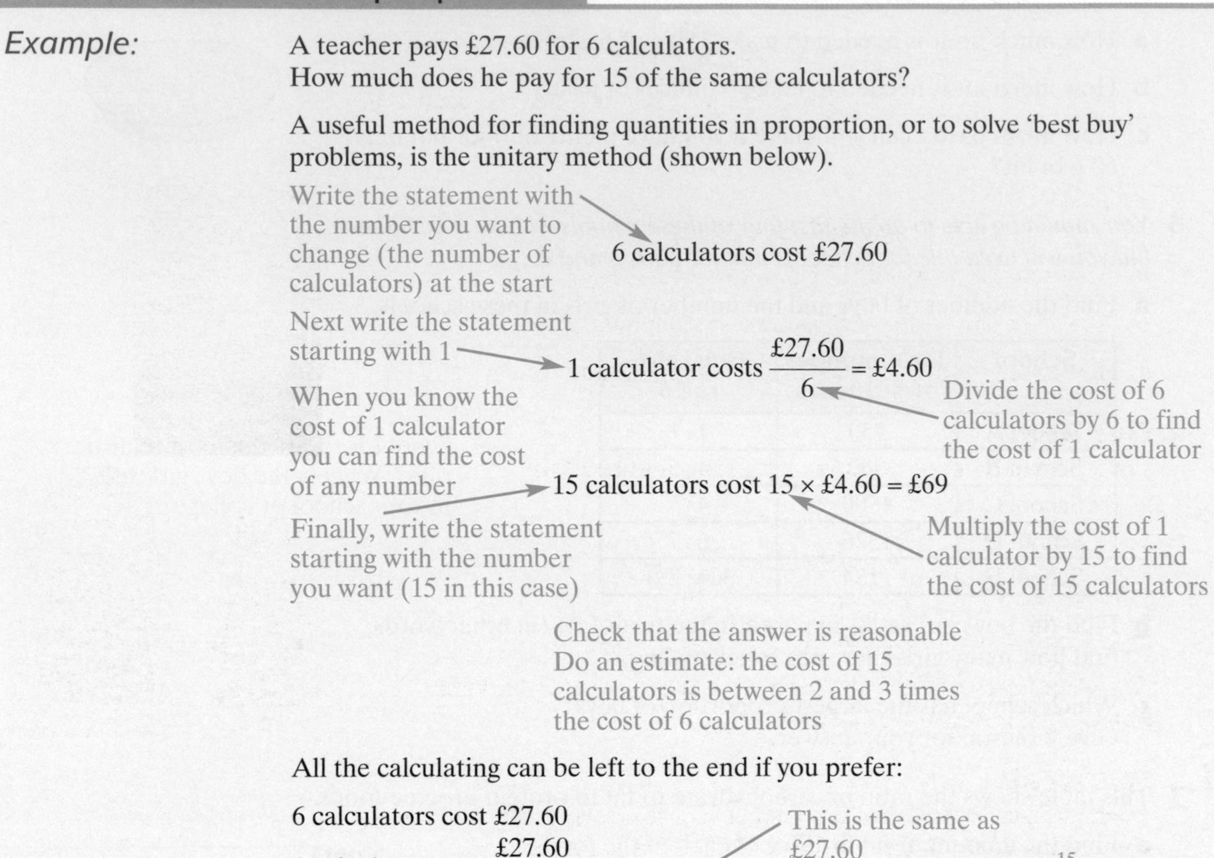

Example:

A teacher pays £27.60 for 6 calculators.
How much does he pay for 15 of the same calculators?

A useful method for finding quantities in proportion, or to solve 'best buy' problems, is the unitary method (shown below).

Write the statement with the number you want to change (the number of calculators) at the start → 6 calculators cost £27.60

Next write the statement starting with 1 → 1 calculator costs $\frac{£27.60}{6} = £4.60$

Divide the cost of 6 calculators by 6 to find the cost of 1 calculator

When you know the cost of 1 calculator you can find the cost of any number

Finally, write the statement starting with the number you want (15 in this case) → 15 calculators cost $15 \times £4.60 = £69$

Multiply the cost of 1 calculator by 15 to find the cost of 15 calculators

Check that the answer is reasonable
Do an estimate: the cost of 15 calculators is between 2 and 3 times the cost of 6 calculators

All the calculating can be left to the end if you prefer:

6 calculators cost £27.60

1 calculator costs $\frac{£27.60}{6}$

15 calculators cost $15 \times \frac{£27.60}{6} = £69$

This is the same as $\frac{£27.60}{6} \times 15$ or $£27.60 \times \frac{15}{6}$

If you feel confident with problems like this, you can do them in one step by combining the multiplication and division, but be careful and check that your answer is sensible

Apply 3

1 Get Real!
Check that, in the example above, the ratio 'cost of 1 calculator : cost of 6 calculators : cost of 15 calculators' is 1 : 6 : 15.

2 Get Real!
Sajid worked for 8 hours and was paid £30.

a How much will he be paid for working 10 hours at the same rate of pay?

b Complete a copy of this table. Plot the values in the table as points on a graph, using the numbers of hours worked as the x-coordinates and the money earned as the corresponding y-coordinates.

Number of hours worked	0	2	4	6	8	10
Money earned (£)					30	

c The points should lie in a straight line through (0, 0).

i Explain why.

ii What does the gradient of the line represent?

iii Show how to use the graph to find out how much Sajid earns in 5 hours.

3 Get Real!

50 grams of fish food will feed 8 fish for 1 day.

a How much food would 12 fish require for 1 day?

b How many days can 2 fish survive on 50 grams of food?

c How much food is needed for 10 fish for 7 days?

4 Get Real!

Lovelylocks shampoo is sold in travel size and large size.

	Amount of shampoo	Price
Travel size	40 grams	75p
Large size	125 grams	£2.25

Calculate which of the two sizes gives you better value for money. Show all your working clearly.

5 Get Real!

'Rich and Dark' chocolate is sold in a 55 g size costing 60p and a 100 g size costing £1.05. Which of these is better value for money?

6 Get Real!

On the motorway, Jacob drove a distance of 84 miles in 3 hours.

a How far would Jacob travel in 4 hours at the same average speed?

b How far would he go in three-quarters of an hour at this average speed?

c How long would it take for Jacob to travel 60 miles at this average speed?

7 Get Real!

Notice that the two parts of this question are really the same! Use part **a** to help you work out part **b**.

a 80% of a number is 16. Use the unitary method to find 100% of the number.

b A sweater is reduced by 20% to £16 in the sale. What was the original price of the sweater?

8 **a** Two numbers are in the ratio 1 : 0.75
The first number is 12; what is the second?

b Two numbers are in the ratio 1 : 0.75
The second number is 12; what is the first?

c Three numbers are in the ratio 1.1 : 1 : 0.9
The third number is 36; what are the other two numbers?

9 Get Real!

The weights of objects on other planets are proportional to their weights on Earth. A person weighing 120 pounds on Earth would weigh 20 pounds on the moon and 300 pounds on Jupiter.

a What would a teenager weighing 80 pounds on Earth weigh on Jupiter?

b What would a rock weighing 10 kilograms on the moon weigh on Earth?

c This graph shows the weights of objects on Jupiter compared with their weights on Earth. Copy the graph and sketch a line on it to show the weights of objects on the moon compared with their weights on Earth.

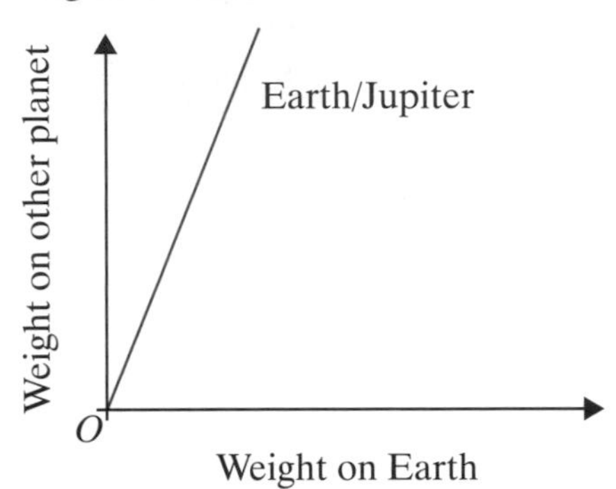

d Express the ratio 'weight of object on Earth : weight of object on moon : weight of object on Jupiter' in its simplest form.

Explore

You may already know something about the Fibonacci sequences

Each term is found by adding together the last two terms

So, starting with 1, 1, the series continues 1, 1, 2, 3, 5, 8, ...

- Carry the sequence on until you have at least 20 terms (would a spreadsheet be useful?)
- Work out, in the form $1 : n$, the ratio of
  term 1 to term 2
  term 2 to term 3
  term 3 to term 4 and so on
- What can you say about the ratios as you go through the series?

Investigate further

Learn 4 Calculating proportional changes

Examples:

In a laboratory experiment, the number of bacteria in a colony increases by 10% every hour. At the start, the number of bacteria is 1000.

a What is the number of bacteria after 1 hour?
b What is the number of bacteria after 2 hours?
c What is the number of bacteria after 10 hours?

You could work out 10% of the number of bacteria and add it on but the one-step method here is more powerful
100% + 10% is 110%

a After one hour, the number of bacteria = 110% of 1000 = $\frac{110}{100} \times 1000 = 1100$

$\frac{110}{100} = 1.1$, so finding 110% of something is the same as multiplying it by 1.1

b After 2 hours, the number of bacteria is 110% of 1100 = $1.1 \times 1100 = 1210$ and so on.

You could do this step by step, multiplying the number each hour by 1.1 to get the number for the next hour, but it takes time! This method does all 10 multiplications in one step

c After 10 hours, the number of bacteria is $1000 \times 1.1^{10} = 2593.74246... \approx 2590$

Apply 4

1 Get Real!
In the example above, find to the nearest 10 the number of bacteria after:

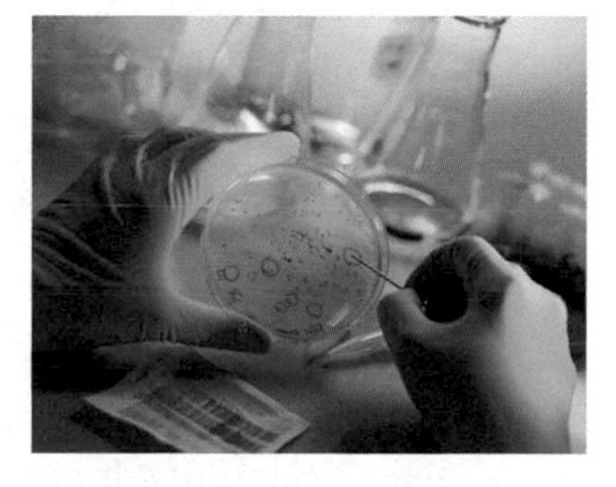

a 4 hours

b two and a half hours

c 45 minutes

d one day.

2 Get Real!
David puts £500 into a savings account that pays 4% per year in interest, added once a year.
The interest is added onto the money in the account.
How much will David have in the account at the end of:

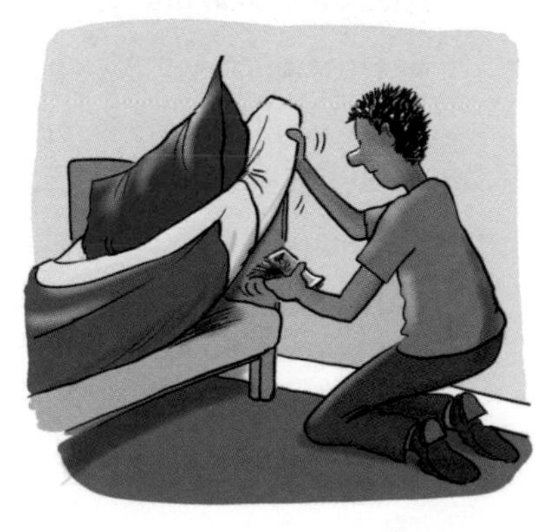

a 1 year

b 2 years

c 5 years?

d How much interest has David gained at the end of 5 years?

Round your answers sensibly.

3 Get Real!

A type of bacteria doubles its numbers every hour.
There are 1 million bacteria at the start.

a Find the number of bacteria after:

i 3 hours

ii 5 hours

iii half an hour.

b Use trial and improvement to estimate, in hours and minutes, the time taken for the number of bacteria to reach 10 million.

4 Get Real!

Gavin is offered a job with a starting salary of £18 000 a year with an annual increase of 2.5%.

a How much will Gavin earn per month after 1 year?

b Will he be earning over £20 000 a year after 5 years?

5 Get Real!

Sue puts £500 into a savings account. After a year, she has £515.

a What annual rate of interest is paid on her account?

b How much money will Sue have after 2 years?

c How many years will it take for Sue's money to exceed £550?

6 Get Real!

In one part of the country, house prices rose by an average rate of 10% a year for the years 2000–2005.

a What did a house that cost £50 000 in 2000 cost in 2005?

b How much did the price of a house costing £78 000 go up in these 5 years?

c What is the price in 2005 of a house that cost £x in 2000?

7 Get Real!

Some banks and building societies add interest on to savings accounts every 6 months instead of every year. So if the annual rate of interest is 4%, 2% interest is added every 6 months.
Alex puts £800 into an account paying 2% every 6 months.

a How much will Alex have in his account at the end of 1 year?

b How much more is this than if 4% interest had been added at the end of the year?

c Compare the difference in the interest after 2 years.

d Explain why, with simple interest the total amount of interest is the same with both methods but with compound interest the amounts of interest are different.

8 Get Real!

The value of a car depreciates by 10% each year from when it is 4 years old to when it is 8 years old. Its value when it is 4 years old is £3500.

What is its value when it is 8 years old?
Give your answer to a sensible degree of accuracy.

9 Get Real!

Research shows that the number of tigers in India has recently been declining at a rate of approximately 1% per year.
The number of tigers in 2002 was estimated as 3836.

a Calculate the number of tigers in India in 2005 at this rate of decrease.

b How many tigers would be lost in the 10 years from 2002 to 2012?

c Explain why working out 10% of the 3836 does not give the correct answer to part **b**.

10 Get Real!

The area of rainforest in Brazil is declining at an estimated rate of 2.3% each year. The current area of rainforest is 1.8 million square kilometres. According to this estimated rate:

a how many square kilometres will be lost next year

b what will the area of rainforest in Brazil be in 5 years' time?

11 The value of a car depreciated from £8000 to £5000 in 2 years.

a Find the percentage depreciation per year.

b What would a car costing £2250 at the end of the 2 years have been worth at the beginning if it had this percentage rate of depreciation?

12 The price of a house went up from £49 000 to £70 000 in 5 years.

a What is the average annual percentage rate of increase?

b Explain why it is not correct to find the total increase as a percentage of the £49 000 divided by 5.

Explore

If someone puts £100 into a bank that pays 100% interest once a year (very unlikely!) the total amount in the bank at the end of year would be £200

- Show that, if the interest was added every 6 months instead of every year, the amount at the end of a year would be £225
- What would the amount be at the end of a year if the interest was added every month?
- Every week?
  Every day?

Investigate further

Learn 5 Direct and inverse proportion

Examples:

The area of a rectangle is fixed. When the length of the rectangle is 15 cm, its width is 7.2 cm.

The product of the length and the width is fixed. This means that the length of the rectangle is inversely proportional to its width

a Find an equation expressing the length of the rectangle in terms of its width.

b Use your equation to find the length of the rectangle when the width is 6 cm.

c Sketch a graph of length against width.

Let the area be A cm^2, the length l cm and the width w cm.

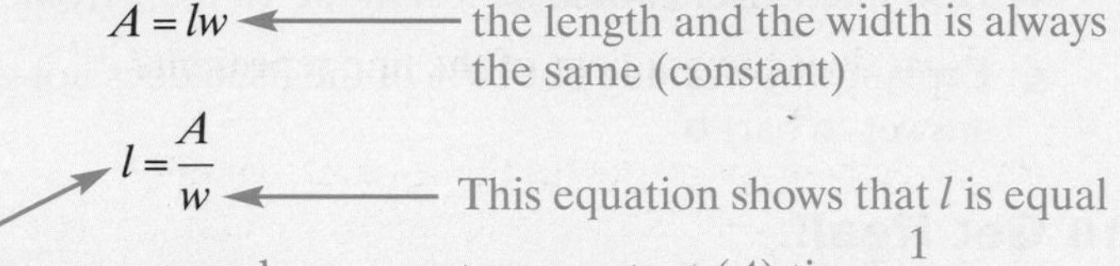

a $$A = lw$$

The area is fixed, so the product of the length and the width is always the same (constant)

$$l = \frac{A}{w}$$

The equation has been rearranged. It is like rearranging the 'distance = speed × time' formula to give speed $= \frac{\text{distance}}{\text{time}}$

This equation shows that l is equal to a constant (A) times $\frac{1}{w}$

In other words, l is **directly proportional** to $\frac{1}{w}$, or **inversely proportional** to w

Another way of expressing this is to write $l \propto \frac{1}{w}$

$\propto$ means 'is proportional to'

Since $l = 15$ when $w = 7.2$,

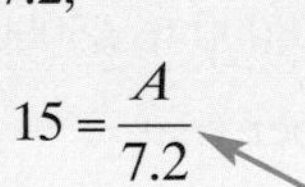

$$15 = \frac{A}{7.2}$$

You could work the area out at the beginning but this shows the process of working with inverse proportion

Multiplying by 7.2 gives

$$A = 15 \times 7.2 = 108$$

So an equation for l in terms of w is

$$l = \frac{108}{w}$$

b When the width is 6 cm, $w = 6$, therefore,

$$l = \frac{108}{6} = 18$$

So, when the width is 6 cm, the length is 18 cm.

c A sketch of length against width is:

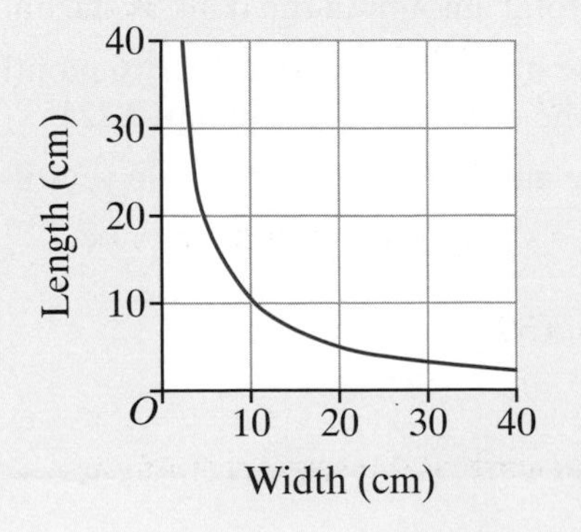

All inverse proportion graphs are this shape. Notice its symmetry and that it never actually meets the x- and y-axes, just gets close to them

Apply 5

1 Get Real!
The amount Emma is paid is directly proportional to the hours she works.
When she works for 6 hours she is paid £33.
Emma is paid £P after working h hours.

a Find a formula expressing P in terms of h and use it to find:

i how much Emma will be paid for 8 hours work

ii how long Emma must work to be paid £110.

b Sketch a graph of P against h.

c What does the gradient of the line represent?

2 P is directly proportional to the square root of Q.
If $P = 6$ when $Q = 100$, find a formula expressing P in terms of Q and use it to find:

a P when $Q = 64$

b Q when $P = 3.6$

c a formula expressing Q in terms of P.

3 The circumference of a circle is directly proportional to its diameter.
A circle with a diameter of 20 cm has a circumference of approximately 62.8 cm.

a Find a formula connecting C and d, where C cm is the circumference and d cm is the diameter.

b Use your formula to find the circumference of a circle with diameter 12 cm.

c Use your formula to find the diameter of a circle with circumference 1 metre.

4 Get Real!
A petrol pump takes 20 seconds to fill an 8-litre can.
How long will it take to fill a 70-litre car fuel tank?

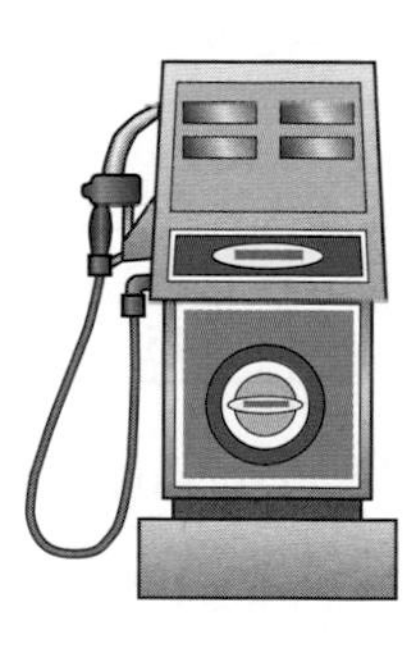

5 Get Real!
£10 is worth €16.20. Write a formula to find the number of pounds, P, you get for E euro.

6 If the weight, w grams, of a sphere is directly proportional to the radius, r cm, cubed, then when the radius is 2 cm the weight is 4 grams.
Find the weight of a sphere with a radius of 5 cm.

7 Get Real!
The amount that Sajid earns, £P, is directly proportional to the number of hours, n, for which he works.
When he worked for 8 hours he was paid £30.

a Find an equation connecting P and n.

b How much will he be paid for 10 hours work at the same rate of pay?

c Sketch a graph of P against n.

d What are the links between this question and Apply **3** question **2**?

8 Get Real!

The number of days a food supply lasts is inversely proportional to the number of people to be fed. You have enough food for 5 people for 7 days.
How long will the food last if there are only 2 people?

9 In each part of this question, the two variables x and y are inversely proportional.
Use the given pair of values of x and y to find the constant k in the equation. Then use the equation to find the missing x- or y-value. Round off sensibly when necessary.

a When $x = 5, y = 12$. Find y when x is 10

b When $x = 2.5, y = 24$. Find x when y is 10

c When $x = 0.1, y = 10$. Find y when x is 0.2

d When $x = \frac{1}{10}, y = 10$. Find x when y is $\frac{1}{5}$

e When $x = 0.53, y = 2.6$. Find y when x is 0.72

10 Two quantities are inversely proportional.
If one quantity is doubled, what happens to the other?
Illustrate your answer with an example.

11 Say whether each pair of variables is directly proportional, inversely proportional, or neither.

a Length and width of a rectangle with fixed perimeter.

b Length and width of a rectangle with fixed area.

c Time taken and distance travelled at constant speed.

d Speed and time taken for a journey of fixed distance.

e Length and perimeter of square.

f Length and area of square.

12 Given that $y \propto x$, copy and complete this table.

x	5		40
y	45	180	

13 Match the following:

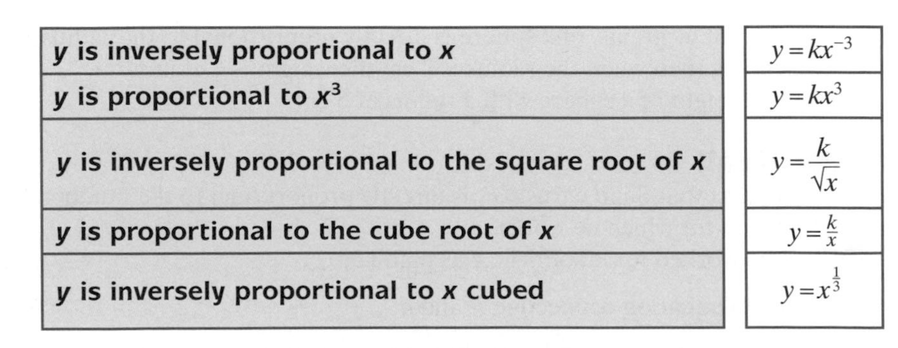

y is inversely proportional to x	$y = kx^{-3}$
y is proportional to x^3	$y = kx^3$
y is inversely proportional to the square root of x	$y = \frac{k}{\sqrt{x}}$
y is proportional to the cube root of x	$y = \frac{k}{x}$
y is inversely proportional to x cubed	$y = x^{\frac{1}{3}}$

14 The diagram below shows these relationships:

i y is proportional to x

ii y is proportional to x^2

iii y is proportional to x^3

iv y is inversely proportional to x

v y is proportional to the square root of x

Which relationship matches which graph?

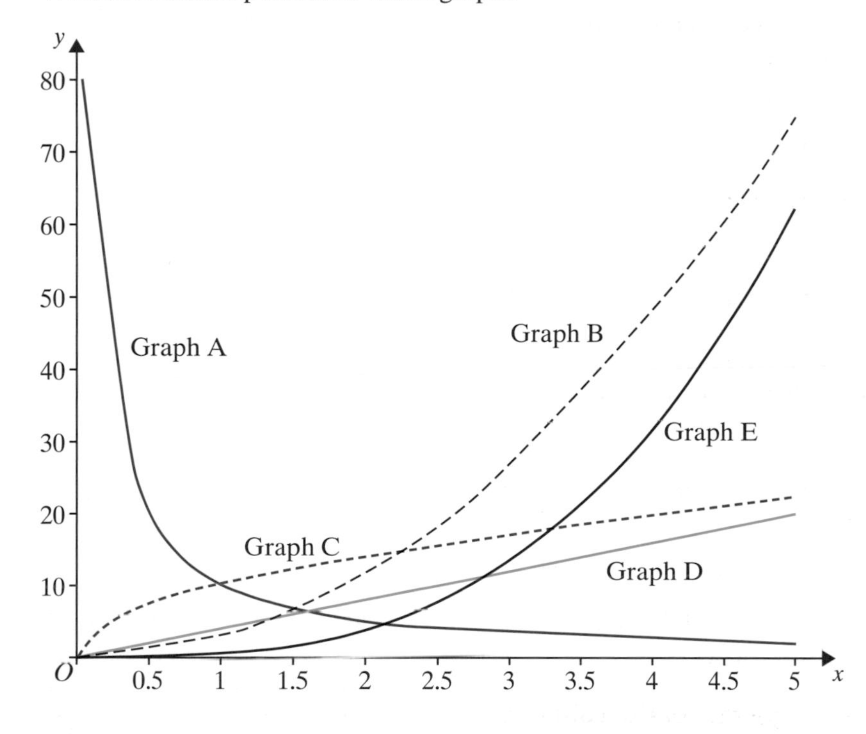

Ratio and proportion

ASSESS

The following exercise tests your understanding of this chapter, with the questions appearing in order of increasing difficulty.

1 **a** In a choir there are 12 boys and 18 girls.
Express this as a ratio in its simplest form.

b Two more boys and two more girls join the choir.
Express the new ratio in its simplest form.

2 **a** 2400 people voted in a local election.
Votes for the three candidates were in the ratio 5 : 6 : 9.
How many votes did each candidate get?

b A drink is made up of water, orange and lemon in the ratio 5 : 1 : 2.
Find the amount of water, orange and lemon in a 1 litre bottle.

3 a Jamie is cooking omelettes.
To make omelettes for 4 people he uses 6 eggs.
How many eggs does Jamie need to make omelettes for 10 people?

b The supermarkets 'Lessprice' and 'Lowerpay' both sell packs of pens.
'Lessprice' sells a pack of 5 pens for £1.25
'Lowerpay' sells a pack of 6 of the same pens for £1.44
Which supermarket gives the greater value?

c Two circles have radii of 5 cm and 6 cm respectively.
What is the ratio of:

i their circumferences

ii their areas?

d It takes Kelly 25 seconds to run 200 m.
At the same pace, how long will it take her to run:

i 56 m

ii 128 m?

4 a Due to illness, a man increases his weight each year in the ratio 10 : 11.
When diagnosed he weighed 13 stones.
How heavy was he at the end of the third year?

b Mr Clumsy has dropped oil onto his carpet.
The area of the stain is increasing in the ratio of 5 : 6 each minute.
The original stain was 50 cm^2.
How big is the stain after 3 minutes?

c A car, originally bought for £10 000, depreciates at a rate of 30% yearly.
How much is it worth after 4 years?

5 a If $a = kb^2$ you say that a is proportional to b^2.
Express the following equations in a similar way:

i $A = \pi r^2$ **iii** $y = \dfrac{5}{x^2}$

ii $v = \frac{4}{3}\pi r^3$ **iv** $T = \frac{3}{4}\sqrt{l}$

b Christopher Columbus is h m above sea level.
The distance, d km, that he can see to the horizon is proportional to the square root of the height.
When Christopher is 100 m above sea level the horizon is 36 km away.
Calculate:

i the distance to the horizon when Christopher is 150 m above sea level

ii how far up Christopher must go to be able to see 40 km out to sea

iii how much further Christopher can see when he climbs from a height of 81 m to 121 m above sea level.

c Boyle's law states that, under certain conditions, the pressure exerted by a particular mass of gas is inversely proportional to the volume it occupies. In these conditions a volume of 150 cm^3 exerts a pressure of 6×10^4 Nm^{-2}. The volume is reduced to 80 cm^3.
What is the new pressure?

6 These five sketch graphs show proportional relationships.

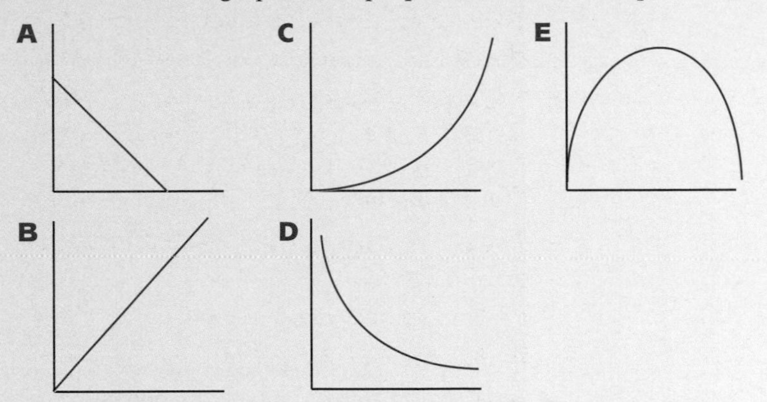

Match each of these statements with an appropriate graph.

i The distance a stone falls down a well is proportional to the square of the time it takes.

ii The price of a piece of wood is proportional to its length.

iii The time taken to drive a set distance is inversely proportional to the average speed.

Glossary

Algebraic expression – a collection of terms separated by + and – signs such as $x + 2y$ or $a^2 + 2ab + b^2$

Amount – the total you will have in the bank or the total you will owe the bank, at the end of the period of time

Arc (of a circle) – part of the circumference of a circle; a minor arc is less than half the circumference and a major arc is greater than half the circumference

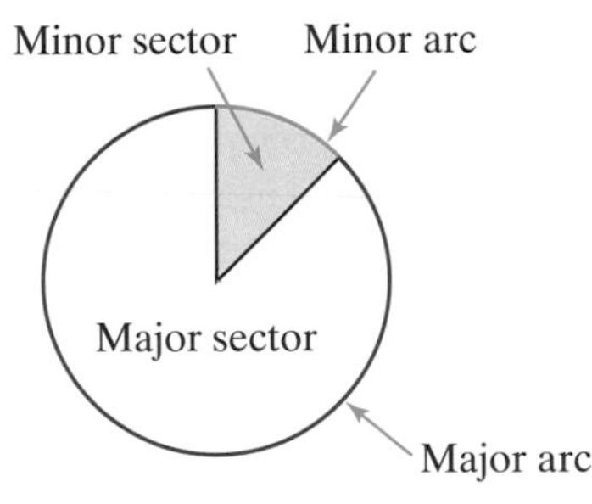

Area – the amount of enclosed space inside a shape

Average – a single value that is used to represent a set of data

Axis (pl. **axes**) – the lines used to locate a point in the coordinates system; in two dimensions, the x-axis is horizontal, and the y-axis is vertical. This system of Cartesian coordinates was devised by the French mathematician and philosopher, René Descartes

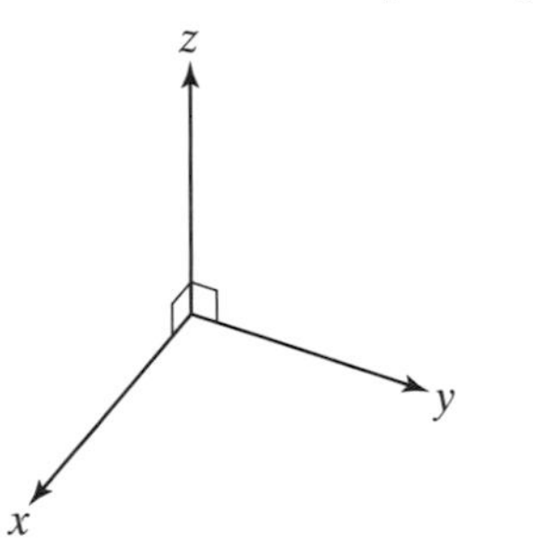

In three dimensions, the x- and y-axes are horizontal and at right angles to each other and the z-axis is vertical

Axis of symmetry – the mirror line in a reflection

Back-to-back stem-and-leaf diagram – a stem-and-leaf diagram used to represent two sets of data

Number of minutes to complete a task

Leaf (units) Girls	Stem (tens)	Leaf (units) Boys
7 7 6 5 4 2 2	1	1 6 7 8 9
7 6 4 3 2 1	2	2 2 7 7 7 7 8 9
7 0	3	1 4 6

Key: 3|2 represents 23 minutes   Key: 3|4 represents 34 minutes

Balance – the amount of money you have in your bank account or the amount of money you owe after you have paid a deposit

Box plot or **box and whisker plot** – used to show how the data is distributed

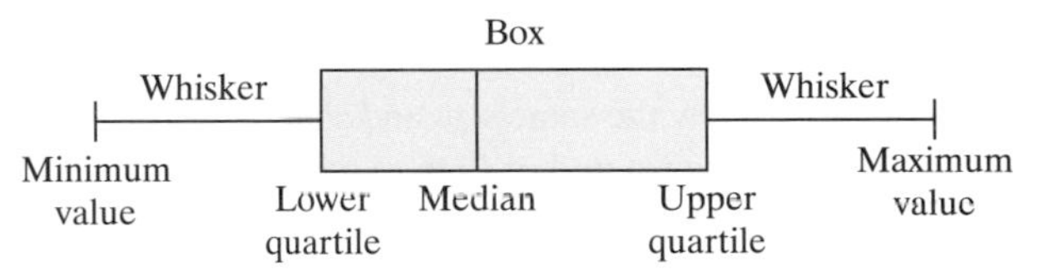

Brackets – these show that the terms inside should be treated alike, for example,

$2(3x + 5) = 2 \times 3x + 2 \times 5 = 6x + 10$

Capacity – the amount of liquid a hollow container can hold, commonly measured in litres (1 litre = 1000 cm^3)

Categorical data – see qualitative data

Centre of rotation – the fixed point around which the object is rotated

Chord – a straight line joining two points on the circumference of a circle

Circle – a shape formed by a set of points that are all the same distance from a fixed point (the centre of the circle)

Circumference – the perimeter of a circle

Cluster sampling – this is useful where the population is large and it is possible to split the population into smaller groups or clusters

Coefficient – the number (with its sign) in front of the letter representing the unknown, for example:

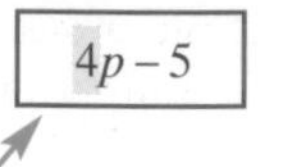

$2-3p^2$

4 is the coefficient of p $\quad$ -3 is the coefficient of p^2

Collect like terms – to group together terms of the same variable, for example, $2x + 4x + 3y = 6x + 3y$

Common factor – factors that are in common for two or more numbers, for example,

the factors of 6 are 1, 2, 3, 6
the factors of 9 are 1, 3, 9
the common factors are 1 and 3

Common fraction – see fraction

Compound interest – pays interest on both the original sum and the interest already earned

Concave polygon – a polygon with at least one interior reflex angle

Cone – a pyramid with a circular base and a curved surface rising to a vertex

Congruent – exactly the same size and shape; one of the shapes might be rotated or flipped over

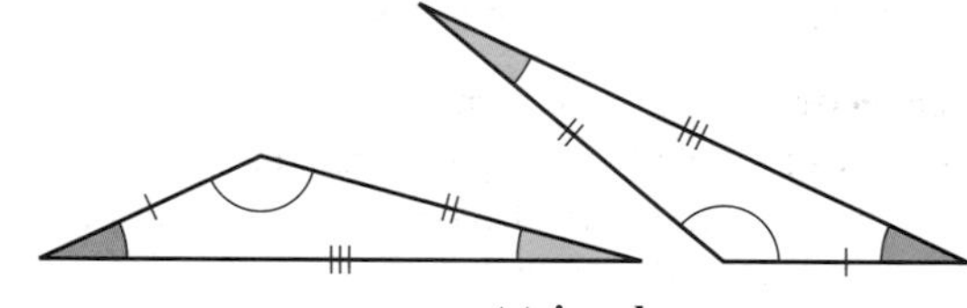

congruent triangles

Consecutive – in sequence

Constant – a number that does not change, for example, the formula $P = 4l$ states that the perimeter of a square is always four times the length of one side; 4 is a constant and P and l are variables

Continuous data – data that can be measured and take any value; length, weight and temperature are all examples of continuous data

Convenience or **opportunity sampling** – a survey that is conducted using the first people who come along, or those who are convenient to sample (such as friends and family)

Convex polygon – a polygon with no interior reflex angles

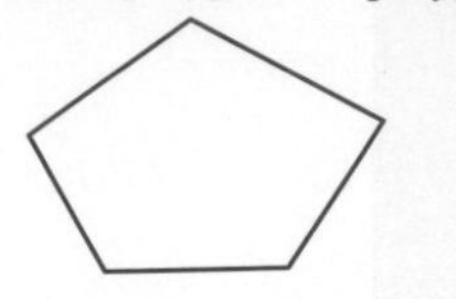

Coordinates – a system used to identify a point; an x-coordinate and a y-coordinate give the horizontal and vertical positions

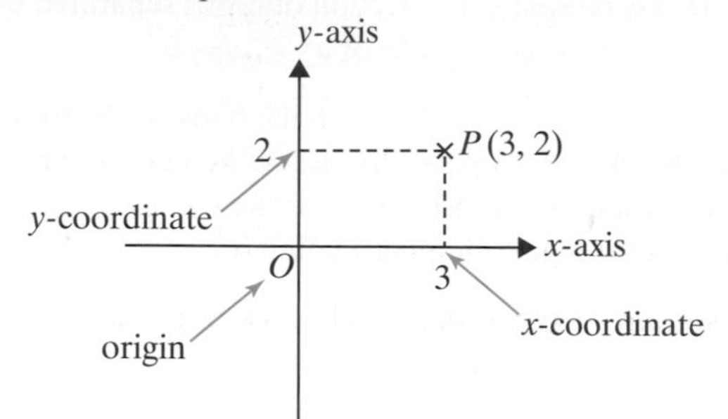

Correlation – a measure of the relationship between two sets of data; correlation is measured in terms of type and strength

Strength of correlation

The strength of correlation is an indication of how close the points lie to a straight line (perfect correlation)

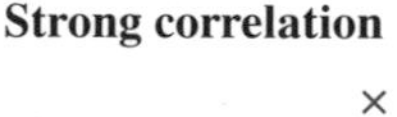

Strong correlation

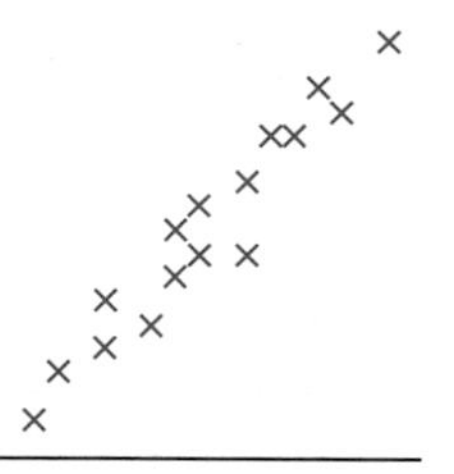

Weak correlation

Correlation is usually described in terms of strong correlation, weak correlation or no correlation

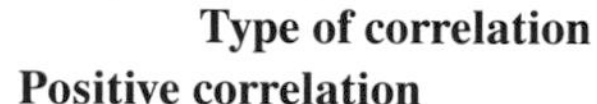

Type of correlation

Positive correlation

In positive correlation an increase in one set of variables results in an increase in the other set of variables

Negative correlation

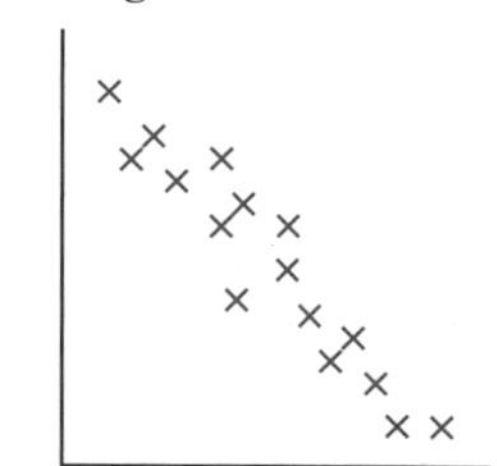

In negative correlation an increase in one set of variables results in a decrease in the other set of variables

Zero or no correlation

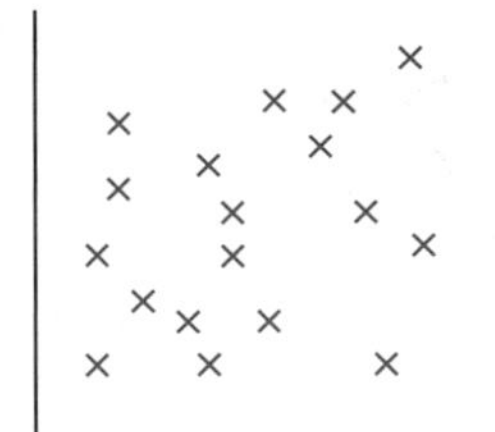

Zero or no correlation is where there is no obvious relationship between the two sets of data

Credit – when you buy goods 'on credit' you do not pay all the cost at once; instead you make a number of payments at regular intervals, often once a month

Cross-section – a cut at right angles to a face and usually at right angles to the length of a prism

Cube – a solid with six identical square faces

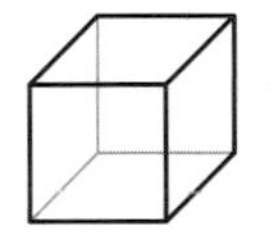

Cube number – a cube number is the outcome when a whole number is multiplied by itself then multiplied by itself again; cube numbers are 1, 8, 27, 64, 125, ...

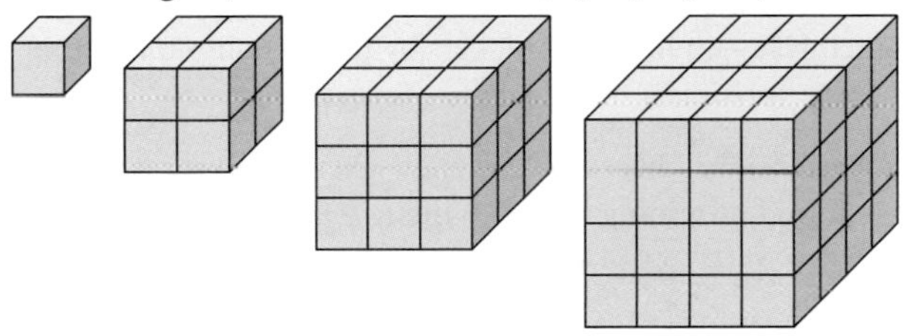

The rule for the *n*th term of the cube numbers is n^3

Cube root – the cube root of a number such as 125 is a number whose outcome is 125 when multiplied by itself then multiplied by itself again

Cuboid – a solid with six rectangular faces (two or four of the faces can be squares)

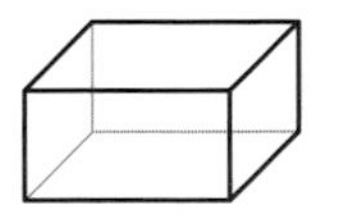

Cumulative frequency diagram – a cumulative frequency diagram can be used to find an estimate for the mean and quartiles of a set of data; find the cumulative frequency by adding the frequencies in turn to give a 'running total'

Cumulative frequency diagram

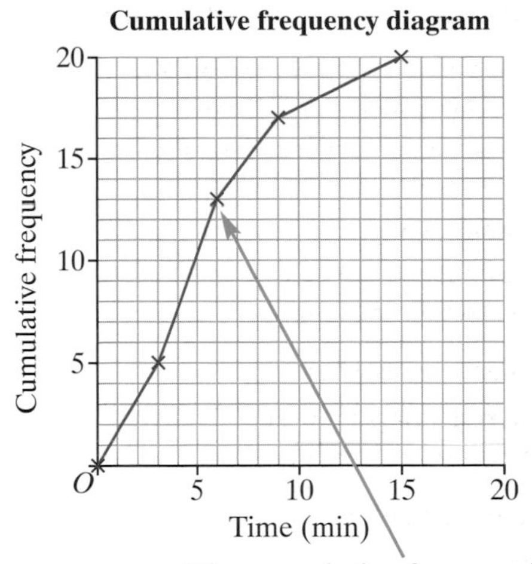

The cumulative frequencies are plotted at the end of the interval to which they relate

Cylinder – a prism with a circle as a cross-sectional face

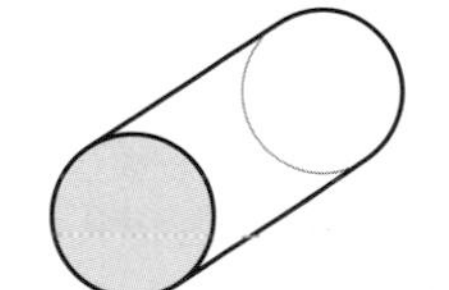

Data – information that has been collected

Data collection sheets – these are used to record the responses to the different questions on a questionnaire; they can also be used with computers to load data onto a database

Decagon – a polygon with ten sides

Decimal – a number in which a decimal point separates the whole number part from the decimal part, for example, 24.8

Decimal fraction – a fraction consisting of tenths, hundredths, thousandths, and so on, expressed in a decimal form, for example, 0.65 (6 tenths and 5 hundredths)

Decimal places – the digits to the right of a decimal point in a number, for example, in the number 23.657, the number 6 is the first decimal place (worth $\frac{6}{10}$), the number 5 is the second decimal place (worth $\frac{5}{100}$) and 7 is the third decimal place (worth $\frac{7}{1000}$); the number 23.657 has 3 decimal places

Degree of accuracy – the accuracy to which a measurement or a number is given, for example, to the nearest 1000, nearest 100, nearest 10, nearest integer, 2 significant figures, 3 decimal places

Denominator – the number on the bottom of a fraction

Deposit – an amount of money you pay towards the cost of an item, with the rest of the cost to be paid later

Depreciation – a reduction in value, for example, due to age or condition

Diameter – a chord passing through the centre of a circle; the diameter is twice the length of the radius

Digit – any of the numerals from 0 to 9

Direct observation – collecting data first-hand, for example, counting cars at a motorway junction or observing someone shopping

Direct proportion – if two variables are in direct proportion, one is equal to a constant multiple of the other, so that if one increases, the other increases and if one decreases then the other decreases

In general $x \propto y$ and $x = kx$

Discount – a reduction in the price, perhaps for paying in cash or paying early

Discrete data – data that can only be counted and take certain values, for example, number of cars (you can have 3 cars or 4 cars but nothing in between, so $3\frac{1}{2}$ cars is not possible)

Edge – a line segment that joins two vertices of a solid

Equation – a statement showing that two expressions are equal, for example, $2y - 7 = 15$

Equidistant – the same distance; if A is equidistant from B and C, then AB and AC are the same length

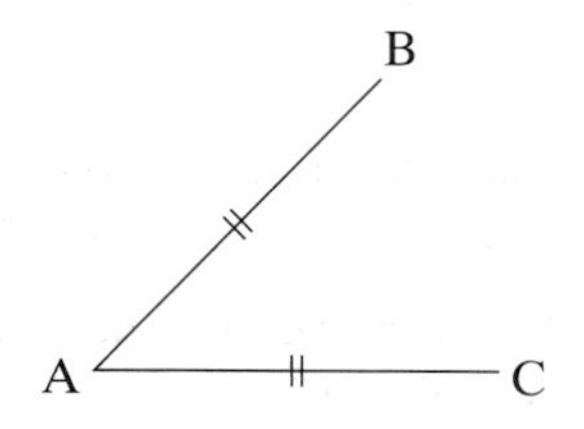

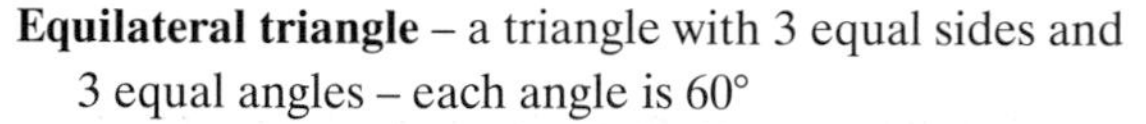

Equilateral triangle – a triangle with 3 equal sides and 3 equal angles – each angle is 60°

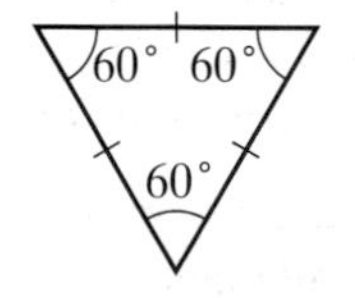

Equivalent fraction – a fraction that has the same value as another, for example, $\frac{3}{5}$ is equivalent to $\frac{30}{50}, \frac{6}{10}, \frac{60}{100}, \frac{15}{25}, \frac{1.5}{2.5}, \ldots$

Estimate – find an approximate value of a calculation; this is usually found by rounding all of the numbers to one significant figure, for example,

$\frac{20.4 \times 4.3}{5.2}$ is approximately $\frac{20 \times 4}{5}$ where each number is rounded to 1 s.f.; the answer can be worked out in your head to give 16

Expand – to remove brackets to create an equivalent expression (expanding is the opposite of factorising)

Exponent – see index

Exterior angle – the angle between one side of a polygon and the extension of the adjacent side

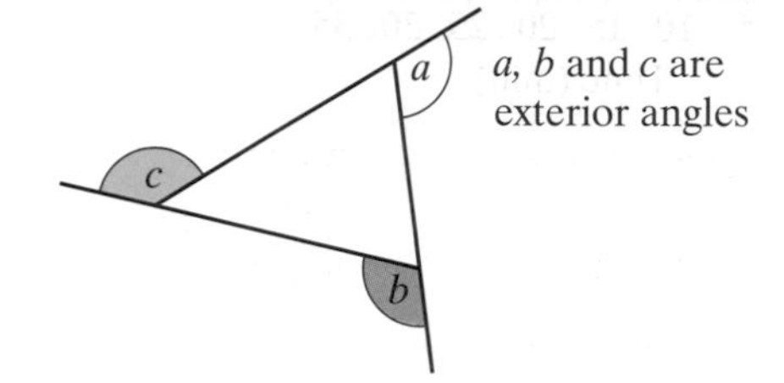

Face – one of the flat surfaces of a solid

Factor – a natural number which divides exactly into another number (no remainder), for example, the factors of 12 are 1, 2, 3, 4, 6 and 12

Factorise – to include brackets by taking common factors (factorising is the opposite of expanding)

Fibonacci numbers – a sequence where each term is found by adding together the two previous terms

1, 1, 2, 3, 5, 8, 13, 21, …

(2 = 1+1, 3 = 1+2, 5 = 2+3, 8 = 3+5, 13 = 5+8, 21 = 8+13)

Formula – an equation showing the relationship between two or more variables, for example, $E = mc^2$

Fraction or **simple fraction** or **common fraction** or **vulgar fraction** – a number written as one whole number over another, for example, $\frac{3}{8}$ (three eighths), which has the same value as $3 \div 8$

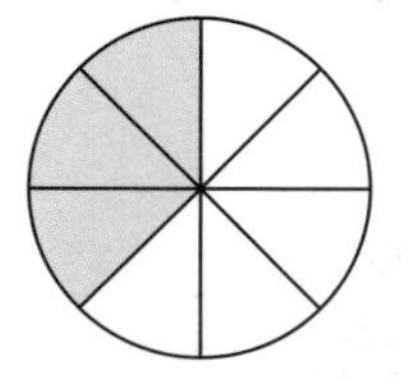

Frequency density – in a histogram, the area of the bars represents the frequency and the height represents the frequency density

Frequency density (bar height) = $\frac{\text{frequency}}{\text{class width}}$

Frequency diagram – a graphical method of showing how many results or observations fall into each category in a survey or experiment

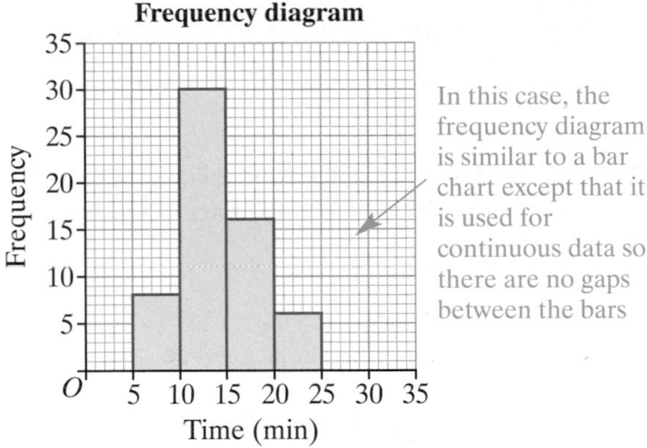

Frequency polygon – this is drawn from a histogram (or bar chart) by joining the midpoints of the tops of the bars with straight lines to form a polygon

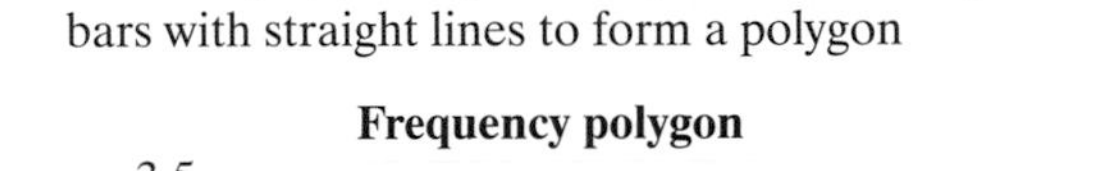
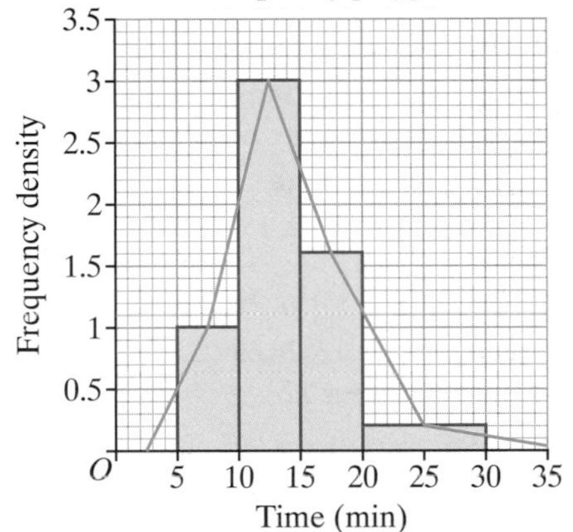

Frequency table or **frequency distribution** – a table showing how many times each quantity occurs, for example:

Number in family	2	3	4	5	6	7	8
Frequency	2	3	8	4	2	0	1

Frustum (of a cone) – a cone with the top part cut off

Gradient – a measure of how steep a line is

$$\text{Gradient} = \frac{\text{change in vertical distance}}{\text{change in horizontal distance}} = \frac{y}{x}$$

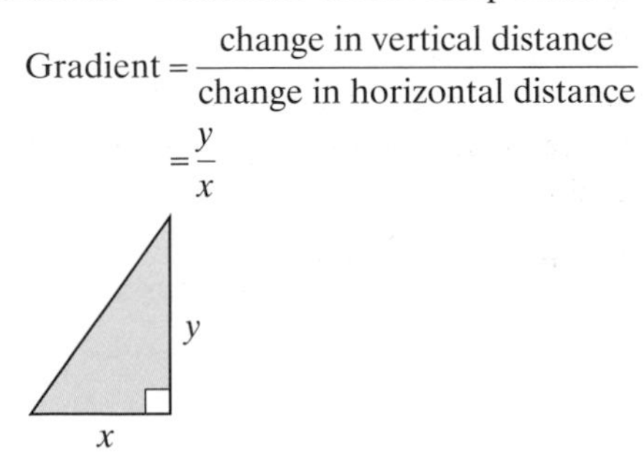

Grouped data – data that has been grouped into specific intervals

Hemisphere – a half sphere

Heptagon – a polygon with seven sides

Hexagon – a polygon with six sides

Highest common factor (HCF) – the highest factor that two or more numbers have in common, for example,

the factors of 16 are 1, 2, 4, 8, 16
the factors of 24 are 1, 2, 3, 4, 6, 8, 12, 24
the common factors are 1, 2, 4, 8
the highest common factor is 8

Histogram – a histogram is similar to a bar chart except that the *area* of the bar represents the frequency

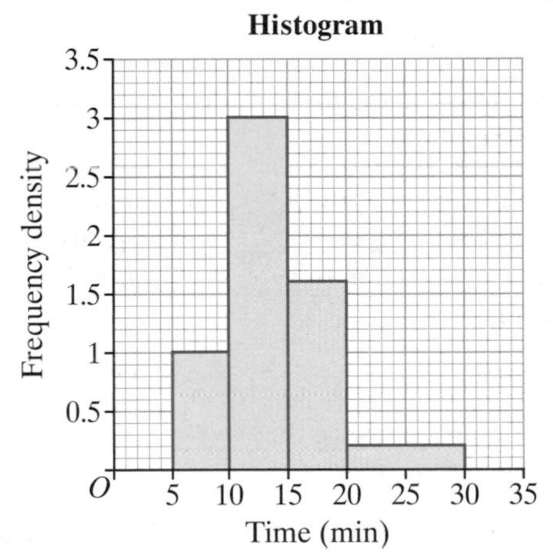

Horizontal – from left to right; parallel to the horizon

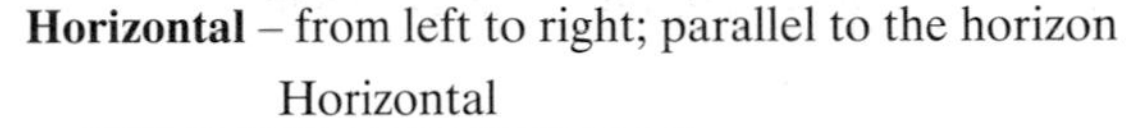

Identity – two expressions linked by the ≡ sign are true for all values of the variable, for example, $3x + 3 \equiv 3(x + 1)$

Image – the shape after it undergoes a transformation, for example, reflection, rotation, translation or enlargement

Improper fraction or **top-heavy fraction** – a fraction in which the numerator is bigger than the denominator, for example, $\frac{13}{5}$, which is equal to the mixed number $2\frac{3}{5}$

Index or **power** or **exponent** – the index tells you how many times the base number is to be multiplied by itself

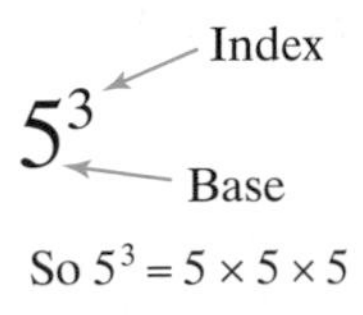

So $5^3 = 5 \times 5 \times 5$

Index notation – when a product such as $2 \times 2 \times 2 \times 2$ is written as 2^4, the 4 is the index (plural **indices**)

Index number – an index number is used to compare measurements over a period of time. A common index is the price index which compares prices over a period of time. The base year is given a value of 100 (representing 100%) and the price index shows the increase (or decrease) since the base year. A price index of 120 represents a 20% increase and a price index of 90 represents a 10% decrease.

Indices – the plural of index

Integer – any positive or negative whole number or zero, for example, $-2, -1, 0, 1, 2 \ldots$

Intercept – the y-coordinate of the point at which the line crosses the y-axis

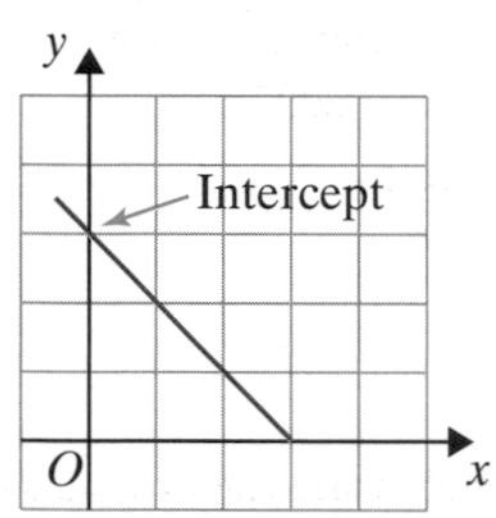

Interest – money paid to you by a bank, building society or other financial institution if you put your money in an account or the money you pay for borrowing from a bank

Interior angle – an angle inside a polygon

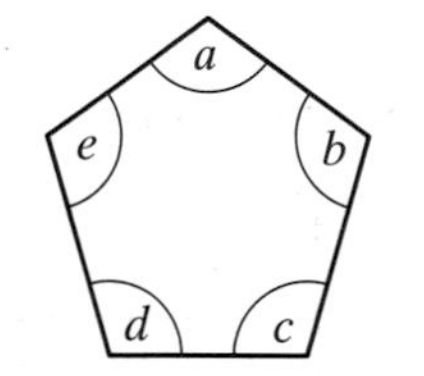

a, b, c, d and e are interior angles

Interquartile range – the difference between the upper quartile and the lower quartile

Interquartile range = $\begin{matrix}\text{upper}\\\text{quartile}\end{matrix} - \begin{matrix}\text{lower}\\\text{quartile}\end{matrix}$

Inverse operation – the operation that undoes or reverses a previous operation, for example, subtract is the inverse of add:

$15 + 8 = 23$ Add 8

$23 - 8 = 15$ Subtract 8 to return to the starting number 15

Inverse proportion – if two variables are in inverse proportion, their product is a constant; so that if one increases, the other decreases and vice versa

In general $x \propto \frac{1}{y}$ and $x = k\frac{1}{y}$ and $xy = k$

Irrational number – a number that is not an integer and cannot be written as a fraction, for example, $\sqrt{2}, \sqrt{3}, \sqrt{5}$ and π; irrational numbers, when expressed as decimals, are infinite, non-recurring decimals

Irregular polygon – a polygon whose sides and angles are not all equal (they do not all have to be different)

Isosceles trapezium – a quadrilateral with one pair of parallel sides. Non-parallel sides are equal

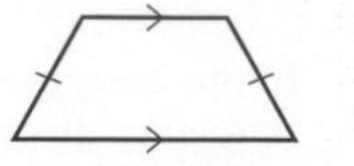

Isosceles triangle – a triangle with 2 equal sides and 2 equal angles; the equal angles are called **base angles**

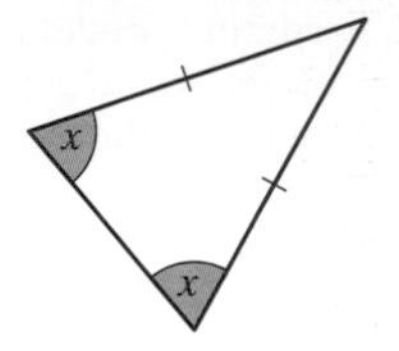

Kite – a quadrilateral with two pairs of equal adjacent sides

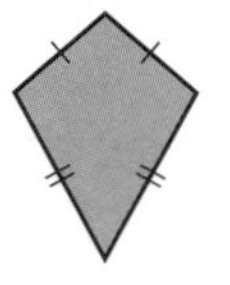

Least common multiple (LCM) – the lowest multiple which is common to two or more numbers, for example,

the multiples of 3 are 3, 6, 9, 12, 15, 18, 24, 27, 30, 33, 36 ...

the multiples of 4 are 4, 8, 12, 16, 20, 24, 28, 32, 36 ...

the common multiples are 12, 24, 36 ...

the least common multiple is 12

Linear equation – an equation where the highest power of the variable is 1; for example, $3x + 2 = 7$ is a linear equation but $3x^2 + 2 = 7$ is not

Linear expression – a combination of terms where the highest power of the variable is 1

Linear expressions	**Non-linear expressions**
x	x^2
$x + 2$	$\frac{1}{x}$
$3x + 2$	$3x^2 + 2$
$3x + 4y$	$(x + 1)(x + 2)$
$2a + 3b + 4c + \ldots$	x^3

Line graph – a line graph is a series of points joined with straight lines

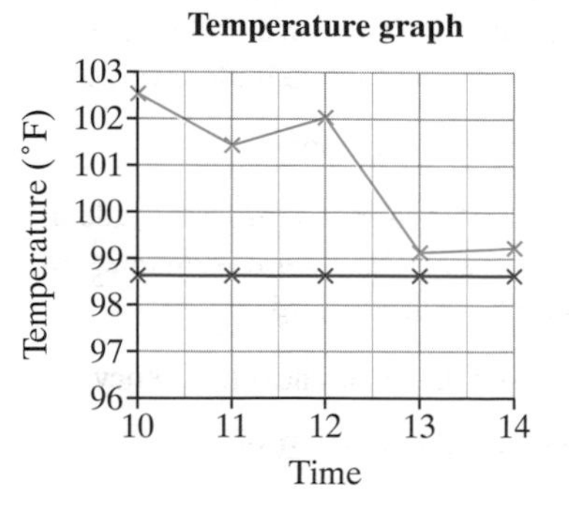

Line of best fit – a line drawn to represent the relationship between two sets of data. Ideally it should only be drawn where the correlation is strong, for example,

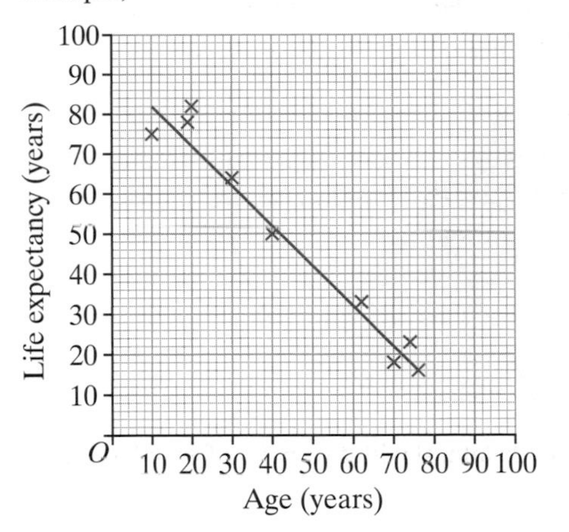

Line of symmetry – a shape has reflection symmetry about a line through its centre if reflecting it in that line gives an identical-looking shape

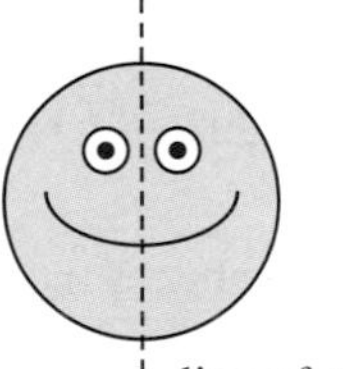

Line segment – the part of a line joining two points

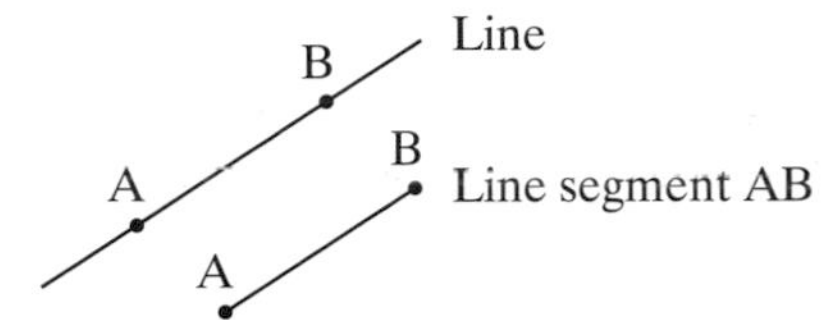

Lower bound – this is the minimum possible value of a measurement, for example, if a length is measured as 37 cm correct to the nearest centimetre, the lower bound of the length is 36.5 cm

Lower quartile – the value 25% of the way through the data

Mean – found by calculating $\frac{\text{the total of all the values}}{\text{the number of values}}$

Median – the middle value when all the values have been arranged in order of size; for an even set of numbers, the median is the mean of the two middle values

Midpoint – the middle point of a line

Mixed number or **mixed fraction** – a number made up of a whole number and a fraction, for example, $2\frac{3}{5}$, which is equal to the improper fraction $\frac{13}{5}$

Modal class – the class with the highest frequency

Mode – the value that occurs most often

Moving average – used to smooth out the fluctuations in a time series, for example, a four-point moving average is found by averaging successive groups of four readings

The four-point moving averages can be plotted on the graph as shown

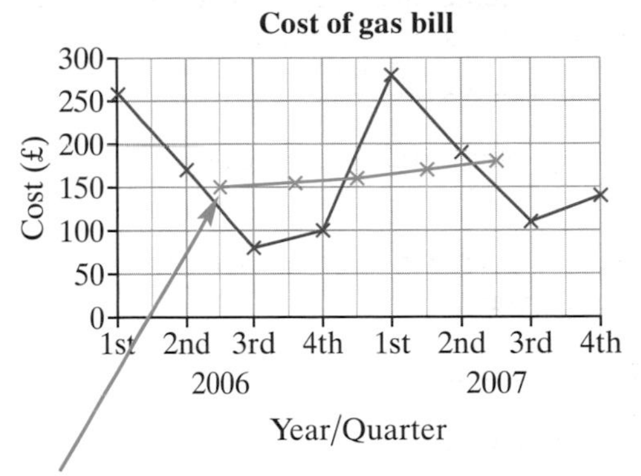

The first four-point moving average is plotted in the 'middle' of the first four points, and so on

Multiple – the multiples of a number are the products of its multiplication table, for example, the multiples of 3 are 3, 6, 9, 12, 15 ...

Multiplier – a number used to multiply an amount

Net – a two-dimensional shape made of polygons that can be folded to make a three-dimensional solid, for example,

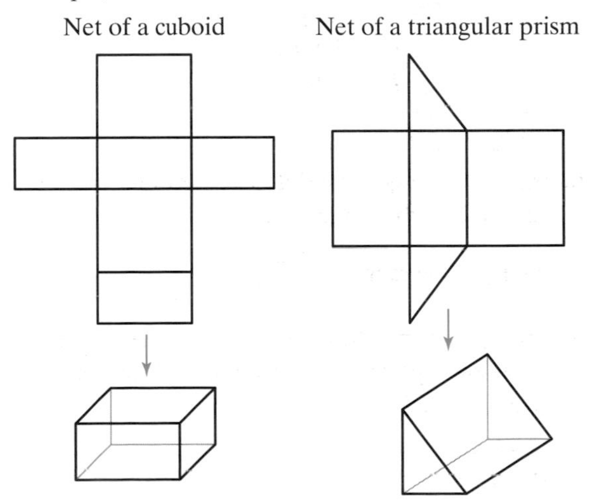

Nonagon – a polygon with nine sides

***n*th term** – this phrase is often used to describe a 'general' term in a sequence; if you are given the *n*th term, you can use this to find the terms of a sequence

Numerator – the number on the top of a fraction

Numerator → $\frac{3}{8}$ ← Denominator

Octagon – a polygon with eight sides

Operation – a rule for combining two numbers or variables, such as add, subtract, multiply or divide

Opportunity sampling – see convenience sampling

Ordered stem-and-leaf diagram – a stem-and-leaf diagram where the data is placed in order

Number of minutes to complete a task

Stem (tens)	Leaf (units)
1	1 6 7 8 9
2	2 2 7 7 7 8 9
3	1 4 6

Key: 3|4 represents 34 minutes

Order of rotation symmetry – the number of ways a shape would fit on top of itself as it is rotated through 360°

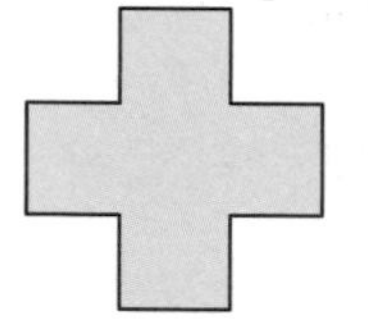

rotation symmetry order 4

(Shapes that are not symmetrical have rotation symmetry of order 1 because a rotation of 360° always produces an identical-looking shape)

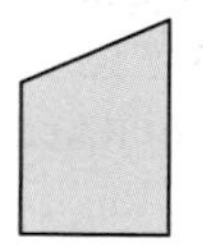

rotation symmetry order 1 (i.e. not symmetrical)

Origin – the point (0, 0) on a coordinate grid

Outlier – a value that does not fit the general trend, for example,

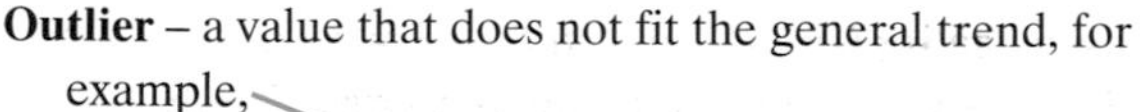

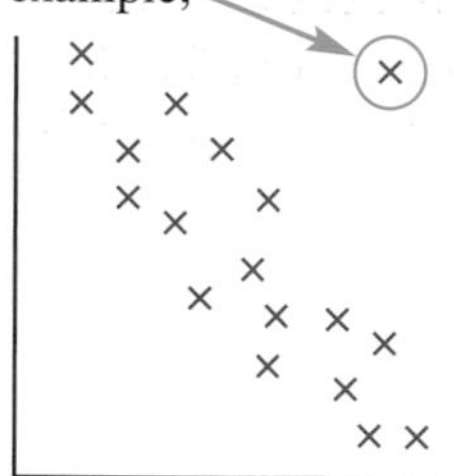

Parallelogram – a quadrilateral with opposite sides equal and parallel

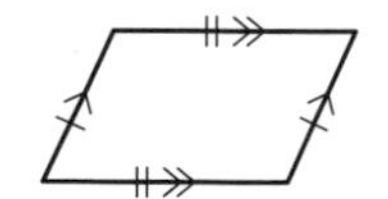

Pentagon – a polygon with five sides

Percentage – a number of parts per hundred, for example, 15% means $\frac{15}{100}$

Perimeter – the distance around an enclosed shape

Pie chart – in a pie chart, frequency is shown by the angles (or areas) of the sectors of a circle

Mobile phone sales

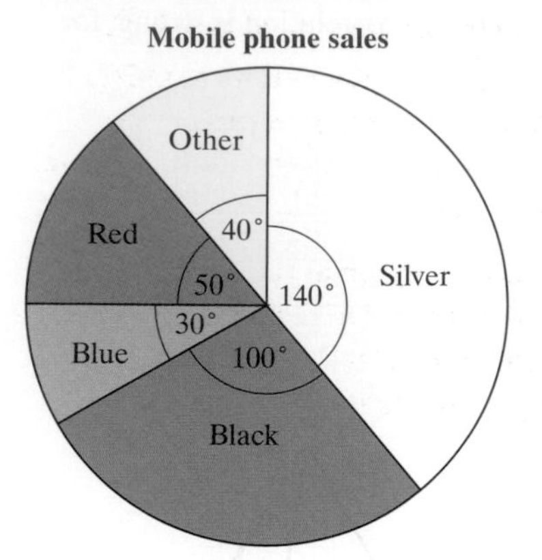

Pilot survey – a small-scale survey to check for any unforeseen problems with the main survey

Polygon – a closed two-dimensional shape made from straight lines

Power – see index

Primary data – data that you collect yourself; this is new data and is usually gathered for the purpose of a task or project

Prime factor decomposition – writing a number as the product of its prime factors, for example, $12 = 2^2 \times 3$

Prime number – a natural number with exactly two factors, for example, 2 (factors are 1 and 2), 3 (factors are 1 and 3), 5 (factors are 1 and 5), 7, 11, 13, 17, 23, ... 59 ...

Principal – the money put into the bank or borrowed from the bank

Prism – a three-dimensional solid with two cross-sectional faces that are identical polygons, parallel to each other; all other faces are either parallelograms or rectangles

Prisms are named according to the cross-sectional face; for example,

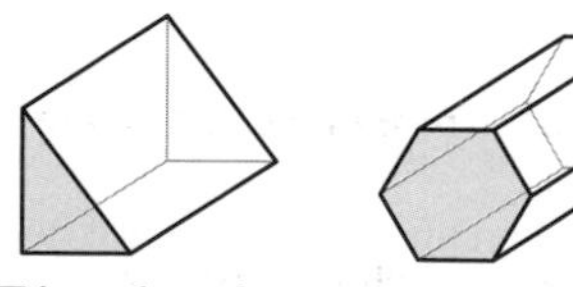

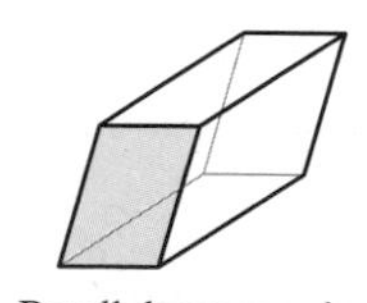

Triangular prism Hexagonal prism Parallelogram prism

Product – the result of multiplying together two (or more) numbers, variables, terms or expressions

Proper fraction – a fraction in which the numerator is smaller than the denominator, for example, $\frac{5}{13}$

Proportion – if a class has 12 boys and 18 girls, the proportion of boys in the class is $\frac{12}{30}$, which simplifies to $\frac{2}{5}$, and the proportion of girls is $\frac{18}{30}$, which simplifies to $\frac{3}{5}$ (the **ratio** of boys to girls is 12 : 18, which simplifies to 2 : 3) – a proportion compares one part with the whole; a ratio compares parts with one another

Pyramid – a solid with a polygon as the base and one other vertex; all the vertices of the base are joined to this vertex forming triangular faces. Pyramids are named according to their base, for example,

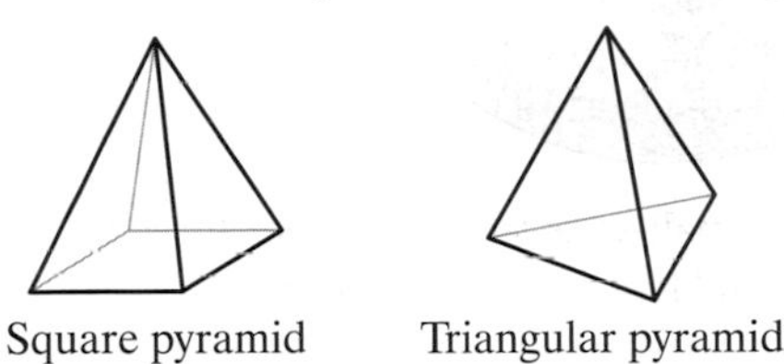

Square pyramid Triangular pyramid

Quadrant (of a circle) – one quarter of a circle

Quadratic expression – an expression containing terms where the highest power of the variable is 2

Quadratic expressions	Non-quadratic expressions
x^2	x
$x^2 + 2$	$2x$
$3x^2 + 2$	$\frac{1}{x}$
$4 + 4y^2$	$3x^2 + 5x^3$
$(x + 1)(x + 2)$	$x(x + 1)(x + 2)$

Quadrilateral – a polygon with four sides

Qualitative or **categorical data** – data that cannot be measured using numbers, for example, type of pet, car colour, taste, people's opinions/feelings, etc.

Quantitative data – data that can be counted or measured using numbers, for example, number of pets, height, weight, temperature, age, shoe size etc.

Quota sampling – this method involves choosing a sample with certain characteristics, for example, select 20 adult men, 20 adult women, 10 teenage girls and 10 teenage boys to conduct a survey about shopping habits

Radius – the distance from the centre of a circle to any point on the circumference

Random sampling – this requires each member of the population to be assigned a number; the numbers are then chosen at random

Range – a measure of spread found by calculating the difference between the largest and smallest values in the data, for example, the range of 1, 2, 3, 4, 5 is $5 - 1 = 4$

Rate – the percentage at which interest is added, usually expressed as per cent per annum (year)

Ratio – the ratio of two or more numbers or quantities is a way of comparing their sizes, for example, if a school has 25 teachers and 500 students, the ratio of teachers to students is 25 to 500, or 25 : 500 (read as 25 to 500)

Rational expression – a fraction, for example,

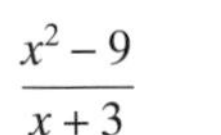

$$\frac{x^2 - 9}{x + 3}$$

Rational number – a number that can be expressed in the form $\frac{p}{q}$ where p and q are both integers, for example, $1(=\frac{1}{1})$, $2\frac{1}{3}\ (=\frac{7}{3})$, $\frac{3}{5}$, $0.\dot{1}(=\frac{1}{9})$; rational numbers, when written as decimals, are terminating decimals or recurring decimals

Reciprocal – any number multiplied by its reciprocal equals one; one divided by a number will give its reciprocal, for example, the reciprocal of 3 is $\frac{1}{3}$ because $3 \times \frac{1}{3} = 1$

Rectangle – a quadrilateral with four right angles, and opposite sides equal in length

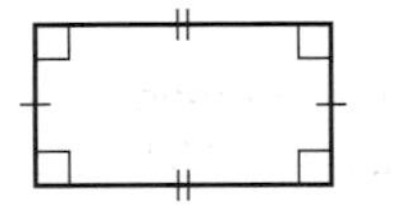

Recurring decimal – a decimal with a repeating digit or group of digits, for example, 0.33333333333 ... (written as $0.\dot{3}$) or 0.25678678678678 ... (written as $0.25\dot{6}7\dot{8}$)

Reflection – a transformation involving a mirror line (or axis of symmetry), in which the line from the shape to its image is perpendicular to the mirror line. To describe a reflection fully, you must describe the position or give the equation of its mirror line, for example, the triangle A is reflected in the mirror line $y = 1$ to give the image B

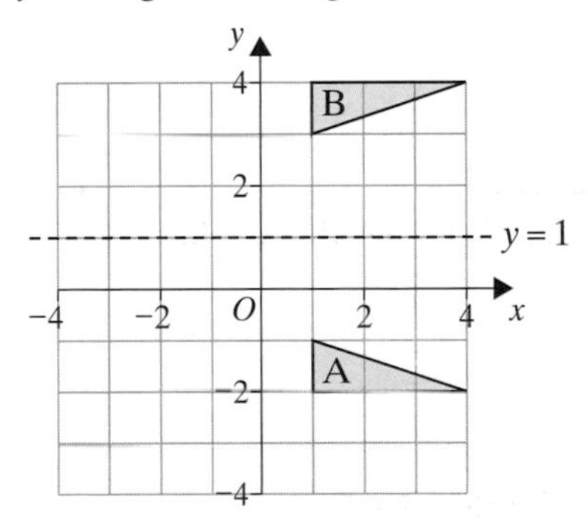

Regular polygon – a polygon with all sides and all angles equal

Respondent – the person who answers the questionnaire

Rhombus – a quadrilateral with four equal sides and opposite sides parallel

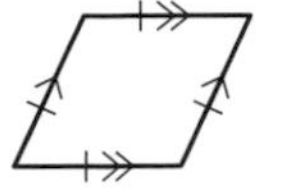

Rotation – a transformation in which the shape is turned about a fixed point called the centre of rotation. To describe a rotation fully, you must give the centre, angle and direction (a *positive angle* is *anticlockwise* and a *negative angle* is *clockwise*), for example, the triangle A is rotated about the origin through 90° anticlockwise to give the image C

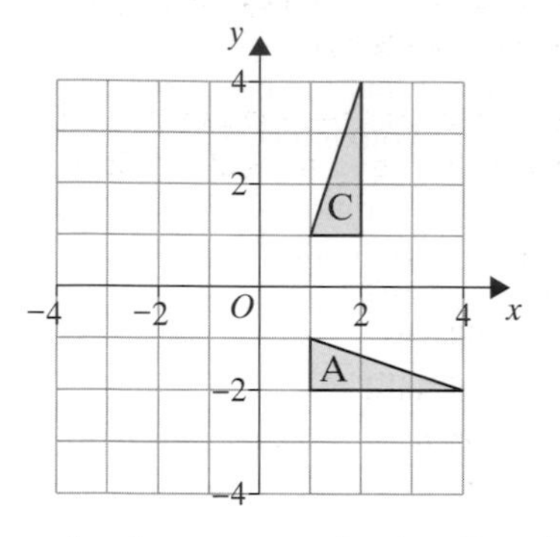

Round – give an approximate value of a number. Numbers can be rounded to the nearest 1000, nearest 100, nearest 10, nearest integer, significant figures, decimal places ... etc

Scatter graph – a graph used to show the relationship between two sets of variables, for example, temperature and ice cream sales

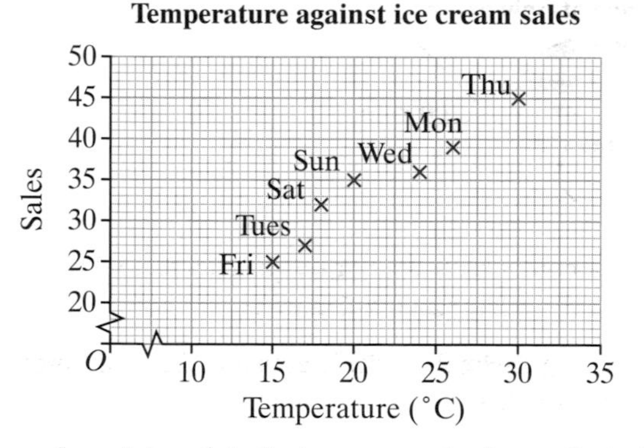

Secondary data – data that someone else has collected; this might include data in books, newspapers, magazines, etc. or data that has been loaded onto a database

Sector (of a circle) – a region in a circle bounded by two radii and an arc

Segment – the region bounded by an arc and a chord

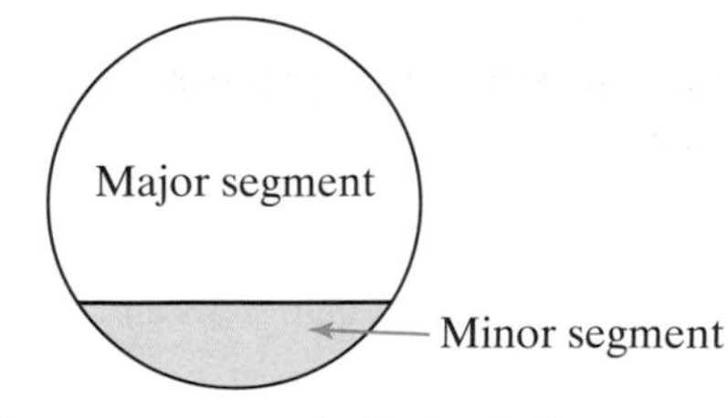

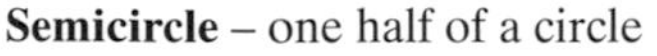

Semicircle – one half of a circle

Sequence – a list of numbers or diagrams that are connected in some way

In this sequence of diagrams, the number of squares is increased by one each time:

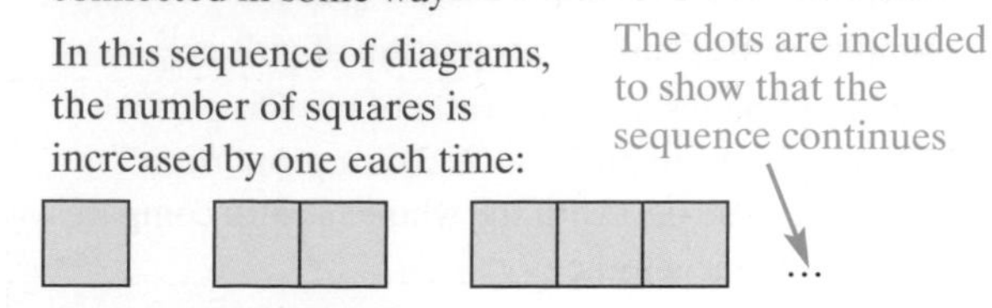

Shape – an enclosed space

Significant figures – the digits in a number; the closer a digit is to the beginning of a number then the more important or significant it is; for example, in the number 23.657, 2 is the most significant digit and is worth 20, 7 is the least significant digit and is worth $\frac{7}{1000}$; the number 23.657 has 5 significant digits

Simple fraction – see fraction

Simple interest – pays interest only on the sum of money originally invested

Simplify – to make simpler by collecting like terms

Simplify a fraction or **express a fraction in its simplest form** – to change a fraction to the simplest equivalent fraction; to do this divide the numerator and the denominator by a common factor (this process is called cancelling or reducing or simplifying the fraction)

Solid – a three-dimensional shape

Solution – the value of the unknown in an equation, for example the solution of the equation $3y = 6$ is $y = 2$

Solve – when you solve an equation you find the solution, in an equation or expression, for example the coefficient of q^2 in the expression $2 - 3q^2$ is -3

Sphere – a solid in which all the points on the surface are the same distance from the centre

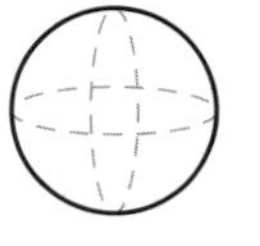

Square – a quadrilateral with four equal sides and four right angles

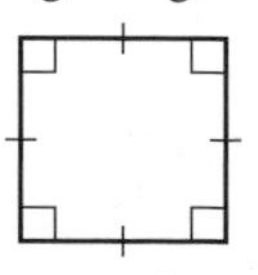

Square number – a square number is the outcome when a whole number is multiplied by itself; square numbers are 1, 4, 9, 16, 25, ...

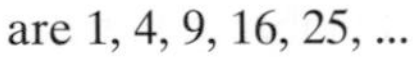

The rule for the *n*th term of the square numbers is n^2

Square root – the square root of a number such as 16 is a number whose outcome is 16 when multiplied by itself

Standard form – standard form is a shorthand way of writing very large and very small numbers; standard form numbers are always written as:

$$A \times 10^n$$

10^n: A power of 10

A: A must be at least 1 but less than 10

Stem-and-leaf diagram – a way of arranging data using a key to explain the 'stem' and 'leaf' so that 3|4 represents 34

Number of minutes to complete a task

Stem (tens)	Leaf (units)
1	6 8 1 9 7
2	7 8 2 7 7 2 9
3	4 1 6

Key: 3|4 represents 34 minutes

Stratified sampling – this involves dividing the population into a series of groups or 'strata' and ensuring that the sample is representative of the population as a whole, for example, if the population has twice as many boys as girls, then the sample should have twice as many boys as girls

Substitute – find the value of an expression when the variable is given a value, for example when $x = 4$, the expression $3x + 2 = 3 \times 4 + 2 = 14$

Surd – a number containing an irrational root, for example, $\sqrt{2}$ or $3 + 2\sqrt{7}$

Survey – a way of collecting data; there are a variety of ways of doing this, including face-to-face, or via telephone, e-mail or post using questionnaires

Symmetrical – a shape that has symmetry

Symmetry (reflection) – a shape has (reflection) symmetry if a reflection through a line passing through its centre produces an identical-looking shape. The shape is said to be symmetrical

Symmetry (rotation) – a shape has (rotation) symmetry if a rotation about its centre through an angle greater than 0° and less than 360° gives an identical-looking shape

Systematic sampling – this is similar to random sampling except that it involves every *n*th member of the population; the number *n* is chosen by dividing the population size by the sample size

Tally chart – a useful way to organise the raw data; the chart can be used to answer questions about the data, for example,

Number of pets	Tally
0	~~IIII~~ IIII
1	~~IIII~~ ~~IIII~~ II
2	~~IIII~~ II
3	III
4	II

The tallies are grouped into five so that

IIII = 4
~~IIII~~ = 5
~~IIII~~ I = 6

This makes the tallies easier to read

Tangent (to a circle) – a straight line that touches the circle at only one point

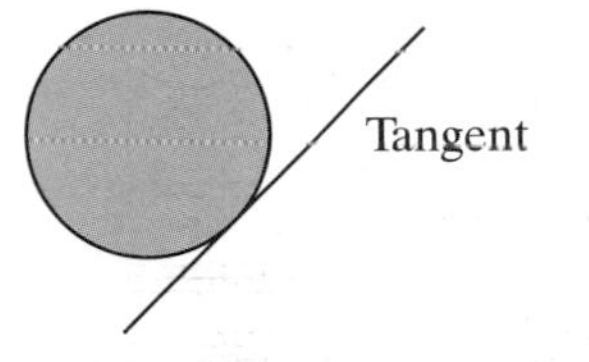

Term – a number, variable or the product of a number and a variable(s) such as 3, x or $3x$

Terminating decimal – a decimal that ends, for example, 0.3, 0.33 or 0.3333

Tessellation – a pattern where one or more shapes are fitted together repeatedly leaving no gaps

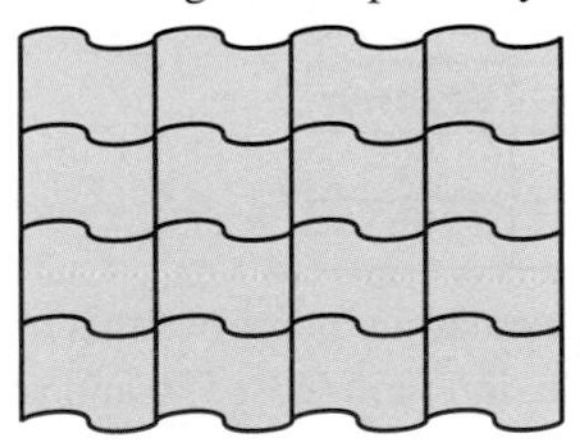

Time – usually measured in years for the purpose of working out interest

Time series – a graph of data recorded at regular intervals

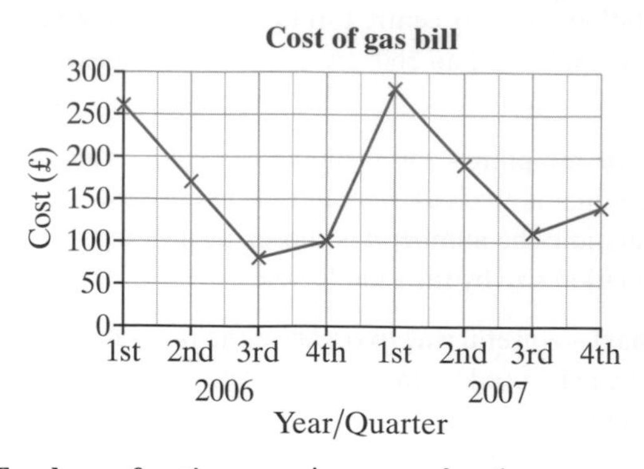

Top-heavy fraction – see improper fraction

Transformation – reflections, rotations, translations and enlargements are examples of transformations as they transform one shape onto another

Trapezium (pl. **trapezia**) – a quadrilateral with one pair of parallel sides

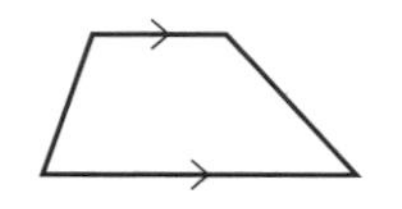

Triangle – a polygon with three sides

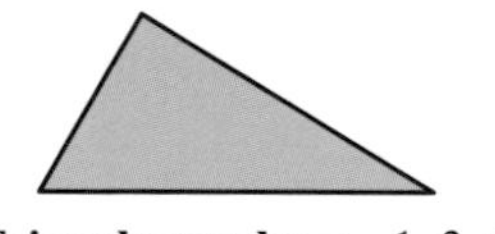

Triangle numbers – 1, 3, 6, 10, 15, ...

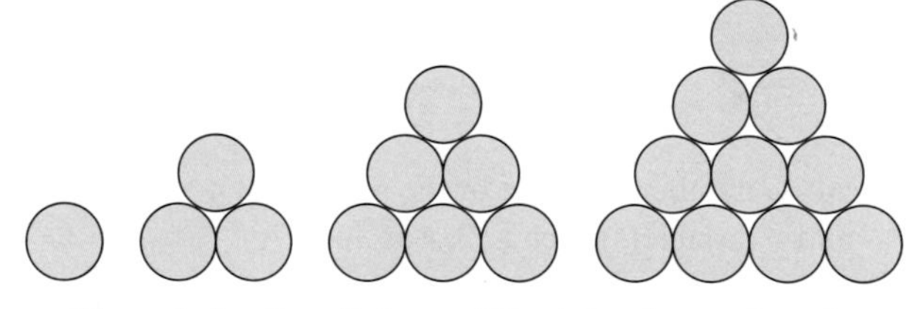

The rule for the nth term of the triangle numbers is $\frac{1}{2}n(n + 1)$

Two-way table – a combination of two sets of data presented in a table form, for example,

	Men	Women
Left-handed	7	6
Right-handed	20	17

Unitary method – a way of calculating quantities that are in proportion, for example, if 6 items cost £30 and you want to know the cost of 10 items, you can first find the cost of one item by dividing by 6, then find the cost of 10 by multiplying by 10

6 items cost £30

1 item costs $\frac{£30}{6} = £5$

10 items cost $10 \times £5 = £50$

Unitary ratio – a ratio in the form $1 : n$ or $n : 1$; for example, for every 100 female babies born, 105 male babies are born. The ratio of the number of females to the number of males is 100 : 105; as a unitary ratio, this is 1 : 1.05, which means that, for every female born, 1.05 males are born

Unit fraction – a fraction with a numerator of 1, for example, $\frac{1}{5}$

Unknown – the letter in an equation such as x or y

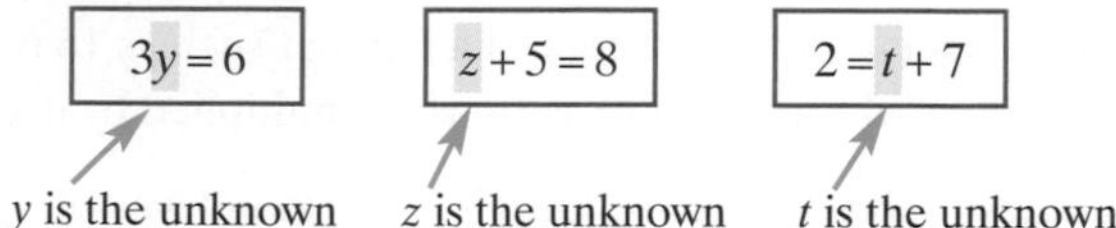

Upper bound – this is the maximum possible value of a measurement, for example, if a length is measured as 37 cm correct to the nearest centimetre, the upper bound of the length is 37.5 cm

Upper quartile – the value 75% of the way through the data

Variable – a symbol representing a quantity that can take different values such as x, y or z

VAT (Value Added Tax) – a tax that has to be added on to the price of goods or services

Vertex (pl. **vertices**) – the point where two or more edges meet

Vertical – directly up and down; perpendicular to the horizontal

Vertical

Volume – a measure of how much space fills a solid, commonly measured in cubic centimetres (cm^3) or cubic metres (m^3)

Vulgar fraction – see fraction